Ai6X6
RESEARCH LAB

Transformation Blueprint Series

Published by EDGE Publish, www.EdgePublish.com

Customer Transformation for the AI Era, First Edition, 2026
ISBN-13: 978-1-971148-03-8, Printed in the United States of America.

Ai6X6 RESEARCH LAB

Transformation Blueprint Series

CUSTOMER TRANSFORMATION FOR THE AI ERA

Designing Experiences That Earn Trust and Loyalty.
The 6x6 Blueprint for Personalization and Journey Coherence.

SUBRATA MUKHERJEE

EDGE PUBLISH
www.EdgePublish.com

"For the grace of God, who lights the path.
And, in loving memory of my mother Rupali and my father Dilip, who taught me the importance of kindness and perseverance. This work is born from the love you gave, the sacrifices you made, and the quiet wisdom that still echoes in my life. This one is for you."

Personal message from our Founder & Executive Director

SUBRATA MUKHERJEE

For over twenty-five years, I have sat at the heart of global digital transformations—from shifting a 180-year-old institution like *The Economist* into a digital powerhouse, to architecting hyper-automation for Fortune 500 leaders. I have seen the rise of the internet, the mobile revolution, and the cloud. But none of these shifts compare to the era we have just entered.

The Ai6x6 Research Lab was born out of a singular, urgent observation: Most organizations are trying to solve tomorrow's problems with yesterday's mental models. Traditional transformation is often treated as a "project"—a linear path with a start and an end. But in an era of Agentic AI and self-directing systems, transformation is a living organism. It is a system of behaviors, decisions, and capabilities that must evolve in unison. I founded this Lab to move beyond the "hype" and the "tools." My motivation is to provide leaders with the Cohesion, Schema, and Custody required to navigate a world where the boundary between human intuition and machine intelligence is disappearing. People often ask why "6×6." Throughout history, the number six has been the sacred geometry of our existence—from the sexagesimal systems of ancient Sumerian astronomy to the hexagonal patterns found in the farthest reaches of our cosmos. Mathematically, six is the first "perfect number" because its divisors (1, 2, and 3) add up to itself, making it uniquely balanced. In my journey across the U.S., Europe, and Asia, I realized that this ancient symmetry holds the key to modern enterprise. By distilling organizational complexity into six fundamental pillars and viewing them through six transformative lenses, we activate a collective "sixth sense" for the organization—the ability to perceive and act on what others miss. This architecture forms the foundation of our three flagship methodologies—the "*guided navigation system every transformation needs*".

My journey to this blueprint was forged in the engine rooms of the world's most complex organizations. I have led wholesale business transformations for *The Economist*, reinvented Bloomberg's digital presence following the *Businessweek* acquisition, revitalized the iconic *Dummies* brand within Wiley, and optimized the global ratings processes within S&P Global and McGraw-Hill. This strategic work was balanced by deep technical execution: implementing RPA, AI, IoT, and Digital Twins within Raytheon, and architecting CRM, BPM, and digital workplace automations for the global clients of Cognizant and Unisys.

This perspective was further refined through rigorous academic pursuit, drawing on my MBA in Marketing from Jadavpur University, my Master's in Technology Management from Columbia University, and executive certifications from Harvard University and the Indian Institute of Management (IIM-C) to bridge the gap between high-level theory and industrial-scale execution.

Today, through the Ai6x6 Research Lab, I lead the development of the frameworks that are now being adopted by executive teams globally. At the center of this work is three methodologies: 6X6 Blueprint, IGNITE Methodology and REVEAL Methodology our multidimensional navigation system that unifies architecture, execution, and evaluation across the enterprise.

However, my background as a Tech Journalist for *The Telegraph* for over a decade and a columnist for *CIO Magazine* has taught me that true insight requires total independence. I have built this Lab as a space where my team and I can express our opinions freely, without the interference of corporate sponsors. Our goal is not just to report the news, but to "fold" it into the Ai6x6 Blueprints—giving you the critical "so-what" and the "what-next" to make global shifts actionable for your organization.

We are democratizing this knowledge because I believe that the ability to thrive in the AI era should not be reserved only for those who can afford expensive global consulting firms. Our mission is to empower businesses of all sizes with the targeted, practical, and battle-tested information that I often wish I had at my disposal during my own leadership assignments. Through the Ai6x6 Transformation Blueprints and our Executive Mentor Network, we are scaling AI literacy to ensure that as systems become more autonomous, leaders become more intentional. We are not just adopting new tools; we are reimagining the nature of the enterprise itself.

Ai6X6 RESEARCH LAB

NEW YORK | LONDON | KOLKATA

To our esteemed Industry Thought Leaders: Thank you!

Your generous time and profound insights have been the bedrock of this book. Each interview has been converted into sections and one-page cards that have been sprinkled across six books. You have provided a vital perspective, transforming our work from theory into a deeply practical guide rooted in real-world leadership in the AI-Era. Your willingness to share your hard-won wisdom, navigating complex challenges and revealing the future of your industries, enriches every page. You all have successfully navigated the AI transition in your careers, your companies and your industries. This book is a testament to your expertise. We are immensely grateful for your unwavering support and belief in this project. Thanks a lot, from me and my team at Ai6x6 and Executive Mentor Network!

TABLE OF CONTENTS

guided navigation for transformation

BLUEPRINT SERIES

SECTION A

Introduction

Ai6x6™ Transformation Blueprint

Ai6x6 Research Lab is a global intellectual institution dedicated to advancing the science and practice of organizational transformation. We develop original frameworks, diagnostic systems, capability playbooks, and research models that help leaders navigate complexity with clarity, precision, and confidence. Our work spans **Leadership, Workforce, Data, Customer, Product, and Operations**—the six pillars that define modern enterprise maturity. We are based in New York, London & Kolkata.

OUR MISSION FOR ORGANIZATIONAL TRANSFORMATION

To build a globally recognized body of knowledge that helps organizations think clearly, act decisively, and transform responsibly in an age of AI acceleration. We essentially play **six roles** for the industry:

THE CATALYST
BRING ALIGNMENT TO LEADERSHIP

THE STRATEGIST
BRING STRUCTURE TO AMBIGUITY

THE ARCHITECT
BRING CLARITY TO COMPLEXITY

THE ENGINEER
BRING DISCIPLINE TO TRANSFORMATION

THE ETHICIST
BRING INTEGRITY TO OUTCOMES

THE GUARDIAN
BRING OVERSIGHT TO AUTONOMY

WHAT WE DO?

1. Build Transformation Frameworks: We architect the Ai6x6 Transformation Blueprints— a structured, multi-lens transformation model that helps orgs decode patterns, risks, behaviors, and decision pathways.
2. Develop Diagnostic Systems: Our maturity diagnostics and radar models translate organizational signals into actionable insights, identifying Growth, Risk, Kill, and Shift zones.
3. Publish Research & Methodologies: We codify best practices in enterprise design through multi-volume series and research playbooks, available in bookstores or on our website ai6x6.com.
4. Create Executive Tools: From RACI matrices to transformation roadmaps, we build high-utility models for immediate executive application. We have many accelerated off-the-shelf offerings on our website.
5. Support Global Organizations: Our frameworks empower enterprises, consulting teams, and innovation labs. Our advisory team worldwide including our Executive Mentor Network is here to help.

Meet the Founder & Executive Director : Subrata Mukherjee

ACCOMPLISHMENTS

Subrata Mukherjee has delivered enterprise-scale transformation across the Americas, Europe, and Asia, shaping modernization programs for some of the world's most influential organizations. His work spans AI/ML, automation, digital product strategy, enterprise architecture, and large-scale organizational transformation. He has led digital reinvention in global publishing, driven automation and AI adoption in aerospace and defense, modernized ratings and compliance processes in financial services, and guided product and data transformation across multiple industries. In 2016, he was recognized as one of Folio's Top 100 People in Publishing for his impact on digital innovation.

INDUSTRIES

His career spans aerospace and defense, financial services, publishing, healthcare, consulting, education, manufacturing, and technology services. He has delivered transformation in highly regulated environments, mission-critical engineering contexts, global media organizations, and fast-moving digital businesses. This cross-industry exposure gives him a multidimensional perspective on how AI reshapes strategy, operations, and organizational identity.

EXPERTISE

Subrata's expertise spans the full arc of enterprise transformation, beginning with early work in CAD/CAM for steel plants and smart-city infrastructure, which shaped his systems-thinking foundation. He then became a specialist in enterprise CRM and BPM transformation, leading large-scale modernization programs that re-architected customer operations, workflow orchestration, and enterprise process design. Over time, he transitioned into a strategist, product thinker, and transformation architect, guiding organizations on designing, adopting, and scaling next-generation digital capabilities. Today, he is recognized as a leading voice in AI and Emerging Technologies, with deep expertise across AI/ML, Generative AI, LLMs, RAG architectures, intelligent automation, hyper-automation, conversational AI, chatbots, agentic systems, and decision-automation frameworks. His work is rooted in strategy, product design, operating-model transformation, and enterprise-scale adoption—not coding or engineering execution—and focuses on translating emerging AI capabilities into practical, scalable, enterprise-grade systems that reshape how organizations operate and compete.

EDUCATION

Subrata holds a Master's in Technology Management from Columbia University, an MBA in Marketing from Jadavpur University, and a Business Communications certification from Harvard University. He also holds a Management by Values certification from the Indian Institute of Management, Calcutta, deepening his grounding in leadership behavior, organizational culture, and values-driven transformation. He is a Certified Product Owner. His academic foundation spans technology, strategy, communication, and organizational leadership, informing his multidisciplinary approach to AI-era transformation.

COMPANIES

Across his career, Subrata has delivered transformation programs for organizations including The Economist, Raytheon Technologies, S&P Global, Bloomberg, Wiley, Cognizant, Unisys, and Gujarat Ambuja Cement. His work has ranged from leading The Economist's digital transformation and modernizing Bloomberg's subscription and data infrastructure, to launching Raytheon's enterprise-wide Citizen Development program for Automation and AI, transforming the Dummies brand at Wiley, modernizing S&P's Ratings Process after the sub-prime mortgage crisis, scaling CRM and Automation/AI consulting portfolios at Cognizant and Unisys, and establishing the technology backbone and smart-building systems for the Gujarat Ambuja conglomerate.

THOUGHT LEADERSHIP

Subrata is a prolific speaker who has presented at more than 100 global conferences and events. His writing has appeared in CIO Magazine, and he served as a weekly technology columnist for The Telegraph. Through the Ai6x6 Research Lab, he publishes frameworks, diagnostics, and transformation blueprints that help organizations navigate the AI economy. In 2026–27, he will launch the Executive Mentor Network to develop a global community of AI Strategy–certified coaches and expand AI literacy among senior leaders. Beyond his professional work, he is an avid traveler who has explored more than 40 countries and an accomplished portrait artist.

www.linkedin.com/in/sm360

1. GIVE **VALUE**, REPEATEDLY, BECOME A GOTO BOOK

   The 18-book series brings in a lot of content from a practitioner perspective. The advice we are providing have worked for us in our past projects across industries; so we are preaching what we practiced before

2. BE **BRIEF**, PROVIDE CONTENT IN SMALL NUGGETS

   I have never read a book cover to cover. I have short attention span, and more and more people are suffering from the same syndrome. Hence each micro-topic we have introduced is covered in two to three pages

3. BE **COMPREHENSIVE** FOR REAL WORLD IMPACT

   Transformation is not easy. In the real world there are a lot of inter-connected challenges. We have proposed detailed frameworks that will connect the dots and provide comprehensive solutions to your team/client

4. BE **PRACTICAL**, PROVIDE EXAMPLES TO DEMYSTIFY

   Most companies and leaders know they and their companies need to transform, but don't know where to start. I wanted to provide a step-by-step guide or a MOSCOW priority list on getting started

5. PUSH **BOUNDARIES** AND DARE LEADERS TO DREAM

   Real transformation is not tinkering around the edges and playing safe; its often about self disrupting and pushing boundaries and reinventing oneself. Applies to individuals, teams and organizations.

6. PUSH LEADERS TO **INTROSPECT** & ASK QUESTIONS

   Since the world around us is moving so fast, leaders often have a fear of been exposed about their lack of knowledge and ask "stupid" questions. These book educates you enough, so you can ask intelligent questions

7. BE **EXCITING**, DON'T BE BORING

   I have tried to keep the book interesting, small text, lot of structured tables, quite a few images, bold formatting, case studies, Tips and Tricks, Interviews, RACI matrix, buying guides, assignment questions etc.

8. BRING **OTHERS**, HEAR FROM MY COMMUNITY

   Tech sector worldwide is very collaborative. We learn from each other every day and we use the knowledge to impact lives positively. This book attempts to bring a few voices from my network and hear their perspective

9. PROMOTE **AI ECONOMY**, USE AI TOOLS YOURSELF

   Generative AI has been used extensively for formatting content in a repeatable format & page size control. Without GenAI such an extensive book series would not have been possible in a short time.

10. BE **DIFFERENT**, CHARTER YOUR OWN COURSE

    Most management self-help books are pages and pages of long text with guidelines provided by publishers that are a bin antiquated. Modern learning happens byte sized and is multi-modal and multi-channel..

The **Guided Navigation for Transformation** framework is a comprehensive, end-to-end methodology designed to de-risk and accelerate AI adoption within an organization. It moves away from "random acts of digital" by providing a disciplined, three-stage approach that balances human capability with technical execution. The core of the system is the **6x6 Lens Sequence**, which ensures that every strategic move is evaluated through six critical dimensions: **Leadership, Workforce, Data, Customer, Product, and Operations**.

By applying these lenses consistently across the Blueprint, Reveal, and Ignite phases, the framework creates a "Golden Thread" that links high-level strategy to on-the-ground deliverables. This is not just a consulting model, but an educational ecosystem with an 18-book series that ensures that AI transformation is both measurable and sustainable.

"Guided Navigation for Transformation" in the AI Era

We provide guided navigation through three distinct phases: First, we establish your **Transformation Blueprint** to align skills. Second, we run the **Reveal Diagnostic** to assess your current maturity. Finally, we execute the **Consulting Playbook** to turn the 'Art of the Possible' into a 'Science of Deliverables'.

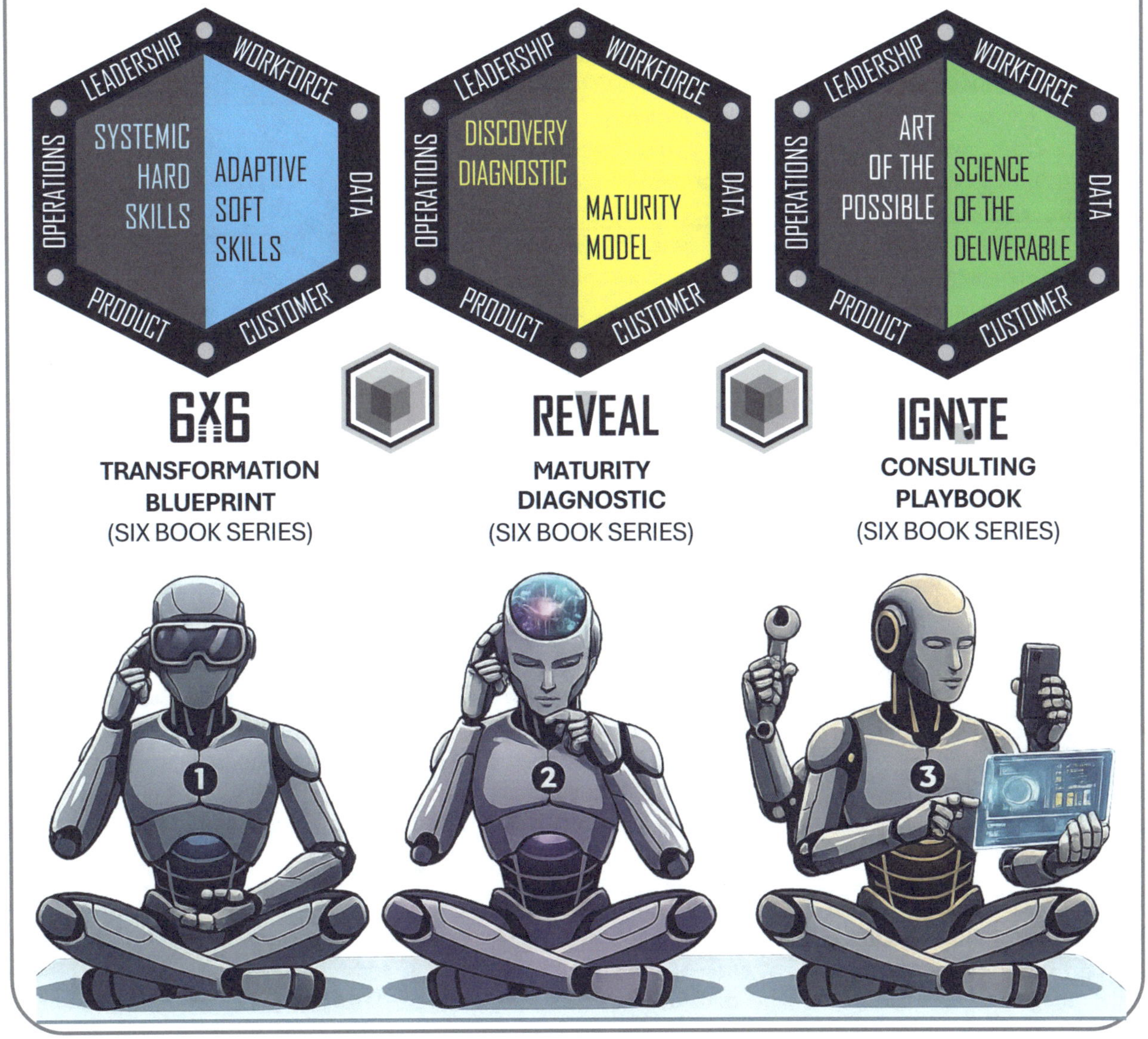

The **AI6x6 Transformation Engine™** is the flagship construct of the AI6x6 Research Lab—a precision-engineered ecosystem that serves as the definitive "Operating System" for the modern enterprise. Built on a proprietary 6×6 geometry, it unifies architecture, execution, and calibration into a single, high-performance unit. The Engine provides leaders with the high-torque machinery and **Guided Navigation** required to convert organizational inertia into scalable momentum in the era of autonomous systems.

The **Transformation Blueprint Series** defines the engine's architecture and establishes the structural logic of the entire system. By introducing the **Six Pillars** and **Six Lenses**, it forms the foundational geometry of enterprise modernization. Written for CEOs, Boards, and Strategy Leaders, this series acts as the **Navigation Deck**, providing the "Architectural North Star." It ensures that every move is strategically calibrated and ethically anchored before the gears are engaged, offering a shared mental model for aligning leadership teams. This intelligence layer removes the guesswork from transformation, providing a rigid yet adaptive framework that allows organizations to visualize their future state with absolute clarity. It is the vital starting point where complex, high-stakes ambition is codified into a disciplined, institutional reality that can withstand the pressures of rapid technological shift.

The **Maturity Diagnostic Series** provides the engine's accountability and calibration system, ensuring peak performance through proprietary **REVEAL ™** methodology. . Acting as the **Sensor-Driven Intake**, it utilizes 36 diagnostic cylinders to quantify organizational readiness and prioritize investments based on evidence rather than intuition. This series allows executive teams to identify hidden gaps and measure progress with scientific clarity, functioning as the **Measuring Layer** of the entire ecosystem. It ensures the engine remains tuned for resilience through a data-driven maturity model that tracks evolution over time. By providing rigorous assessment instruments and scoring logic, it offers the accountability required to sustain a long-term transformation journey. This layer acts as the engine's nervous system, constantly scanning the internal and external environment to calibrate the Blueprint's trajectory and ensure that every unit of energy spent drives the organization toward sustained excellence and institutional stability.

The **Consulting Playbook Series** translates architecture into action by operationalizing the engine through the proprietary **IGNITE™** methodology. It serves as the **Heavy-Duty Gearbox**, providing the repeatable consulting patterns, execution models, and directional force required to activate transformation across diverse domains. Designed for PMOs and delivery leads, these playbooks convert high-level strategy into tangible enterprise-scale outcomes, ensuring that the "torque" of the engine is effectively applied to the road. They function as the essential doing layer, reducing internal friction and shifting the organization from abstract concept to deep-seated conviction. By providing step-by-step pathways and intervention models, the series bridges the gap between executive intent and operational reality. This transmission layer ensures that momentum is not just created, but sustained and scaled, allowing the enterprise to accelerate through complex transitions with precision and measurable impact.

The **Transformation Engine** is built for scale across the global economy. Future **Industry Evolution** volumes will apply this 6×6 engine to specific sectors—such as Healthcare, Banking, and Manufacturing—revealing how the model drives specialized, high-velocity transformation. By adapting the Pillars and Lenses to sector-specific demands, these extensions ensure the Engine remains the definitive standard for navigating the complexities of the age of intelligent systems.

SERIES 1 6 BOOKS

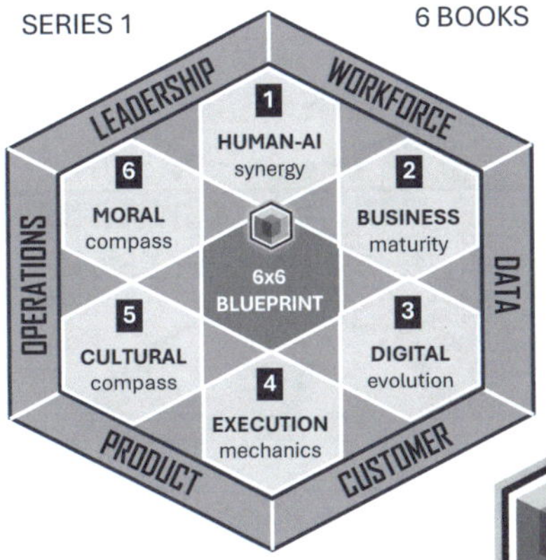

SERIES 2 6 BOOKS

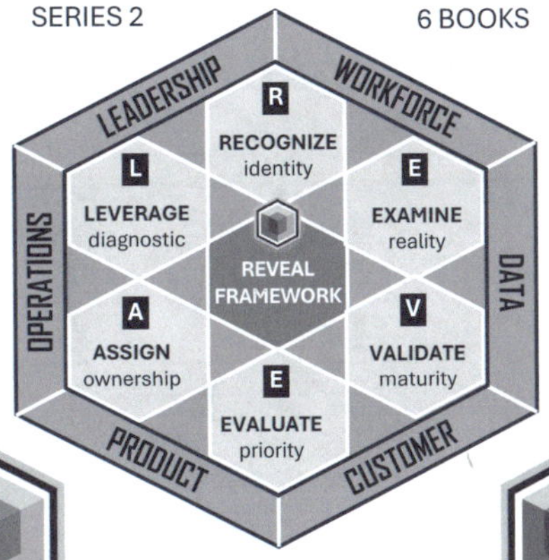

SERIES 3 6 BOOKS

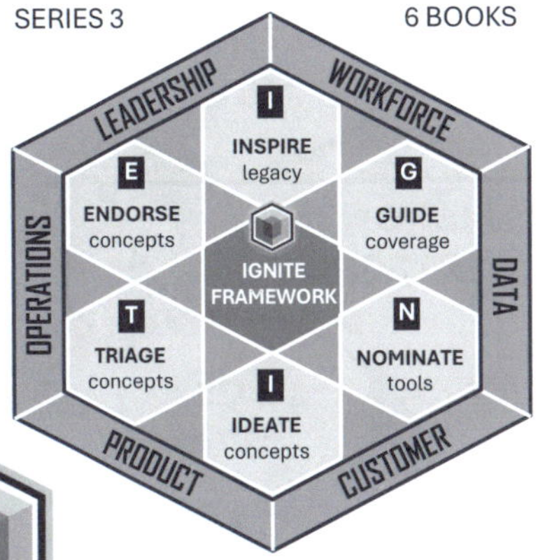

6X6 TRANSFORMATION BLUEPRINT

REVEAL MATURITY DIAGNOSTIC

IGNITE CONSULTING PLAYBOOK

see **think** **act**

The Ai6x6 palette is a semantic system, not a decorative choice. Each color encodes the emotional and operational signature of its lens: **pink** captures the human spark that ignites potential; **green** reflects the strategic discipline required for maturity; **purple** signals the depth and computational power of digital evolution; **orange** embodies the kinetic urgency of execution; **blue** anchors the cultural soul in trust and cohesion; and **yellow** illuminates the ethical clarity of the moral audit. Together, these colors form a coherent visual language that guides the reader through the transformation cycle with intuitive, emotionally aligned cues.

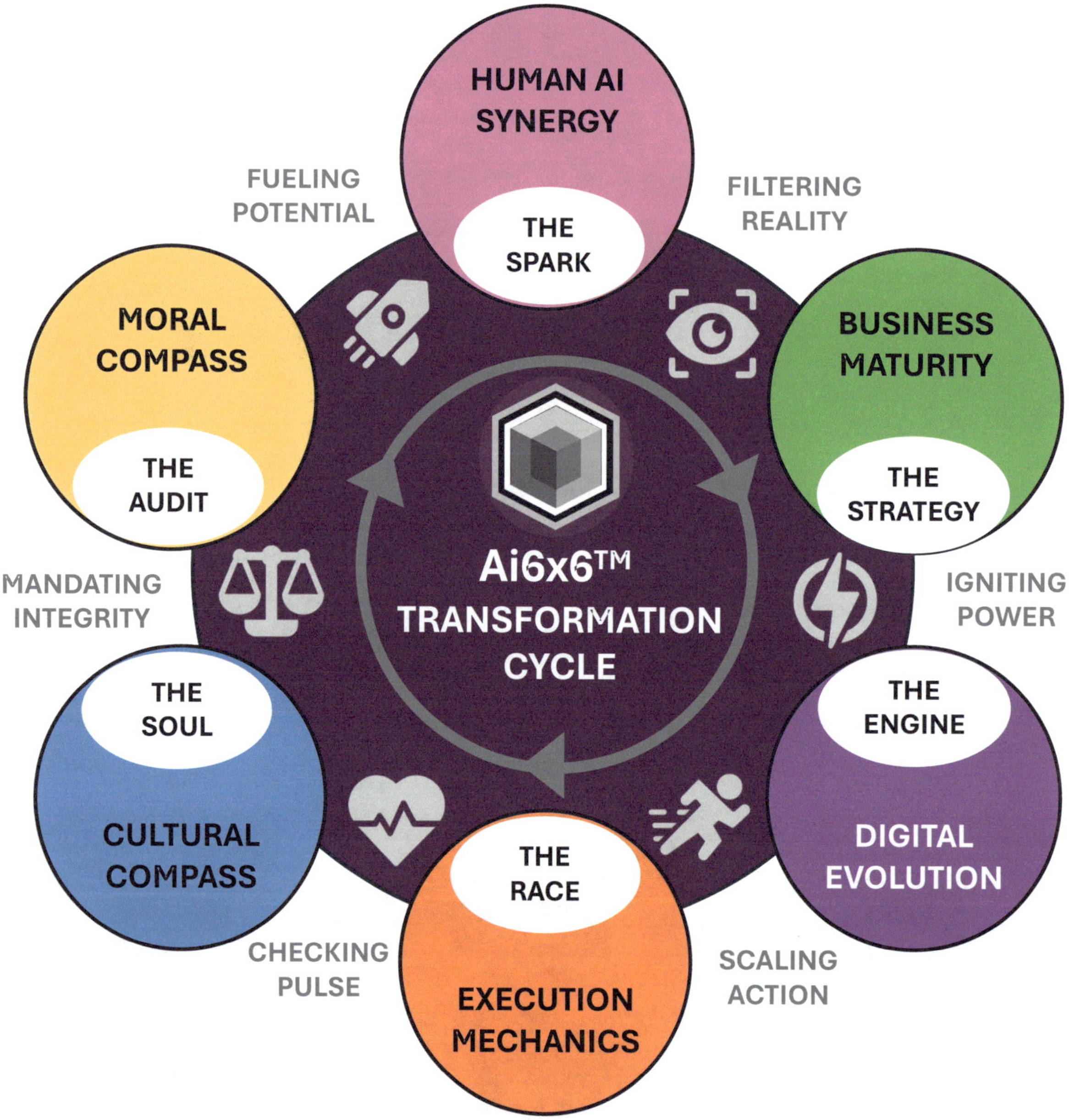

The **COLOR TABS ALONG THE PAGE EDGES IN THE MIDDLE** serve as the physical navigation system of the Ai6x6 series. Each tab mirrors the chromatic identity of its lens, allowing readers to locate themselves instantly within the transformation cycle. This creates a spatial reading experience where orientation is continuous and intuitive. The tabs turn our book into a modular interface—one where Human–AI Synergy, Business Maturity, Digital Evolution, Execution Mechanics, Cultural Compass, and Moral Compass become distinct territories with their own visual signatures. The result is a playbook that reads like a designed system, not a linear text.

6x6™ Transformation Blueprint

The Prologue to a Purpose

"The only way to go forward is to let go of the past."
The Myth of Orpheus and Eurydice

The tragic lesson of this myth is a cautionary tale for business. Orpheus's failure to trust his journey and looking back results in losing everything, underscoring the importance of fully committing to a forward-looking transformation.

The mission of this book series is simple: to transform leaders from passive observers of the AI Era into its active pioneers. We will ignite your thinking with exciting, highly practical content that pushes boundaries and provides the essential tools and frameworks you need to build real-world impact. We won't just tell you what to do; we'll show you how, using a diverse community's experiences and real-world examples that bring the abstract to life. This isn't a passive read; it's an active partnership and a dynamic dialogue designed to turn your natural curiosity into a definitive competitive advantage. This book is the first of a six-part series : Leadership Transformation; which governs your journey on how you should lead from the top and how you should inspire others to take on the AI Era and excel in it.

The **AI6x6 Transformation Engine** is a precision-engineered ecosystem designed to deliver **Guided Navigation** for enterprises facing high-stakes digital evolution. By integrating three proprietary systems—the **Transformation Blueprint**, the **Consulting Playbook**, and the **Maturity Diagnostic**—the Engine provides the robust "Operating System" required to dominate the AI era.

The architecture begins with the **Transformation Blueprint**, the Engine's intelligence layer. Here, the system utilizes **Six Lenses**—including the Cultural and Moral Compasses—to calibrate your trajectory and codify a master architectural plan. This strategic map then engages the **Consulting Playbook**, where the gears of the **IGNITE** methodology translate executive intent into immediate directional force through rapid-response execution cycles. Finally, the **Maturity Diagnostic** acts as a sensor-driven intake, utilizing six diagnostic cylinders to compress raw organizational data into high-pressure strategic insights that validate the transformation's impact. Bolted to a rigid **Chassis of Six Pillars**, the Engine ensures the enterprise remains resilient, balanced, and unshakeable even at peak acceleration. It is more than a strategy; it is a proprietary architecture for sustained institutional excellence.

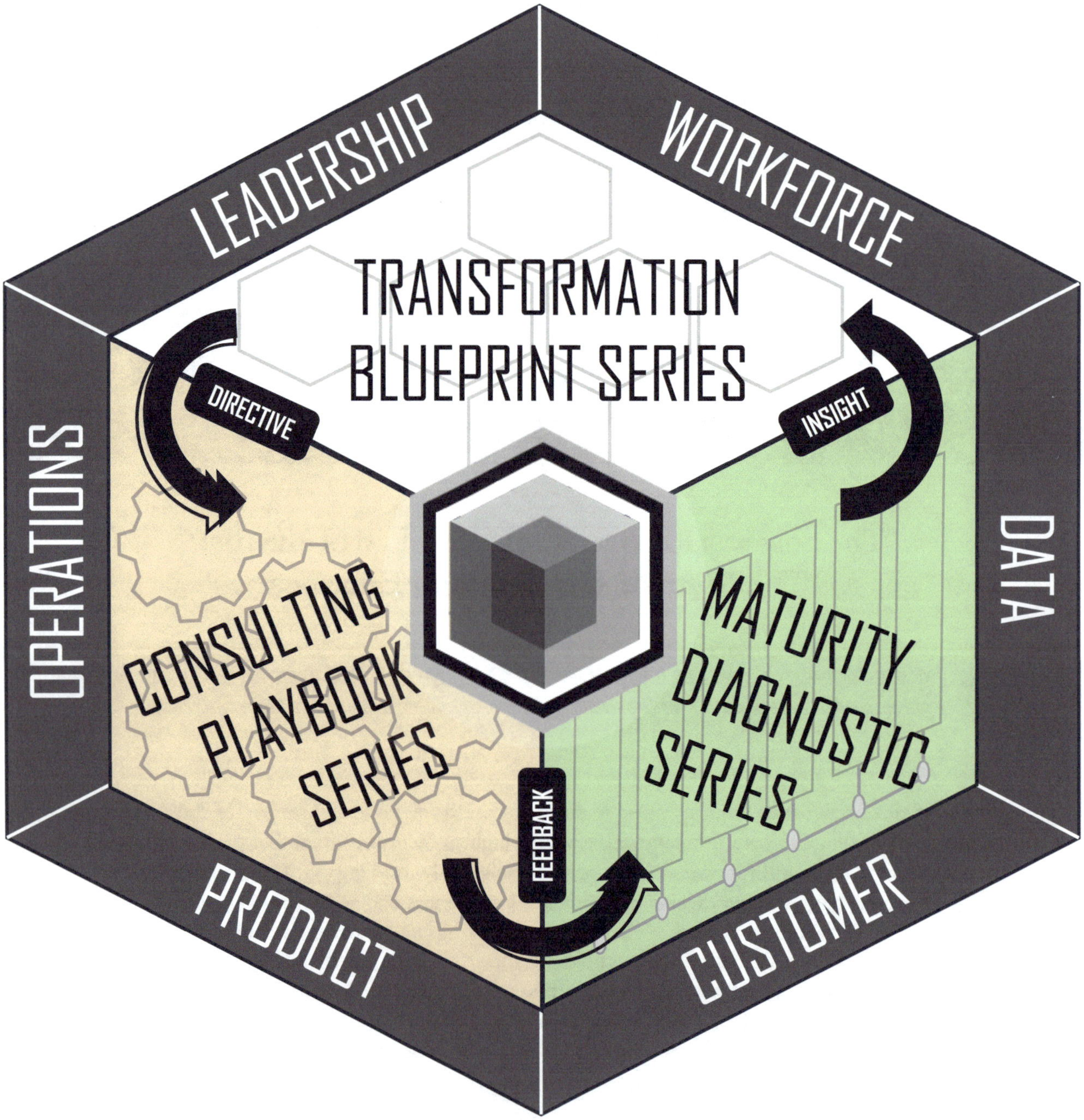

The AI6x6 Transformation Engine is a precision-engineered ecosystem that replaces guesswork with **Guided Navigation**. It is designed to convert organizational inertia into scalable momentum through a disciplined, mechanical approach to digital evolution.

CHASSIS (SIX PILLARS)
The entire system is anchored to a rigid structural frame. This chassis ensures that even as you reach peak acceleration, your enterprise remains resilient and unshakeable.

NAVIGATION DECK (SIX LENSES):
This is the intelligence layer. By applying our **Six Lenses**, we provide the **Guided Navigation** required to calibrate your trajectory, preventing costly drifts in strategy.

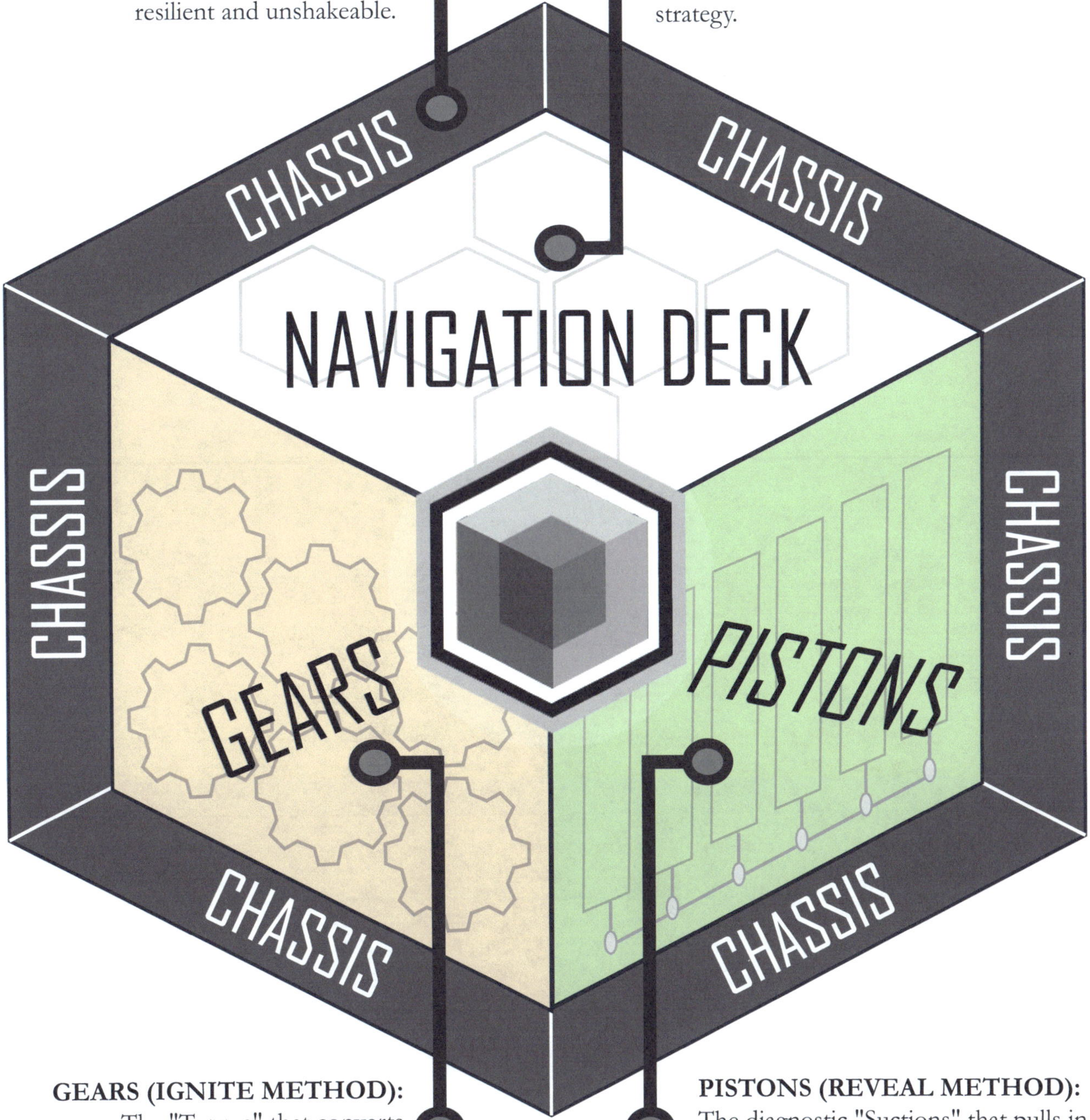

GEARS (IGNITE METHOD):
The "Torque" that converts diagnostic energy into immediate directional force through pilot programs and playbooks.

PISTONS (REVEAL METHOD):
The diagnostic "Suctions" that pulls in raw organizational data and compresses it into high-pressure actionable insights.

A car works best when its actions are coordinated through systems that generate power, transmit motion, and guide direction. Each component plays a precise mechanical role that keeps the vehicle stable, responsive, and moving with purpose.

CHASSIS
Supports vehicle structure, absorbs road stress, anchors mechanical systems, maintains stability, and ensures safe, balanced handling across varying speeds, terrains, and driving conditions.

NAVIGATION DECK
Displays real-time driving information, maps, routes, and alerts, helping the driver maintain direction, anticipate conditions, and make confident, informed decisions.

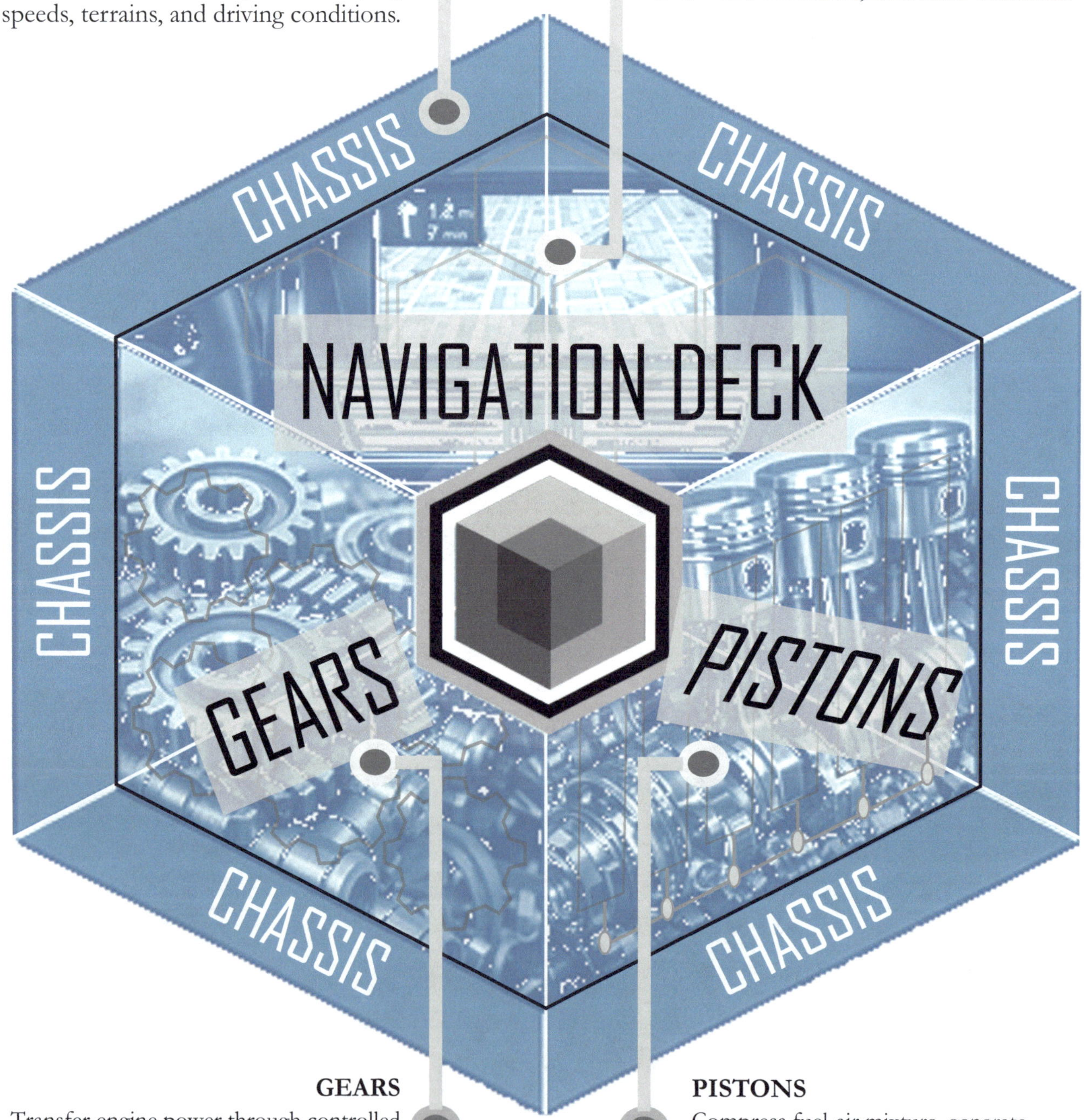

GEARS
Transfer engine power through controlled ratios, regulating speed and torque, ensuring smooth acceleration, efficient performance, and seamless coordination between mechanical components.

PISTONS
Compress fuel-air mixture, generate combustion force, drive the crankshaft, & convert explosive energy into continuous motion powering the vehicle's entire drivetrain system.

The **AI6x6 Transformation Engine** replaces the ambiguity of traditional consulting with mechanical certainty, converting organizational inertia into scalable momentum. By integrating three proprietary systems—the **REVEAL** diagnostic intake, the **IGNITE** consulting transmission, and the **6x6 Blueprint** navigation deck—the Engine provides the robust "Operating System" required to dominate the AI era. The **REVEAL** methodology acts as a sensor-driven intake, utilizing 6x6 diagnostic cylinders to compress raw organizational data into high-pressure strategic insights. This "data fuel" then powers the **IGNITE** phase, where consulting playbooks serve as a heavy-duty gearbox, translating executive intent into immediate directional force through rapid-response execution cycles. This entire power plant is governed by the **6x6 Blueprint**, the intelligence layer that uses **Six Lenses**—including the Cultural and Moral Compasses—to calibrate your trajectory and ensure ethical, operational alignment. Bolted to a rigid **Chassis** of **Six Pillars**, the Engine ensures the enterprise remains resilient, balanced, and unshakeable even at peak acceleration. It is more than a strategy; it is a proprietary architecture for sustained institutional excellence.

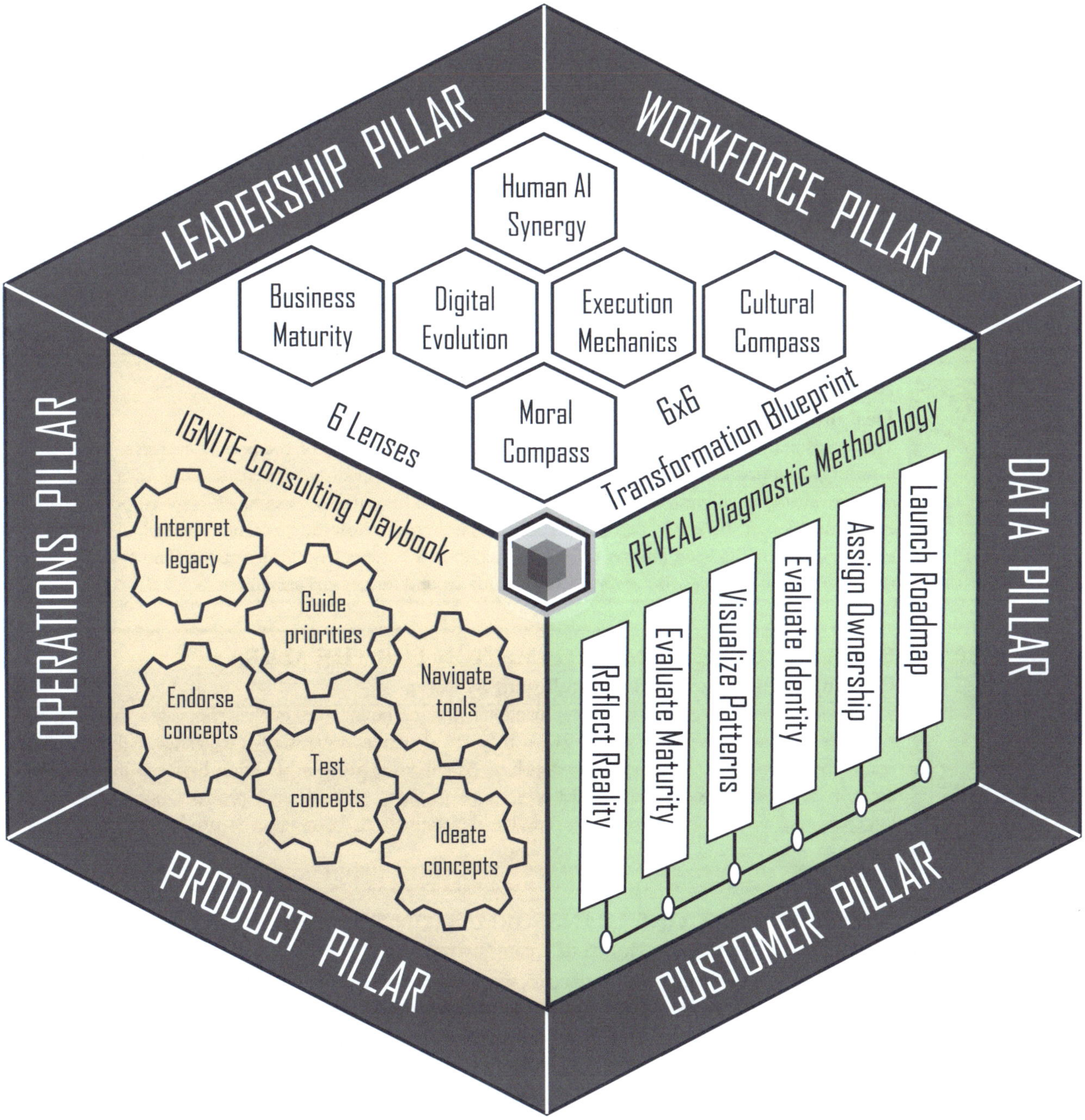

6 BOOKS OF THE Ai6x6 TRANSFORMATION BLUEPRINT SERIES

BOOK 1: LEADERSHIP TRANSFORMATION FOR THE AI ERA

Designing the Vision, Not Just Directing It

The Leadership book reframes what it means to lead in an era defined by intelligence, acceleration, and ambiguity. It explores how leaders think, decide, communicate, and shape meaning across the enterprise, emphasizing that leadership is not a role but an architecture. The narrative examines how leaders make decisions systemically, exercise judgment in an AI-augmented world, create alignment through narrative clarity, and uphold ethical stewardship as the foundation of trust. This book sets the tone for the entire series by positioning leadership as the integrator of all transformation forces.

BOOK 2: WORKFORCE TRANSFORMATION FOR THE AI ERA

Elevating Human Capability in a Technology-Driven Era

The Workforce book examines the human system of the enterprise — the skills, roles, mindsets, structures, and experiences that determine how people show up to the work of transformation. It explores how organizations support people as AI reshapes every function, and how leaders build environments where individuals feel empowered rather than displaced. The narrative dives into the evolution of adaptive talent ecosystems, the emergence of AI-augmented roles, the importance of team cohesion and psychological safety, and the recognition of human capability as a strategic asset.

BOOK 3: DATA TRANSFORMATION FOR THE AI ERA

Building the Intelligence Fabric of the Enterprise

The Data book establishes the informational backbone of the organization. It explores how data is collected, governed, interpreted, and transformed into intelligence, emphasizing that data is no longer a technical asset but the raw material of decision-making, personalization, automation, and innovation. The narrative examines how trustworthy data governance is built, how intelligence flows across systems, how data architectures are designed, and how data becomes a source of enterprise-level advantage. This book gives leaders the clarity to build an intelligent, insight-driven organization.

BOOK 4: CUSTOMER TRANSFORMATION FOR THE AI ERA

Designing Experiences That Earn Trust and Loyalty

The Customer book anchors the enterprise in the lived experience of the people it serves. It explores how organizations understand, anticipate, and respond to customer needs — not as a marketing exercise but as a cultural and operational discipline. The narrative examines how emotion shapes experience design, how AI enables personalization at scale, how transparency strengthens customer trust, and how end-to-end journey coherence becomes a competitive differentiator. This book ensures the organization evolves around the customer, not around its internal structures.

BOOK 5: PRODUCT TRANSFORMATION FOR THE AI ERA

Creating Value in a World of Intelligent Systems

The Product book explores how organizations design, build, and evolve the value they deliver. Products are no longer static offerings — they are dynamic systems of intelligence, behavior, and continuous adaptation. This book examines how AI enhances product behavior, how continuous value creation replaces episodic feature delivery, how product strategy and product operations must integrate, and how digital ecosystems redefine differentiation. It transforms product thinking from output to outcome, and from features to value.

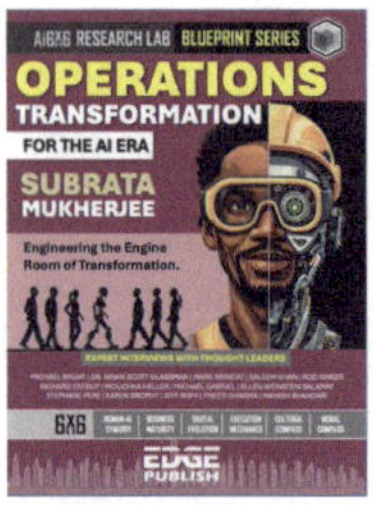

BOOK 6: OPERATIONS TRANSFORMATION FOR THE AI ERA

Engineering the Engine Room of Transformation

The Operations book addresses the systems, workflows, and processes that determine whether strategy becomes reality. It explores how organizations orchestrate complexity, manage risk, and maintain resilience in environments defined by acceleration and uncertainty. The narrative examines how intelligent workflows are designed, how operational resilience is built, how AI optimizes processes, and how execution scales with precision. This is the book where transformation becomes tangible.

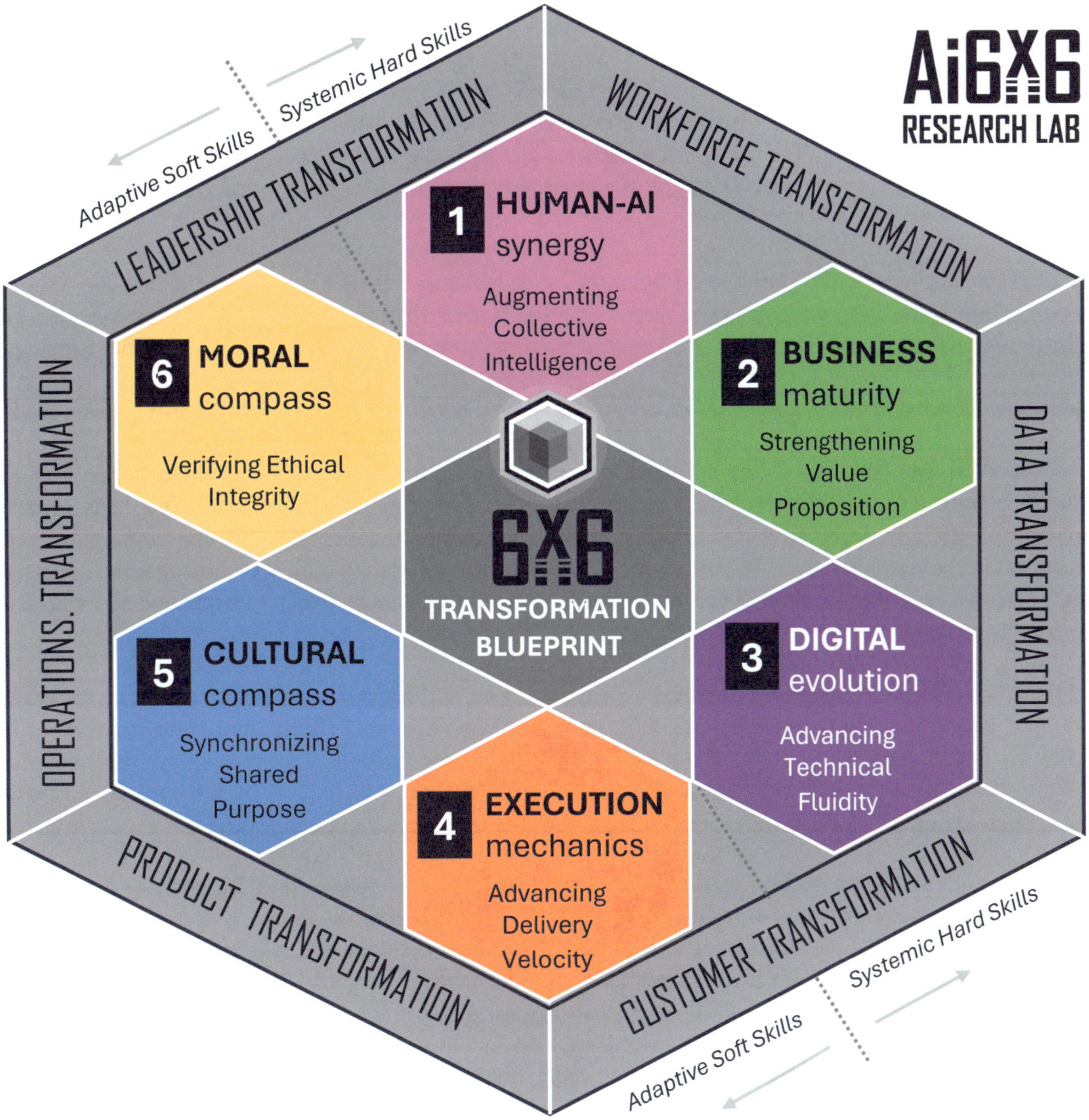

The **Ai6x6 TRANFORMATION BLUEPRINT™** provides the multidimensional structural integrity required for sustainable, high-stakes evolution. Serving as the **Navigation Deck** and **Chassis** of the Transformation Engine, the BLUEPRINT™ moves beyond episodic change to establish a permanent, scalable "Operating System" for the enterprise. The architecture is defined by the intersection of **Six Pillars** and **Six Lenses**. The Pillars—Leadership, Workforce, Data, Operations, Product, and Customer—provide the horizontal breadth and load-bearing strength of the organization. These are bisected vertically by the Six Lenses to reveal the true depth of the enterprise. Leaders apply **Human–AI Synergy** to augment collective intelligence and **Business Maturity** to instill systemic discipline. The **Digital Evolution** lens ensures technical fluidity, while **Execution Mechanics** masters the art of precision delivery. This is anchored by the **Cultural Compass**, which synchronizes shared purpose, and the **Moral Compass**, which verifies ethical integrity.

By aligning these twelve dimensions, the BLUEPRINT™ eliminates strategic drift and structural fragmentation. It ensures that every initiative is not just an isolated project, but a calibrated component of a unified, resilient, and unshakeable corporate architecture. Also remember to follow the Tab colors on the edges to navigate quickly.

The **Ai6x6 Transformation Blueprint™** gives leaders a multidimensional, architecture-grade view of how transformation behaves inside their enterprise. Where REVEAL™ diagnoses the present and IGNITE™ accelerates the future, the Blueprint defines the *structural truth* of the organization. Built on six proprietary Lenses, it converts fragmented observations, cultural noise, and operational ambiguity into a coherent, systems-level understanding. The Blueprint equips leaders with the clarity, depth, and structural intelligence required before designing any transformation roadmap, ensuring decisions are grounded in reality rather than aspiration.

The Blueprint begins with the **Human–AI Synergy Lens**, which examines how people and intelligent systems interact across workflows, decisions, and responsibilities. Rather than simply measuring adoption, this Lens analyzes augmentation, displacement risk, trust dynamics, and readiness for human-machine teaming. It distinguishes where AI enhances capability, where it creates friction, and where new operating models are required. Leaders begin with a grounded understanding of how human and machine roles must evolve together.

Next, the **Business Maturity Lens** evaluates the enterprise's structural readiness for transformation. This includes governance, financial discipline, decision velocity, and leadership alignment. The Lens separates foundational strength from structural fragility, revealing whether the organization can support sustained transformation without collapse. It clarifies which capabilities are stable, which require reinforcement, and which must be redesigned before acceleration. Business Maturity ensures leaders understand the chassis on which transformation will operate.

The third Lens, **Digital Evolution**, maps the depth, coherence, and scalability of the digital ecosystem. Leaders often face fragmented platforms, technical debt, and integration gaps that silently undermine transformation. Digital Evolution cuts through this complexity by assessing architecture, interoperability, data flow, and platform extensibility. It clarifies which digital foundations can scale, which must be modernized, and which create systemic risk. This Lens ensures technology decisions are grounded in architectural truth.

With structural and digital realities established, leaders move into the **Execution Mechanics Lens**, which examines how strategy becomes movement. This Lens evaluates processes, rhythms, accountability systems, and operational precision. It identifies where execution is consistent, where it is brittle, and where it breaks under pressure. Execution Mechanics ensures leaders understand the organization's true capacity to deliver, adapt, and sustain momentum across complex transformation cycles.

The fifth Lens, **Cultural Compass**, explores the beliefs, behaviors, incentives, and informal power structures that shape how people respond to change. Culture is often the silent accelerator or the invisible brake. This Lens reveals alignment, resistance patterns, psychological safety, and leadership influence networks. It distinguishes cultural strengths from cultural liabilities, giving leaders a clear view of the human forces that determine whether transformation will be embraced or rejected.

The final Lens, **Moral Compass**, evaluates ethics, trust, transparency, and responsible AI principles. As organizations adopt intelligent systems, the stakes of ethical decision-making rise dramatically. This Lens ensures transformation is not only effective but also equitable, accountable, and aligned with long-term societal and organizational values. It clarifies where safeguards are strong, where risks exist, and where governance must evolve.

Together, these Six Lenses form a complete, repeatable, consulting-grade architectural framework. The Ai6x6 Transformation Blueprint™ is not a checklist or maturity model; it is a structural intelligence system that moves leaders from assumption to understanding, from understanding to design, and from design to confident transformation planning. It gives leaders the conviction to move forward — not because the future is predictable, but because their decisions are grounded in multidimensional truth, structural clarity, and a deep understanding of how transformation behaves across the enterprise.

Ai6x6™ Transformation Blueprint

Understand Pillars & Lenses

"We don't see things as they are, we see them as we are." — Anaïs Nin

The **Ai6x6 TRANSFORMATION BLUEPRINT** is fundamentally structured around two core concepts: **Pillars and Lenses**. Together, they create a dual-axis architecture that brings both breadth and depth to enterprise transformation. The Pillars provide the WIDTH of the framework, defining the major domains where transformation must take root for the organization to evolve coherently. These domains—Customer, Product, Operations, Data, Workforce, and Leadership—represent the full span of enterprise behavior. Each Pillar receives its own dedicated book, allowing leaders to explore the complexities, tensions, and opportunities of that domain without dilution or compromise.

This ensures that transformation is not treated as a single initiative but as a system of interconnected disciplines that must advance in parallel. Complementing this horizontal structure is the DEPTH created by the Six Lenses, the universal imperatives that cut vertically through every Pillar. These Lenses represent the non-negotiable dimensions of modern transformation—forces that must be activated in every domain, regardless of industry, maturity, or strategic ambition. The six Lenses are: Human–AI Synergy (L1), Business Maturity (L2), Digital Evolution (L3), Execution Mechanics (L4), Cultural Compass (L5), and Moral Compass (L6). Each Lens exposes a different layer of organizational capability, ensuring that no domain is approached superficially, and no critical dimension is overlooked. The goal of the blueprint is to be holistic and very comprehensive so that executives are not blind-sighted by a key aspects being overlooked. Since a lot of ground has to be covered, the books may look very dense, but I have tried my best to keep it brief & engaging.

Transformation has become one of the most OVERUSED words in modern leadership, yet one of the least understood. Organizations declare it, teams attempt it, consultants package it—but few enterprises ever truly achieve it. Not because they lack ambition or intelligence, but because they approach transformation as a sequence of disconnected initiatives rather than a shift in how the organization thinks, behaves, and evolves. I created the **Ai6x6 Transformation Blueprint** because I have seen the same pattern repeat across industries, functions, and eras of technology: transformation fails not at the point of strategy, but at the point of structure. It fails when organizations modernize technology but ignore culture. It fails when they redesign processes but overlook trust. It fails when they invest in AI but fail to prepare people. It fails when they build strategy but cannot execute it. It fails when leaders underestimate the complexity of aligning human capability with machine intelligence.

The Ai6x6 Transformation Blueprint exists to solve this fragmentation. It brings together **six foundational lenses** and **six transformation pillars** into a single, coherent architecture. Each lens represents a dimension of capability that every organization must master. Each pillar represents a domain— Leadership, Workforce, Data, Customer, Product, Operations —where transformation must take root. Together, they form a 6×6 matrix that reveals not only what must change, but how change becomes sustainable, repeatable, and scalable.

This blueprint is intentionally multidimensional. Transformation is not linear. It is not sequential. It is not a checklist. It is a living system—human, technical, cultural, operational, ethical—moving together or not at all. The Ai6x6 model acknowledges this complexity and gives leaders a structure strong enough to hold it.

- It is called a blueprint because it is both **architectural** and **actionable**.
- This blueprint is not a methodology per se. It is not a maturity model in the traditional sense. It is not a checklist as you commonly see it. It is a **HOLISTIC** way of seeing things that are often overlooked.
- It is called Ai6x6 because it integrates **AI-driven intelligence** with **six human-centered disciplines**, applied across **six transformation pillars**.
- It is a system designed for organizations that want to evolve with intention rather than react with urgency.

The Six Lenses emerged from a simple but unavoidable truth: "**no single discipline can carry transformation on its own**". Not AI. Not culture. Not governance. Not execution. Not digital modernization. Not ethics. Each is necessary, but none is sufficient. Real transformation requires all six lenses working together—reinforcing one another, correcting one another, and elevating one another.

It reveals the forces that shape organizational behavior, the tensions that undermine progress, and the capabilities that determine whether transformation becomes sustainable or collapses under its own ambition. It provides leaders with a structural map—one that makes complexity navigable and makes evolution repeatable.

Each lens represents a dimension of organizational life:

1. Human–AI Synergy: how intelligence is created, shared, and amplified.
2. Business Maturity: how discipline, readiness, and structure turn ambition into capability.
3. Digital Evolution: how technology becomes fluid, integrated, and trustworthy.
4. Execution Mechanics: how strategy becomes action and action becomes advantage.
5. Cultural Compass: how beliefs, behaviors, and meaning shape the human system.
6. Moral Compass: how ethics, transparency, and responsibility anchor everything else.

Together, they form a complete operating philosophy for modern transformation—one that is as human as it is technical, as strategic as it is operational, as aspirational as it is grounded. The Ai6x6 Transformation Blueprint is not about predicting the future. It is about preparing organizations to shape it. It is about giving leaders a language for what they instinctively know but rarely articulate. It is about building enterprises that can evolve continuously, coherently, and confidently. Most of all, it is about honoring the truth that **transformation is not something organizations *do*—it is something they *become*.**

The Pillars of the **Ai6x6 Transformation Blueprint** define the WIDTH of enterprise evolution. They represent the six domains where transformation must become real, measurable, and operationally grounded. While the Lenses provide the depth of capability, the Pillars provide the breadth of application—ensuring that transformation is not isolated to a single function but becomes a structural shift across the entire enterprise. Each Pillar receives its own dedicated book, allowing leaders to confront the full complexity of that domain with precision and without compromise.

The Leadership Pillar sits at the top of the architecture because leadership is the force that shapes every other domain. It defines how decisions are made, how priorities are set, how accountability is distributed, and how meaning is communicated. This Pillar examines the evolution of leadership in an era defined by AI, acceleration, and ambiguity. It challenges leaders to move from managing activities to designing systems, from directing work to enabling capability, and from protecting the past to shaping the future. Leadership is not a separate function—it is the integrator of the entire Blueprint.

The Workforce Pillar focuses on the human system that powers the enterprise. It explores how roles evolve, how teams collaborate, how skills are developed, and how people experience their work. In a world where AI augments every function, the Workforce Pillar examines how organizations support human capability, build adaptive talent ecosystems, and create environments where people feel empowered rather than displaced. This Pillar forces leaders to confront the truth that transformation succeeds or fails based on whether people feel equipped, supported, and connected to the mission.

The Data Pillar establishes the informational foundation of the enterprise. It examines how data is collected, governed, interpreted, and transformed into intelligence. In an AI-driven world, data is no longer a byproduct of operations—it is the raw material of decision-making, personalization, automation, and innovation. This Pillar explores how organizations build trustworthy data ecosystems, eliminate silos, and ensure that intelligence flows across the enterprise. It challenges leaders to treat data not as a technical asset but as a strategic capability that shapes every other Pillar.

The Customer Pillar anchors the organization in the lived experience of the people it serves. It forces leaders to confront the reality that customer expectations evolve faster than internal processes, and that loyalty is earned through coherence, trust, and emotional resonance. This Pillar examines how organizations understand, anticipate, and respond to customer needs—not as a marketing exercise, but as a cultural and operational discipline. It challenges leaders to build enterprises where customer experience is not a department, but a shared responsibility embedded in every decision.

The Product Pillar focuses on how organizations design, build, and evolve the value they deliver. Products are no longer static offerings—they are dynamic systems of intelligence, behavior, and continuous adaptation. This Pillar explores how product strategy, product design, and product operations must work together to create experiences that feel alive and responsive. It examines how AI reshapes product behavior, how digital ecosystems redefine differentiation, and how organizations must shift from feature delivery to continuous value creation.

The Operations Pillar addresses the engine room of the enterprise—the systems, workflows, and processes that determine whether strategy becomes reality. It examines how organizations orchestrate complexity, manage risk, and maintain resilience in environments defined by acceleration and uncertainty. This Pillar forces leaders to recognize that operational excellence is not about efficiency alone; it is about adaptability, intelligence, and the ability to respond to change without losing coherence. Operations is where transformation becomes tangible.

Together, these six Pillars form the structural WIDTH of the Ai6x6 Transformation Blueprint. They define the terrain on which the Six Lenses operate, ensuring that transformation is comprehensive, coherent, and enterprise-wide.

4
Earned relationships
CUSTOMER
"Who we serve?"
5
Brand Experience
PRODUCT
"What we serve?"
6
Engine room of service
OPERATIONS
"How we serve?"
2
Human Capability
WORKFORCE
"Human Assets"
3
Information foundation
DATA
"Insights"
1
Vision, mission & agenda
LEADERSHIP
"C-Suite"
Each Pillar = "Width" of the Transformation

4
Discerning Diners
CUSTOMER
"The Patrons"
5
Food, Drinks, Ambiance
PRODUCT
"The Feast"
6
Ingredients, Purchase
OPERATIONS
"The Logistics"
2
Cooks, Stewards, Servers
WORKFORCE
"The other staff"
3
Menu, Recipe & Reviews
DATA
"The Menu"
1
Lead Culinary aspect
LEADERSHIP
"The Exec Chef"

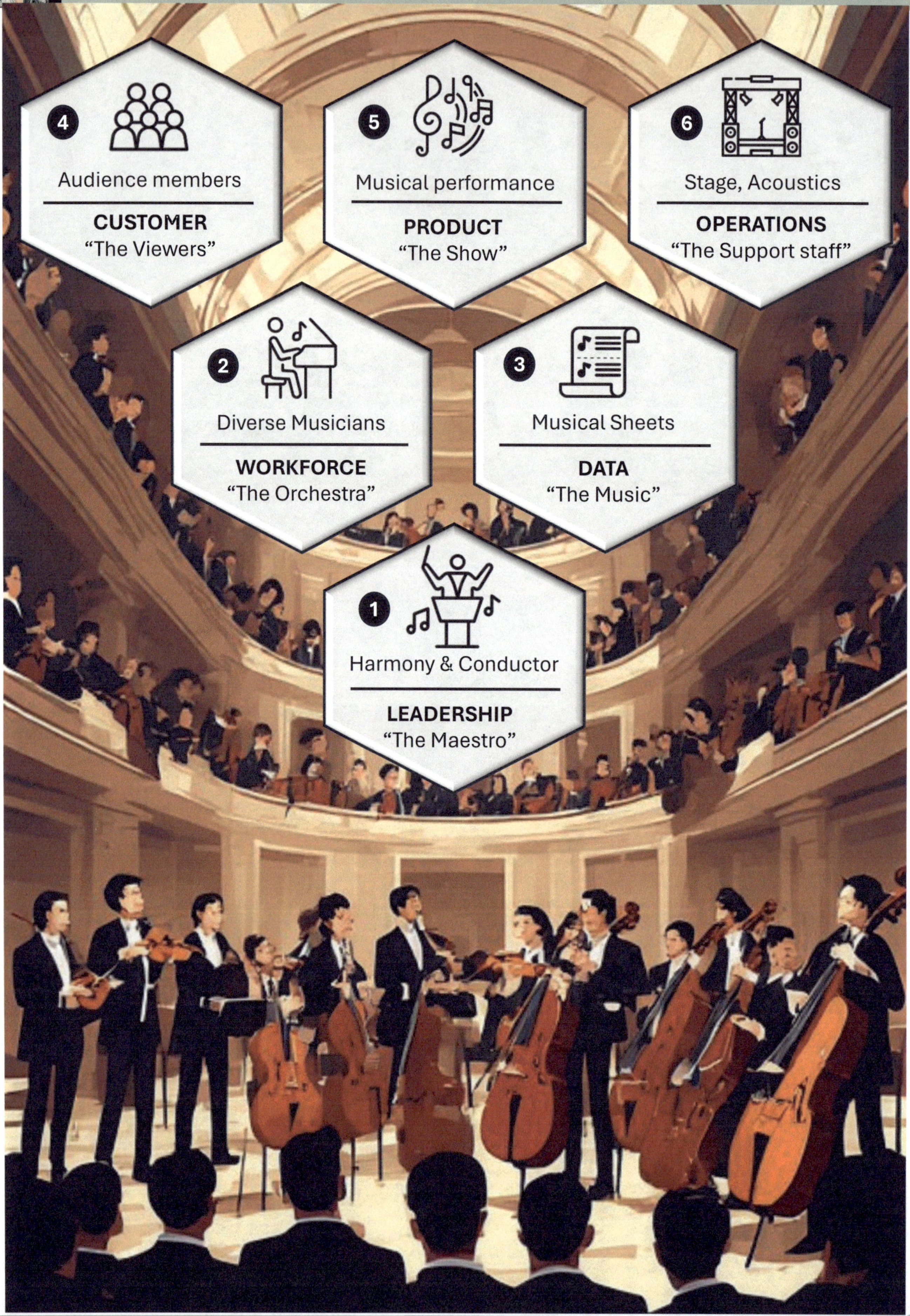
4
Audience members
CUSTOMER
"The Viewers"
5
Musical performance
PRODUCT
"The Show"
6
Stage, Acoustics
OPERATIONS
"The Support staff"
2
Diverse Musicians
WORKFORCE
"The Orchestra"
3
Musical Sheets
DATA
"The Music"
1
Harmony & Conductor
LEADERSHIP
"The Maestro"

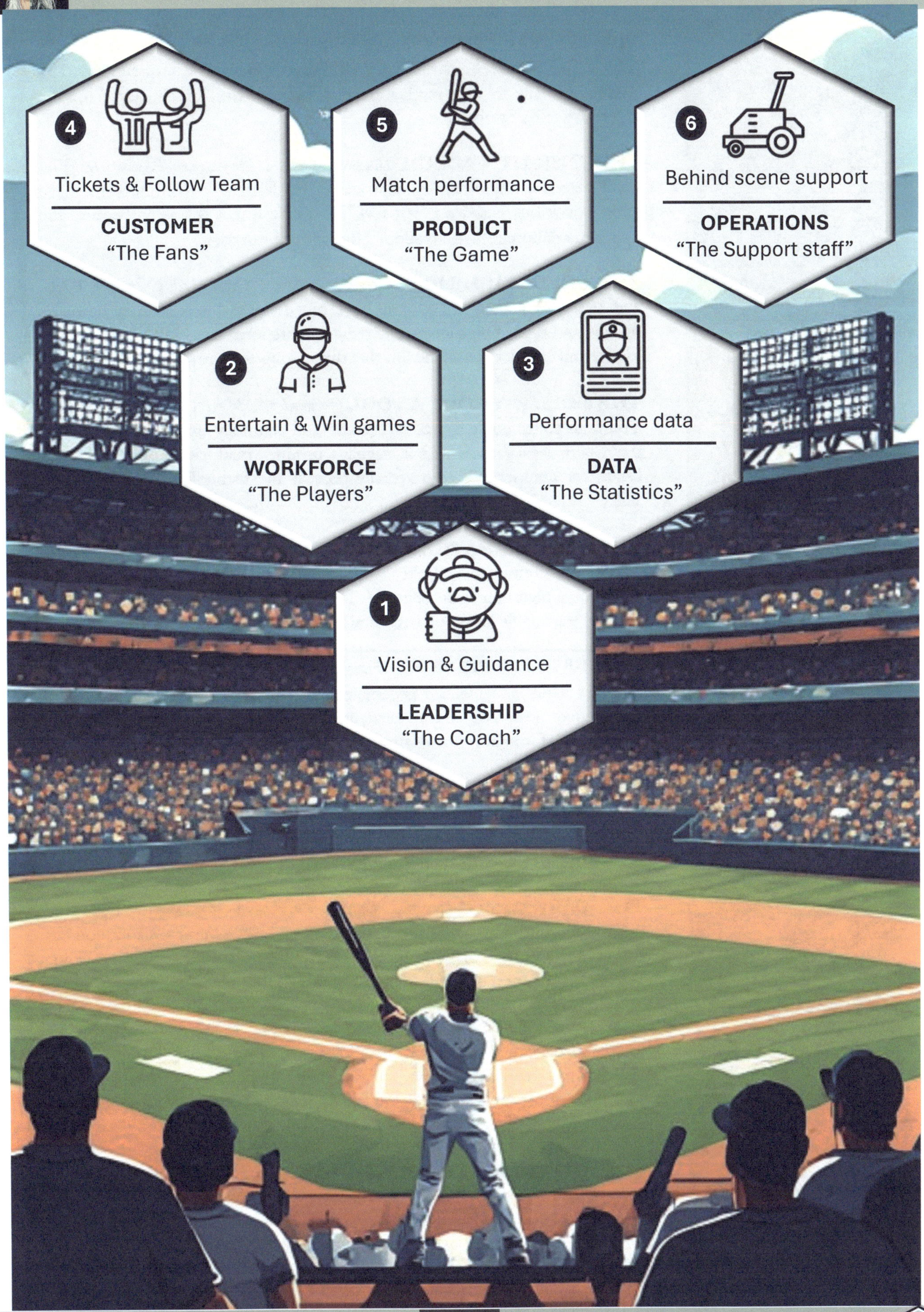
4
Tickets & Follow Team
CUSTOMER
"The Fans"
5
Match performance
PRODUCT
"The Game"
6
Behind scene support
OPERATIONS
"The Support staff"
2
Entertain & Win games
WORKFORCE
"The Players"
3
Performance data
DATA
"The Statistics"
1
Vision & Guidance
LEADERSHIP
"The Coach"

THE LITERACY GAP (*Invested: Product | Overlooked: Workforce*): You launch genius AI-driven features that your staff cannot explain, sell, or support. The technology fails to gain market traction because the "human layer" of the enterprise remains obsolete.

THE INSIGHT VACUUM (*Invested: Data | Overlooked: Leadership*): You build massive, high-integrity data lakes, but leadership continues to make critical decisions based on "gut feel." The investment is wasted because the upper-tier culture refuses to adopt a data-driven mindset.

THE COLD MACHINE (*Invested: Operations | Overlooked: Customer*): Back-end efficiency reaches an all-time high, but the actual user journey becomes robotic and rigid. You successfully reduce operational costs but lose the brand loyalty and emotional connection that drives long-term revenue.

THE SKILL WITHOUT A TOOL (*Invested: Workforce | Overlooked: Product*): You train your people to be "AI-Ready" and culturally agile, but you provide them with legacy tools and a stagnant product roadmap. Your best talent leaves for competitors who actually possess the technology to match their skills.

THE HALLUCINATION (*Invested: Leadership | Overlooked: Operations*): Executives announce a bold "Digital First" vision, but the underlying operational plumbing is decades old. The strategy never leaves the boardroom because the engine lacks the structural integrity to turn the wheels.

THE BROKEN PROMISE (*Invested: Customer | Overlooked: Data*): You market a hyper-personalized experience, but your internal data silos remain a fragmented mess. You end up alienating customers with irrelevant offers and inaccurate information that destroys their trust.

THE PROTOTYPE TRAP (*Invested: Product | Overlooked: Operations*): A brilliant AI-driven tool is developed in a vacuum, but the legacy supply chain and logistics cannot fulfill the demand it creates. The product succeeds in the lab but collapses the moment it hits the real-world operational floor.

THE DARK ASSET (*Invested: Data | Overlooked: Workforce*): You spend millions cleaning and structuring your data, but no one in the organization has the literacy to query or interpret it. The data sits idle and "dark," providing zero ROI because the people aren't equipped to find the value.

THE ROGUE ENGINE (*Invested: Operations | Overlooked: Leadership*): Middle management automates workflows to save time, but leadership has no visibility or governance over these new "black box" processes. The organization moves faster, but the steering wheel is no longer connected to the captain's chair.

THE INTERNAL ECHO (*Invested: Workforce | Overlooked: Customer*): Employees are empowered and "AI-literate," but they become obsessed with internal process improvements rather than user needs. The company becomes a high-performing bubble that is perfectly optimized for a customer that no longer exists.

THE BLIND CAPTAIN (*Invested: Leadership | Overlooked: Data*): A visionary leader attempts to steer the company through a massive pivot based on bold assumptions. Because the data pillar was ignored, they navigate straight into a market reef that the "missing data" would have clearly identified.

THE FANCY WRAPPER (*Invested: Customer | Overlooked: Product*): Your marketing and customer service are world-class, but the actual product has not evolved to keep up with competitors. You win the initial sale through great branding, but the churn rate sky-rockets once the customer realizes the product is hollow.

THE ANALYTICS ISLAND (*Invested: Data | Overlooked: Operations*): The data team produces perfect, real-time reports on every aspect of the business. However, the operational workflows are so slow and manual that by the time the business acts on the data, the opportunity has already vanished.

THE ORPHAN PROJECT (*Invested: Product | Overlooked: Leadership*): Innovation teams build cutting-edge AI tools that solve real problems, but leadership never provides a permanent budget or strategic home for them. These "pilot projects" eventually wither and die because they lack the top-down air cover to survive.

THE FRUSTRATED EXPERT (*Invested: Workforce | Overlooked: Operations*): You hire the best AI researchers and data scientists, but you force them to work within manual, "broken" internal workflows. Your experts spend 90% of their time fighting the system rather than building the future.

THE IVORY TOWER (*Invested: Leadership | Overlooked: Customer*): The board develops a strategy based purely on shareholder value and financial engineering. By ignoring the Customer pillar, they fail to see a massive shift in user sentiment that makes their entire three-year plan irrelevant.

THE HUMAN BOTTLENECK (*Invested: Operations | Overlooked: Workforce*): You automate 80% of your workflows, but the remaining 20% require human intervention that your staff isn't trained for. The entire system grinds to a halt because you optimized the machine but forgot to upskill the operator.

THE REACTIVE PANIC (*Invested: Customer | Overlooked: Leadership*): The company chases every new customer trend and social media "vibe" to stay relevant. Without a strategic anchor from Leadership, the organization becomes a fragmented mess of contradictory ideas that stand for nothing.

THE WASTEFUL HARVEST (*Invested: Data | Overlooked: Product*): You collect trillions of data points and pay massive cloud storage fees, but none of that data is ever used to iterate or improve the actual product. You are "Data Rich" but "Value Poor," hoarding information with no path to monetization.

THE GRASSROOTS REBELLION (*Invested: Workforce | Overlooked: Leadership*): Staff begins using "Shadow AI" and unsanctioned tools because leadership is too slow or fearful to provide guidance. You have an AI-ready workforce that is operating outside of the company's governance, creating a massive security and legal risk.

If the Pillars define the WIDTH of enterprise transformation, the **Six Lenses** define its DEPTH. They are the universal imperatives that cut vertically through every domain—Leadership, Workforce, Data, Customer, Product, and Operations. Each Lens exposes a different dimension of capability, ensuring that transformation is not superficial, fragmented, or episodic, but deeply rooted in the systems that shape how organizations think, behave, and evolve. These Lenses are not optional. They are not enhancements. They are the structural forces that determine whether transformation becomes sustainable or collapses under its own ambition. Each Lens reveals a layer of truth about the enterprise—its intelligence, its discipline, its technology, its execution, its culture, and its integrity.

The Human–AI Synergy Lens (L1) examines how intelligence is created, shared, and amplified across the enterprise. It explores the evolving relationship between human judgment and machine capability, ensuring that AI augments rather than replaces human potential. This Lens forces leaders to confront how decisions are made, how insights flow, and how people interact with intelligent systems. It challenges organizations to build environments where humans and AI operate as partners—each elevating the other. Human–AI Synergy is the starting point because it defines the cognitive foundation of modern transformation.

The Business Maturity Lens (L2) establishes the structural discipline required for transformation to take hold. It examines governance, alignment, prioritization, funding, risk management, and organizational readiness. This Lens reveals whether the enterprise has the scaffolding to support change or whether ambition is outpacing capability. Business Maturity is where leaders confront the operational truth: transformation is not powered by inspiration alone—it is powered by systems, discipline, and repeatable mechanisms that ensure consistency across the enterprise.

The Digital Evolution Lens (L3) focuses on the technological backbone of the organization. It examines how systems integrate, how data flows, how platforms scale, and how digital environments adapt to new demands. This Lens ensures that technology becomes an enabler rather than a bottleneck. It challenges organizations to build digital ecosystems that are fluid, intelligent, secure, and trustworthy. Digital Evolution is not about tools—it is about creating a technological environment capable of supporting continuous change.

The Execution Mechanics Lens (L4) converts strategy into action. It examines the methodologies, operating models, workflows, and performance systems that determine whether the organization can deliver consistently and at scale. This Lens exposes the gap between what leaders intend and what teams experience. It ensures that execution is not left to improvisation but is engineered with precision. Execution Mechanics is where transformation becomes tangible—where ideas become outcomes and outcomes become advantage.

The Cultural Compass Lens (L5) explores the beliefs, behaviors, and shared meaning that shape how people think and collaborate. It examines psychological safety, communication norms, leadership behaviors, and the emotional climate of the enterprise. This Lens reveals whether the culture accelerates transformation or resists it. Cultural Compass ensures that people feel connected, supported, and aligned—because no transformation can succeed in a culture that does not want it.

The **Moral Compass Lens (L6)** anchors the Blueprint in ethics and transparency. Moving beyond a static matrix, it defines the **True North** for AI governance, data privacy, and leadership accountability. This Lens ensures transformation is principled, not just efficient. In the AI era, a Moral Compass is the boundary condition for legitimacy—without it, innovation loses its direction; with it, trust becomes an unbreakable competitive advantage.

Together, these Six Lenses form the DEPTH of the Ai6x6 Transformation Blueprint. They ensure that every Pillar—Leadership, Workforce, Data, Customer, Product, and Operations—is approached with rigor, coherence, and integrity. They transform transformation from a collection of projects into a complete operating philosophy.

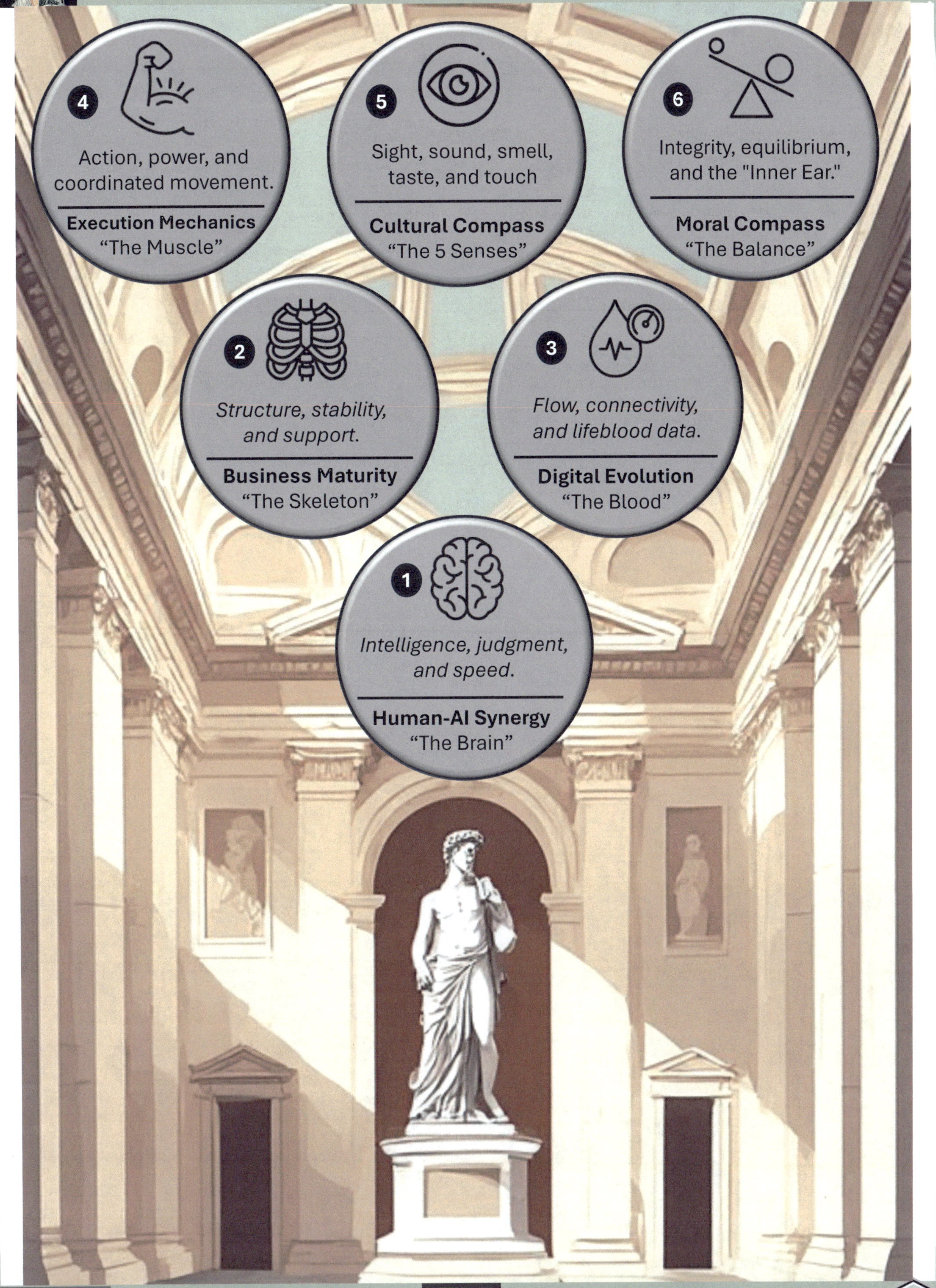
4
Action, power, and coordinated movement.
Execution Mechanics
"The Muscle"
5
Sight, sound, smell, taste, and touch
Cultural Compass
"The 5 Senses"
6
Integrity, equilibrium, and the "Inner Ear."
Moral Compass
"The Balance"
2
Structure, stability, and support.
Business Maturity
"The Skeleton"
3
Flow, connectivity, and lifeblood data.
Digital Evolution
"The Blood"
1
Intelligence, judgment, and speed.
Human-AI Synergy
"The Brain"

4
Gears that Convert energy power to motion
Execution Mechanics
"The Engine"
5
Monitor car health, temperature, audio
Cultural Compass
"The Dashboard"
6
Alignment, stability and stay in 'right lane'
Moral Compass
"The Steering"
2
Structural Frame and body of the car
Business Maturity
"The Chassis"
3
Gasoline, Electric energy & battery
Digital Evolution
"The Fuel"
1
Judgement, GPS, decision-making
Human-AI Synergy
"The Driver"

THE ETHICAL IMPLOSION (*Invested: Digital evolution | Overlooked: Moral compass*): You build a high-speed AI infrastructure with zero ethical guardrails or governance. One biased algorithm creates a public relations nightmare and a massive legal liability that can bankrupt the firm.

THE BUREAUCRATIC FREEZE (*Invested: Business maturity | Overlooked: Cultural compass*): You have established perfect governance and checklists, but you are operating within a terrified culture. People follow the rigid "process" of transformation but stop taking risks, effectively killing all real innovation.

THE THEORETICAL DREAM (*Invested: Human–AI synergy | Overlooked: Execution mechanics*): The organizational "brain" develops brilliant ideas for AI collaboration, but there is no "muscle" to act on them. Innovation stays trapped as a pilot project and never scales to produce enterprise value.

THE AUTOMATED ERROR (*Invested: Execution mechanics | Overlooked: Human–AI synergy*): You build a powerful execution engine but ignore the necessity of human judgment and intervention. The system begins executing "bad" decisions or flawed logic at 100x speed without anyone knowing how to stop it.

THE FUNDED HOBBY (*Invested: Cultural compass | Overlooked: Business maturity*): The internal culture is happy, safe, and collaborative, but there is no structural discipline or funding prioritization. Massive amounts of energy and capital are wasted on uncoordinated "passion projects" that never deliver ROI.

THE PRINCIPLED STANDSTILL (*Invested: Moral compass | Overlooked: Digital evolution*): You have world-class ethics and deep integrity, but you lack the "blood flow" of modern data and systems. You remain a "good" organization, but your technology is too outdated to actually compete with faster rivals.

THE FERRARI IN THE GARAGE (*Invested: Digital evolution | Overlooked: Execution mechanics*): You invest millions in an elite, high-performance tech stack that stays idle because no one has built the workflows to use it. You have purchased the "Engine" but never bothered to build the "Transmission" to turn the wheels.

THE LEGALISTIC LOOPHOLE (*Invested: Business maturity | Overlooked: Moral compass*): You strictly follow every legal rule and regulation while completely violating the spirit of ethics and trust. Your "Governance" is perfect on paper, but your brand is dying because your customers feel exploited by your "technically legal" actions.

THE INTELLIGENT OUTCAST (*Invested: Human–AI synergy | Overlooked: Cultural compass*): You develop AI tools that are technically brilliant and offer high synergy, but the workforce finds them creepy or intrusive. Because you ignored the "Senses" of the culture, the tools are rejected by the very people they were meant to help.

THE EFFICIENT COLLAPSE (*Invested: Execution mechanics | Overlooked: Business maturity*): The organization is moving with incredible power and speed, but it is moving toward a goal that isn't funded or strategically prioritized. You are winning the race, but you've realized too late that you're on the wrong track entirely.

THE NICE LIAR (*Invested: Cultural compass | Overlooked: Moral compass*): The culture is polite, collaborative, and highly "engaged," but it is built on a foundation of hidden ethical skeletons. People are "happy" because they are protecting each other from the truth, leading to an eventual systemic implosion.

THE ANALOG SAINT (*Invested: Moral compass | Overlooked: Human–AI synergy*): The leadership has high integrity but lacks the "AI-IQ" to understand the risks of the modern digital landscape. They are making ethical decisions based on a world that no longer exists, leaving them vulnerable to new forms of disruption.

THE EXPENSIVE CHAOS (*Invested: Digital evolution | Overlooked: Business maturity*): You layer high-tech AI tools on top of disorganized, immature governance and disorganized funding. The tech doesn't solve your problems; it simply makes your existing chaos move much faster and cost much more.

THE STAGNANT FORTRESS (*Invested: Business maturity | Overlooked: Execution mechanics*): The organization has perfect defenses, massive funding reserves, and rigid discipline, but zero kinetic movement. You are a perfectly "safe" business that is being slowly surrounded and suffocated by smaller, faster competitors.

THE DR. FRANKENSTEIN (*Invested: Human–AI synergy | Overlooked: Moral compass*): You achieve brilliant "Brain-to-Machine" innovation that pushes the boundaries of what is possible. However, because you ignored the "Inner Ear" of balance and ethics, you create a system that causes more social harm than organizational good.

THE BURNOUT FACTORY (*Invested: Execution mechanics | Overlooked: Cultural compass*): You optimize your "Muscle" for maximum output but completely ignore your "Senses" and emotional climate. You fail to detect internal fatigue and resentment until your highest-performing talent quits en masse.

THE LUDDITE OASIS (*Invested: Cultural compass | Overlooked: Digital evolution*): You have created the best place in the world to work, with high safety and great vibes. However, because you ignored the "Blood Flow" of digital evolution, your company is being disrupted out of existence by competitors you can't even see.

THE TOOTHLESS POLICY (*Invested: Moral compass | Overlooked: Execution mechanics*): You have beautiful "Values" posters and world-class ethics policies, but no actual workflows or mechanics to enforce them. Your integrity is a theory that falls apart the moment a team is pressured to hit a quarterly target.

THE GHOST SHIP (*Invested: Digital evolution | Overlooked: Human–AI synergy*): You have successfully implemented full automation and "lights out" data systems across the enterprise. However, because there is no human judgment in the loop, the system continues to run on outdated logic that is driving the company into a ditch.

THE PRECISE DINOSAUR (*Invested: Business maturity | Overlooked: Human–AI synergy*): You have spent years optimizing and documenting every single process to a level of "Six Sigma" perfection. Unfortunately, you ignored the "Brain" of AI synergy, and those processes are now perfectly optimized for a market that has already moved on.

Human judgment amplified by AI

Human–AI Synergy defines how intelligence is created, shared, and amplified across the enterprise. It examines the evolving partnership between human judgment and machine capability, ensuring that AI elevates rather than replaces human potential. This Lens forces leaders to confront how decisions are made, how insights flow, and how people interact with intelligent systems. It exposes whether the organization is building a truly augmented workforce or simply layering tools on top of old thinking. Ignoring this Lens leads to fragmented automation, poor adoption, and inconsistent decision-making. Human–AI Synergy is essential because it establishes the cognitive foundation of transformation—how the enterprise thinks, learns, and adapts—and without it, AI becomes a disconnected tool rather than a strategic multiplier.

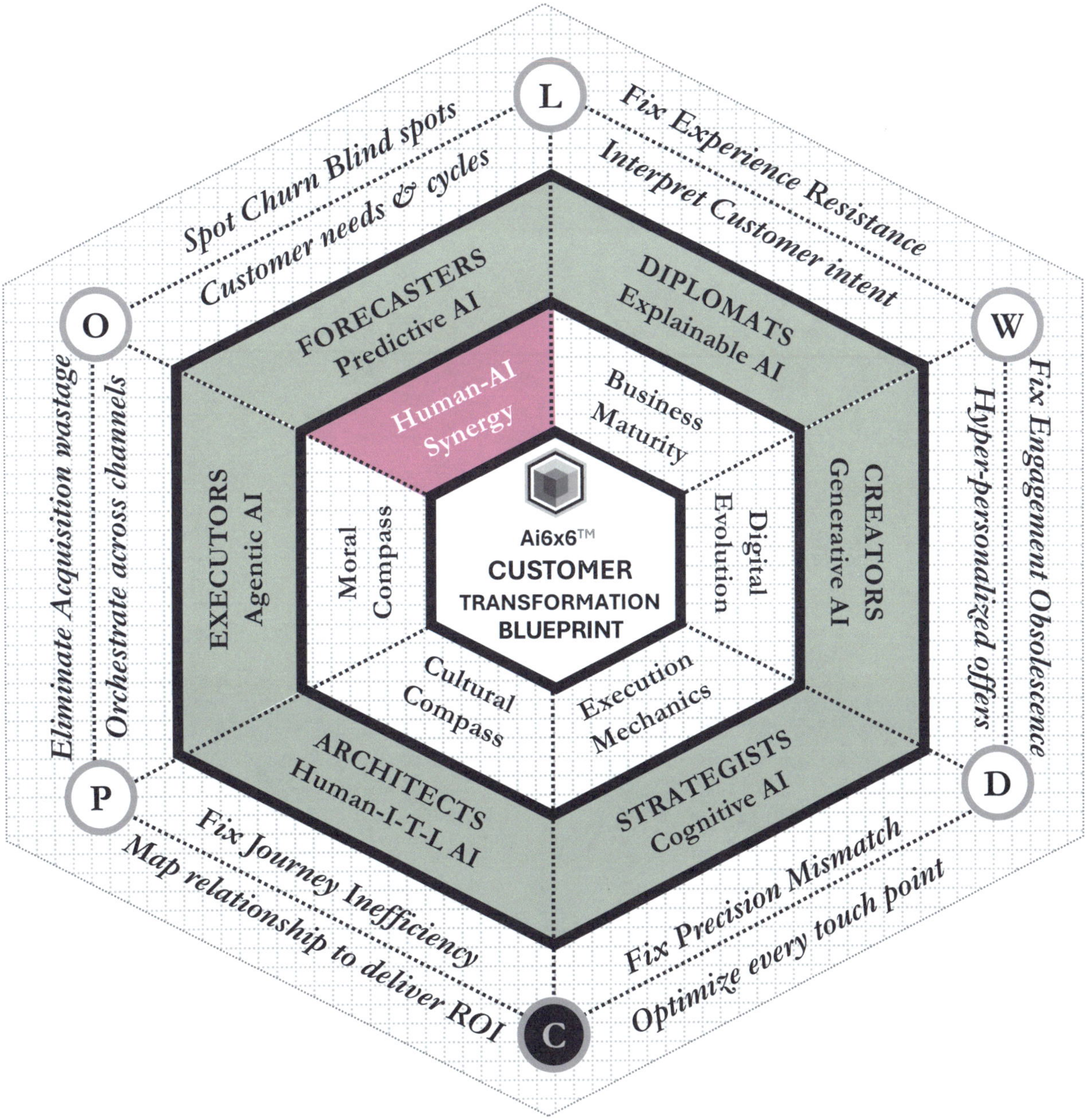

Ai6x6 HUMAN-AI SYNERGY FRAMEWORK FOR **CUSTOMER LIFECYCLE**

Discipline that stabilizes transformation

Business Maturity is the structural discipline that transforms raw ambition into a legacy of Authority. It begins at the Foundation, establishing the governance and funding models that anchor an organization's identity before activation begins. As the enterprise moves through Initiation and Cohesion, this Lens evaluates whether the internal scaffolding is robust enough to support the increasing velocity of the transformation engine. This Lens exposes a fundamental truth: sustained success is engineered through systems, not slogans. It ensures that as Momentum builds, the enterprise maintains the Optimization necessary to prevent collapse under the weight of its own initiatives. Mastering this Lens allows an organization to transcend temporary programs and achieve the permanent market Authority that defines industry leadership.

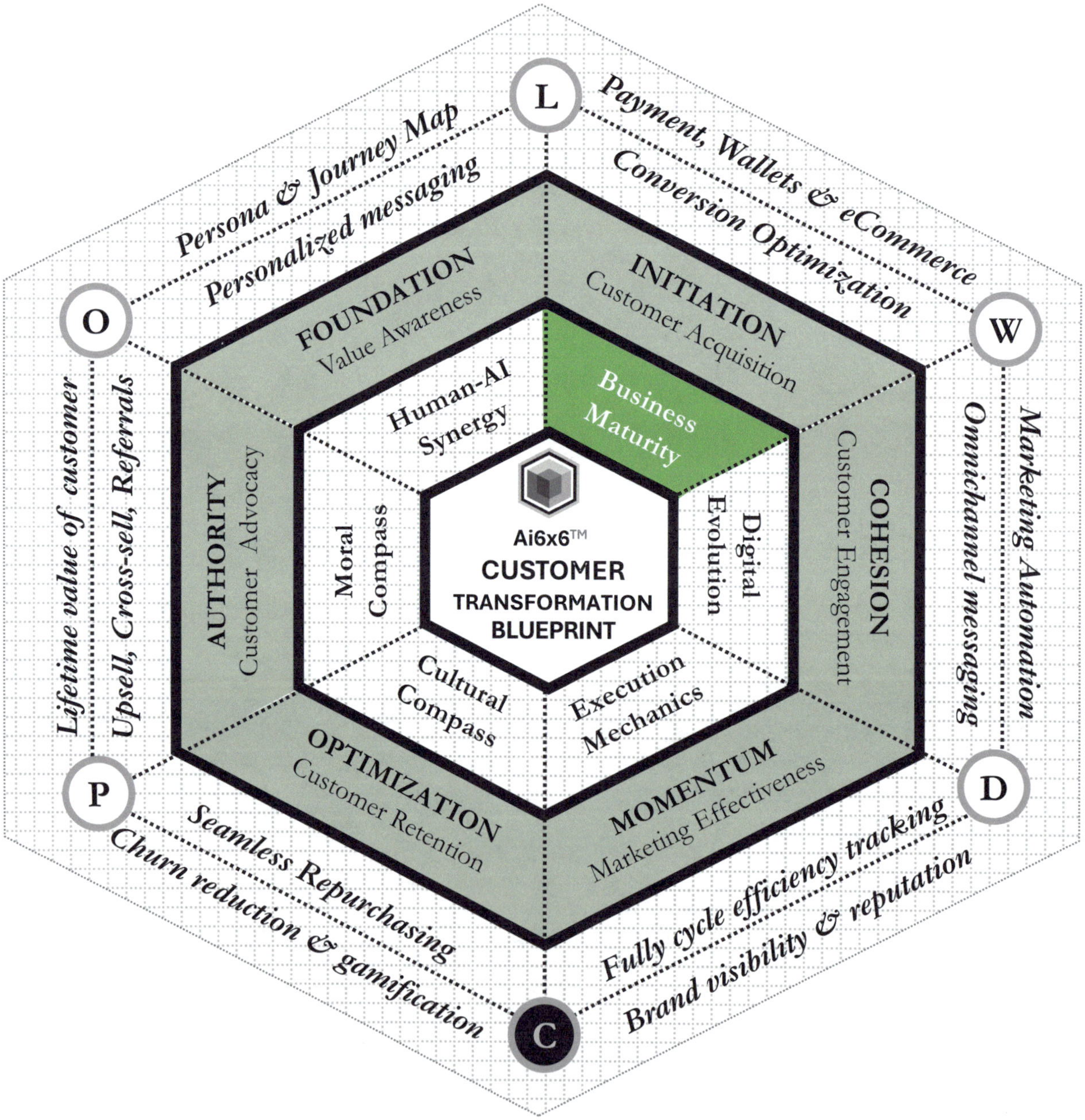

Ai6x6 BUSINESS MATURITY FRAMEWORK FOR **CUSTOMER LIFECYCLE**

Infrastructure built for continuous change

Digital Evolution examines the technological backbone that enables modern transformation. It evaluates system integration, data flow, platform scalability, and the adaptability of digital environments. This Lens ensures technology accelerates rather than obstructs progress. It forces organizations to confront technical debt, architectural fragility, and the true readiness of their digital ecosystem. When ignored, enterprises build brittle systems, slow innovation, and create environments where strategy moves faster than infrastructure. Digital Evolution is not about tools—it is about constructing a digital foundation capable of continuous change. It ensures that technology becomes an enabler of intelligence, speed, and resilience. Without this Lens, even the strongest strategies fail because the underlying environment cannot support enterprise-wide evolution.

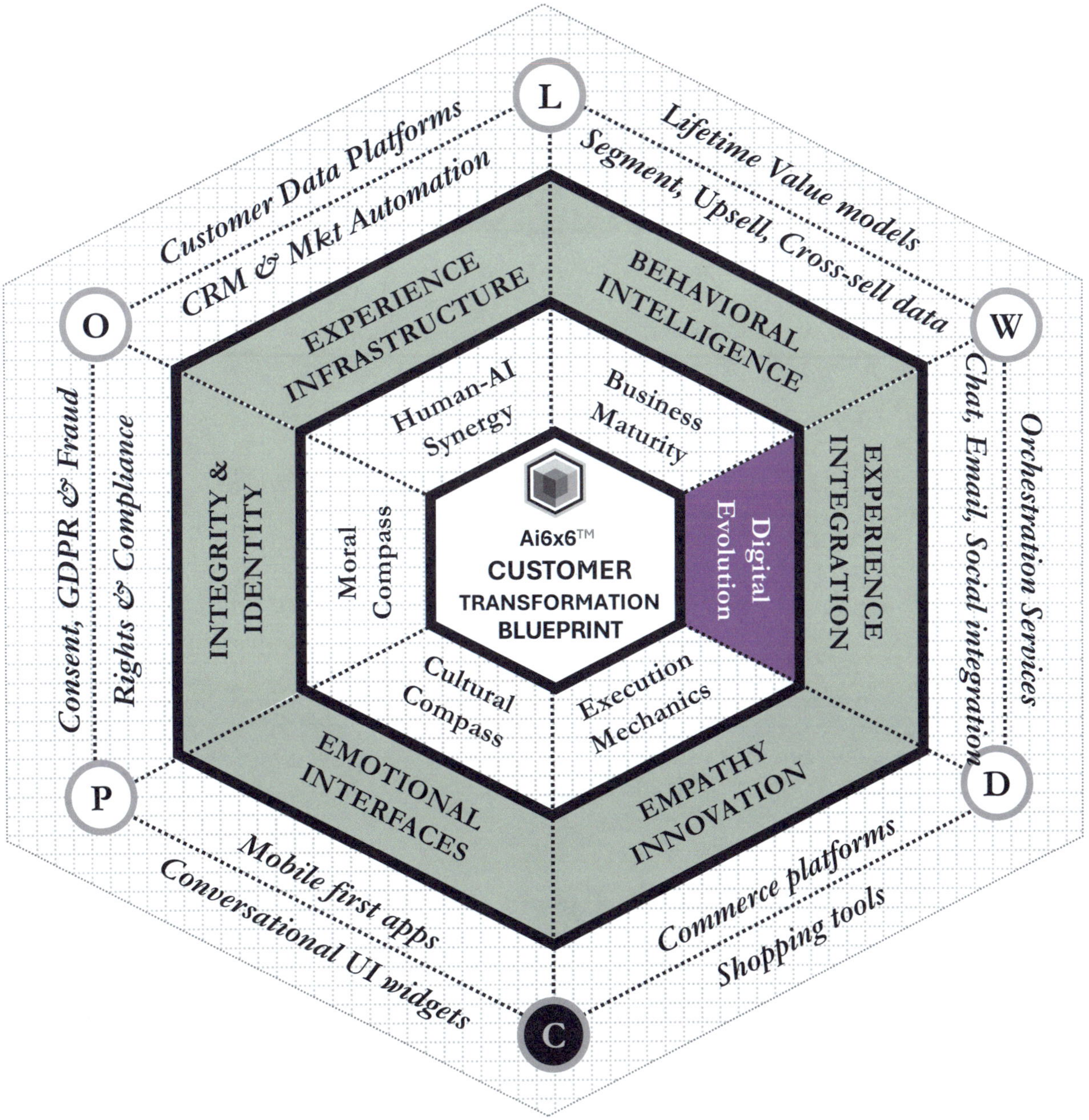

Ai6x6 DIGITAL EVOLUTION FRAMEWORK FOR **CUSTOMER LIFECYCLE**

Strategy delivered with operational precision

Execution Mechanics converts strategy into consistent, scalable action. It examines operating models, workflows, methodologies, and performance systems to determine whether the organization can deliver reliably. This Lens exposes the gap between leadership intent and frontline reality, revealing whether execution is engineered or improvised. It forces leaders to confront bottlenecks, handoff failures, and the true velocity of delivery. Ignoring this Lens results in missed deadlines, inconsistent outcomes, and organizational frustration. Execution Mechanics is essential because it transforms ideas into outcomes and outcomes into advantage. It ensures that transformation is not theoretical but operational—a system of delivery that produces measurable impact at scale.

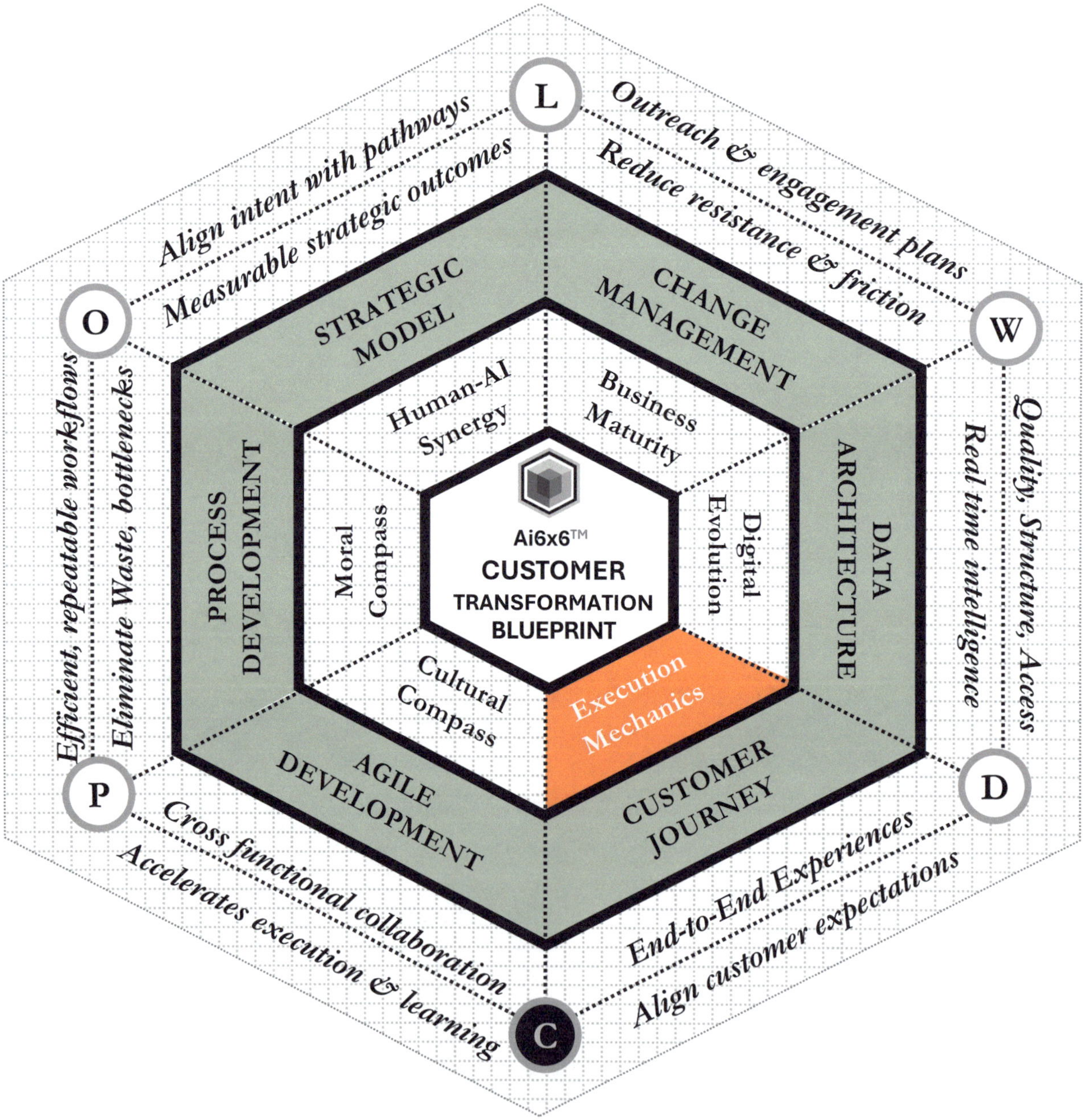

Ai6x6 EXECUTION MECHANICS FRAMEWORK FOR **CUSTOMER LIFECYCLE**

Culture aligned to accelerate change

Cultural Compass explores the beliefs, behaviors, and shared meaning that shape how people think, collaborate, and respond to change. It examines psychological safety, communication norms, leadership behaviors, and emotional climate. This Lens reveals whether culture accelerates transformation or quietly resists it. It forces leaders to confront the invisible forces that determine engagement, trust, and momentum. Ignoring this Lens means ignoring the human system that ultimately determines success. Even the best strategies fail in cultures that are misaligned, fearful, or fatigued. Cultural Compass is essential because transformation is a human endeavor—no system or technology can succeed without cultural readiness. It ensures people are not just compliant but committed, creating the conditions for lasting change.

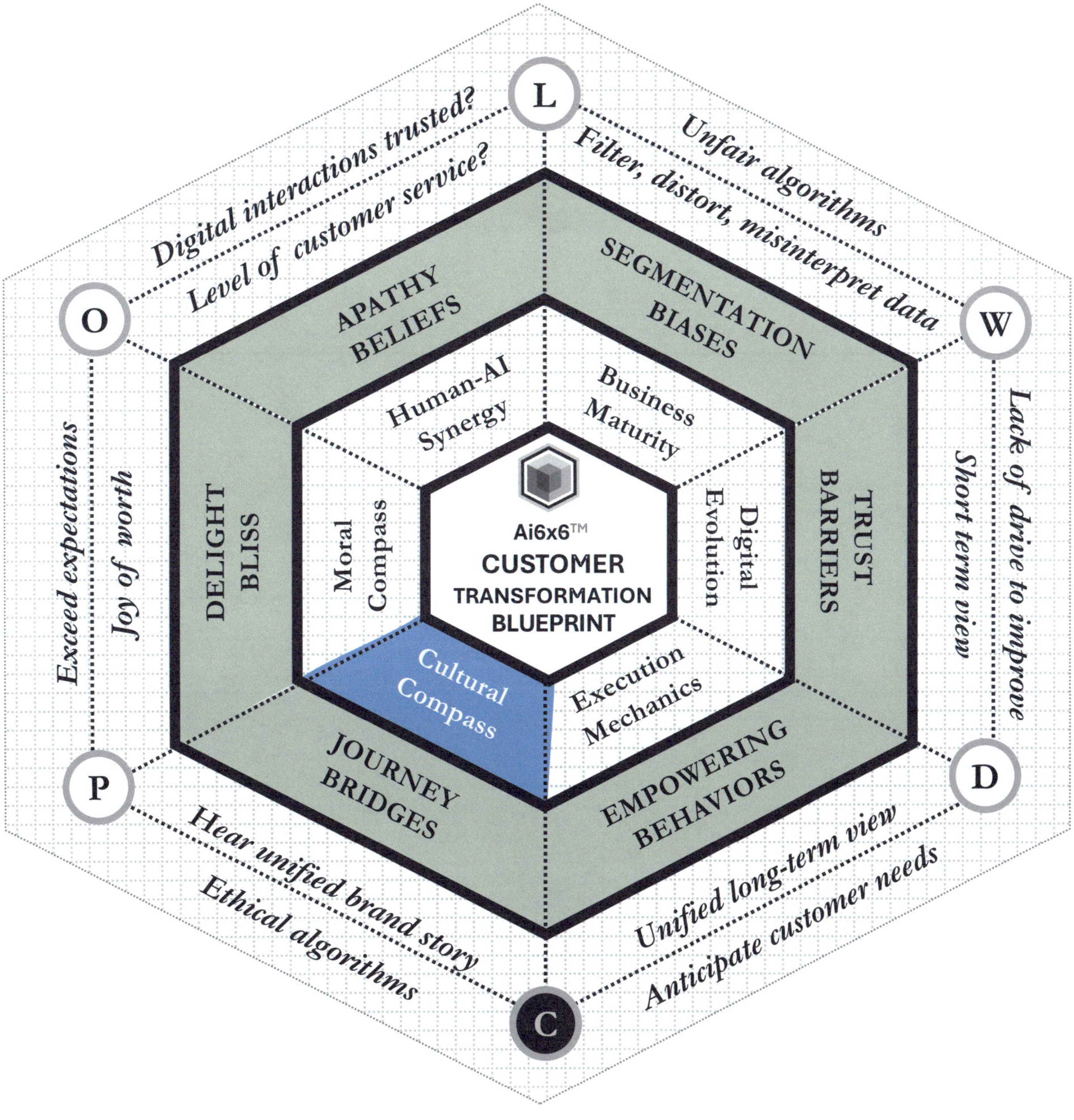

Ai6x6 CULTURAL COMPASS FRAMEWORK FOR **CUSTOMER LIFECYCLE**

Ethics governing intelligent enterprise systems

The **Moral Compass** anchors transformation in ethics, transparency, and responsibility. It evaluates data privacy, AI fairness, accountability structures, and the moral obligations of leadership. This Lens ensures that innovation is principled and that stakeholders feel protected, respected, and informed. It forces leaders to confront the ethical implications of their decisions and the long-term consequences of their systems. Ignoring this Lens exposes organizations to reputational damage, regulatory risk, employee resistance, and customer disengagement. Trust is the boundary condition of modern enterprise—without it, nothing else holds. The Moral Compass is essential because it defines the legitimacy of every decision, system, and outcome, ensuring transformation is not only effective but worthy of confidence.

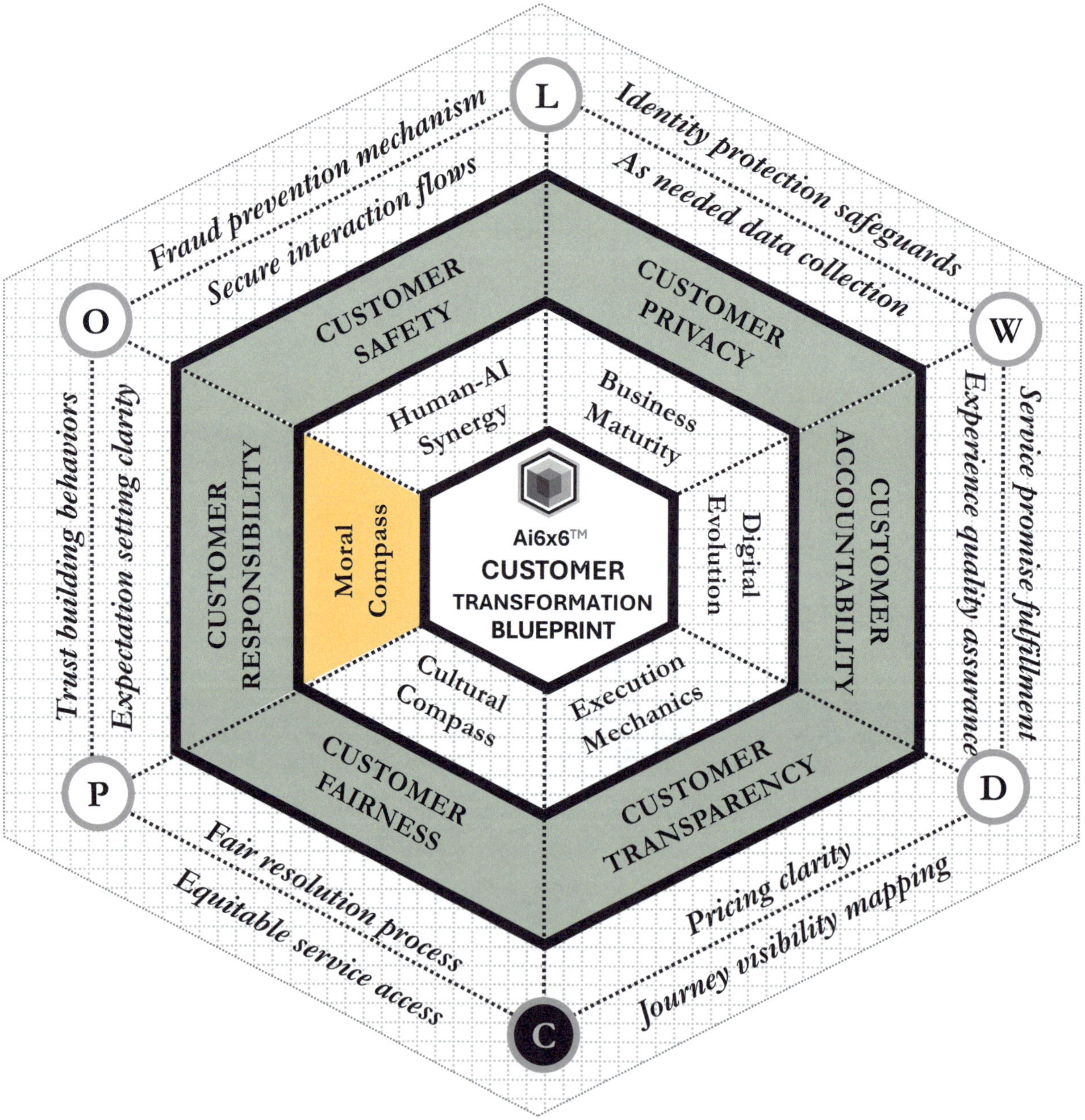

Ai6x6 MORAL COMPASS FRAMEWORK FOR **CUSTOMER LIFECYCLE**

Ai6x6™ Transformation Blueprint

Why Transform Customer?

"You've got to start with the customer experience and work back toward the technology, not the other way around." —Steve Jobs

Steve Jobs was an American entrepreneur, inventor, and co-founder of Apple Inc. He is widely recognized as a pioneer of the personal computer revolution.

Today's customers have more power and choice than ever. A customer transformation is necessary to shift an entire business's mindset from being product-focused to being truly customer-centric. Without this, companies fail to meet rising expectations for personalized, seamless, and meaningful experiences and will lose customers. By prioritizing the customer, an organization can build brand loyalty, create raving fans, and establish a significant competitive advantage that technology alone cannot replicate. The benefit of this transformation is increased customer loyalty, higher retention rates, and a powerful competitive edge.

1. **Hyper-Personalized Customer Journeys:** Systems use a 360-degree, unified customer profile to deliver contextually relevant experiences, recommendations, and services at every digital and physical touchpoint, maximizing conversion and satisfaction. The precise timing and relevance of offers dramatically increases average basket size and the probability of conversion compared to generic marketing campaigns. Every interaction feels relevant, deepening the customer relationship.
2. **Increased Customer Loyalty and Retention:** By consistently delivering friction-free, valuable experiences and being transparent about data usage (Trust Sovereignty), the company solidifies its position, increasing long-term customer lifetime value (CLV). High customer retention rates are the most profitable engine of growth, as the cost of serving an existing loyal customer is significantly lower than acquiring a new one. Loyalty acts as a powerful barrier to competitive entry.
3. **Seamless Omnichannel Experience:** The customer's context and history remain fluid and consistent as they move across channels (mobile, web, service line), eliminating the frustrating need to repeat information or restart their journey. This cohesive experience reduces customer effort, a key driver of satisfaction, and eliminates the service cost associated with redundant data collection.
4. **Higher Net Promoter Score (NPS):** By identifying and eliminating key pain points in the journey, the transformation drives customer satisfaction to a level that encourages organic advocacy and referrals, fueling low-cost growth. The increase in positive word-of-mouth marketing reduces the reliance on expensive paid advertising channels for customer acquisition. Promoters become the most effective and least expensive part of the marketing team.
5. **Enhanced Data Privacy and Ethical Constraints:** The design embeds transparency and user control into the core platform, ensuring customers feel safe and protected, which is vital for building lasting trust in an increasingly data-sensitive market. Demonstrating a commitment to ethical data use turns privacy from a compliance burden into a competitive differentiator and trust signal. Customers who trust the brand are more willing to share the data needed for personalization.
6. **Reduced Customer Service Costs:** Proactive self-service tools, intelligent chatbots, and predictive service models deflect routine inquiries, allowing the service team to focus on high-value issues and significantly lowering the cost per customer interaction. The reallocation of human agents from simple, repetitive issues to complex, emotionally sensitive issues maximizes the ROI of the service payroll. The high-value tasks retained by humans are those that cement loyalty.
7. **Increased Share of Wallet (Cross/Upselling):** Personalization engines deliver highly targeted and timely upgrade or cross-product offers based on predicted needs, leading to a higher volume of successful non-intrusive sales. By using behavioral data to predict the next logical purchase, the organization can organically grow its revenue stream from the existing customer base. Selling to existing customers is the most capital-efficient path to revenue growth.
8. **Faster Feedback Loops for Product Development:** Customer interaction data, journey analytics, and qualitative feedback are channeled directly and immediately to product teams, ensuring development efforts are tightly focused on solving actual user pain. This direct line of sight eliminates the waste of building features nobody uses and ensures the product backlog is always prioritized by confirmed customer value. The product team is constantly validated by real-world behavior.
9. **Shift from Service to Proactive Guidance:** Systems leverage predictive models to anticipate potential customer problems (e.g., service failure, subscription expiration) and resolve them or guide the customer before they realize an issue exists. This pre-emptive issue resolution eliminates the emotional labor of conflict and converts a potential defection into a positive loyalty moment.
10. **Creation of Customer-Led Digital Communities:** Digital platforms are utilized to facilitate customer-to-customer support, co-creation, and knowledge sharing, leveraging network effects for viral growth, peer support, and faster resolution. Community-driven support significantly lowers dependency on the company's own support resources while creating a valuable space for direct product feedback. The customers effectively scale the support function for the brand.

1. From Transactions to Relationships
When Loyalty Became Fragile

Carla had been loyal to the same outdoor gear brand for five years. She'd earned points, wrote glowing reviews, and posted product photos. But when a newer competitor offered a better digital experience and faster service, she left. "They knew my order history," she said, "but not me." The brand had focused on retention metrics—repeat purchases, time between orders—but not relationship signals. They never acknowledged her social engagement or responded to feedback. It wasn't until churn doubled that leaders dug deeper. The insight? Loyalty wasn't transactional—it was relational. The company launched a customer insights hub, integrated behavior-based messaging, and gave service reps autonomy to surprise-and-delight returning customers. Carla came back during a holiday sale and got a handwritten note. It wasn't the price—it was the recognition. Real loyalty doesn't come from points—it comes from feeling seen. Customer transformation means shifting from extracting value to offering it—proactively, personally, and consistently.

2. Service in Silos
When Customers Fell Through the Cracks

Jake returned a package, expecting a refund within days. Three weeks later, he was still waiting—passed between support, billing, and logistics. "Everyone's polite," he told a rep, "but no one owns this." Internally, each team had its own systems, KPIs, and knowledge bases. Leaders considered service "responsive," but customers experienced it as fragmented. The real issue wasn't intent—it was structure. The company connected service data across departments, unified timelines, and empowered agents with resolution authority. They also introduced a shared metric: First Contact Resolution, tied to voice-of-customer scores. When Jake returned for another order, one agent handled the entire interaction—refund, reorder, and apology—within 15 minutes. He wrote a 5-star review. Customers don't experience departments—they experience journeys. Transformation requires breaking down silos, so no one ever says again, "That's not my department." It's always someone's moment to make things right.

3. One-Size-Fits-None
When Personalization Missed the Person

Marissa subscribed to a travel gear brand for years. But every promotion she got felt off. Backpacks she'd never browsed. Ski goggles in August. Gift ideas for "him" when she always shopped for her sister. "They say it's personalized," she said, "but it feels like guesswork." The brand had a rules-based engine built on transaction history, not intent signals. Marissa unsubscribed. Her exit survey sparked action. The CX team invested in contextual intelligence: blending browsing, surveys, and behavior patterns. When Marissa returned for holiday shopping, the site greeted her with curated kits based on past gifting habits and an "edit your preferences" tool. It felt like someone had finally listened. Engagement and AOV rose across the segment. Transformation isn't "more data"—it's the right data, used with empathy. When customers say "recommended for you," they don't want the obvious. They want to feel known—and delighted.

4. Feedback Fatigue
When Customers Were Asked But Ignored

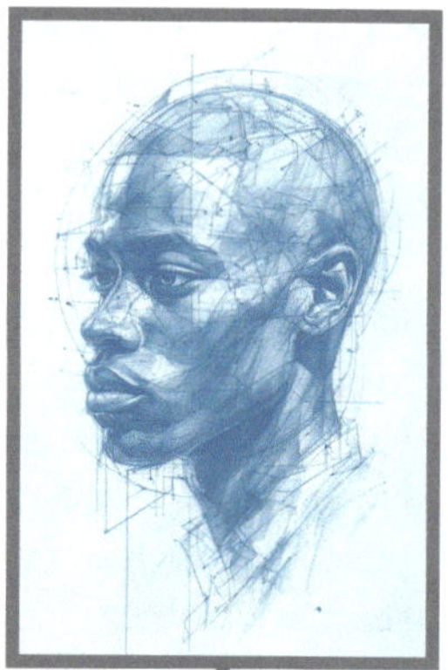

After every order, Sam got a survey. "How'd we do?" But when he flagged a recurring issue with shipping delays—three times—nothing changed. He eventually stopped responding. Internally, the company celebrated high NPS but ignored themes hiding in the comments. When a product leader reviewed 500 open-text responses, a pattern emerged: two issues accounted for over 40% of complaints. Yet no one owned follow-up. The company restructured its Voice of Customer loop. Feedback was now routed in real-time to product and ops leads. Customers who raised concerns received confirmation and resolution updates. Sam got a simple note: "We heard you. The issue's fixed." He replied, "Thanks for listening." Survey response rates improved, but more importantly—trust returned. Feedback without follow-through erodes brand capital. Transformation means treating every data point as a conversation—not a checkbox.

5. The Reactive Brand
Always a Step Behind

A viral TikTok review wiped out inventory of a skincare product overnight. Customers celebrated—then fumed as "in stock" promises turned into waitlists. Support teams apologized. Social media stalled. Meanwhile, marketing launched a campaign... for the same item. It was chaos. Not from bad people—but from disconnected systems. Leadership paused campaigns, integrated supply chain and inventory APIs into marketing workflows, and established a cross-functional "CX command center." Now when demand spikes, alerts trigger proactive communication. Back-in-stock signups are personalized. And when supply runs low, ads pull back automatically. Frustrated tweets became thank-yous. The brand didn't move faster—they moved smarter. Transformation means turning reactivity into readiness. Because the best apology is prevention.

6. The Invisible Customer
When Data Was There, But No One Used It

Alex had been a subscriber for four years—but each time he contacted support, they treated him like a new customer. "Welcome aboard!" one agent said after he upgraded. It stung. Internally, teams had rich data—but it lived across five systems. Support had tickets. Sales had purchase history. Product had usage logs. No one had the whole picture. After customer defection spiked, a cross-functional squad centralized the customer profile. Now every touchpoint pulls from one timeline: engagement, behavior, sentiment. The impact? Agents reference past issues before asking. Sales reps acknowledge milestones. When Alex renewed, he got a thank-you—and an offer tailored to his actual usage. He stayed. Transformation isn't just integration—it's memory. Customers want continuity, not déjà vu. They want to feel like someone, somewhere, actually sees them.

7. Marketing Without Meaning
When Campaigns Became Wallpaper

Flashy banners. Promo blasts. Daily emails. The brand was everywhere—but nowhere meaningful. "They talk at me, not to me," one customer wrote. Engagement plummeted. Then a new hire asked: "What if we stopped pushing products—and told stories?" The team spotlighted real users: a chef using their blender to feed shelters. A teen starting a baking business. Instead of discounts, they shared journeys. Campaigns became content. Loyalty grew. The shift wasn't about tone—it was about trust. Customers didn't want to be targeted. They wanted to be invited.

Transformation means knowing the difference between noise and narrative. Because great brands don't interrupt lives—they enrich them.

8. Voice Disconnected
When Channels Didn't Feel Human

Nina tried chat support. Then email. Then a call. Each time, she re-explained the issue. "I feel like I'm looping through a time warp," she joked to a friend. Internally, each channel had its own ticketing and tagging system. "Omnichannel" meant availability—not memory. Leaders brought in a journey orchestration platform. Now, any interaction triggers a shared case ID, sentiment tagging, and real-time transcript transfer. Agents pick up where the last left off, not from scratch. Nina tried again—and was greeted with, "Hi again—we saw your last message. Let's fix this." She smiled. It wasn't magic. It was design.

In customer transformation, connection matters more than coverage. Because being everywhere means little if it never feels like one experience.

9. Empowered Employees,
Delighted Customers — Or Not

Tom listened empathetically as a loyal customer explained a small late fee mishap—easily waivable under better policies. But the system wouldn't let him. "Policy says no," he explained, watching the customer leave disappointed. Tom felt worse. Later, finance calculated: waived fees cost little—churn cost everything. The company redesigned policies with guardrails and autonomy. Agents could waive up to $50 based on judgment. Customers weren't the only ones smiling—employees felt trusted. Retention went up on both sides. Empowerment isn't lawlessness—it's design with trust. The fastest way to improve CX isn't tech. It's trust at the frontline.

Transformation begins when people closest to problems get the power to solve them.

10. Community, Not Just Commerce
When Customers Built Something Bigger

Without prompting, fans of a wellness brand launched a group chat. They shared hacks, favorite combos, and daily wins. At first, the company hesitated. Should they moderate? Control? Instead, they joined. Quietly. Respectfully. Then generously—with sneak peeks, expert Q&As, and brand spotlights. Engagement tripled. More than buyers, they now had believers. Transformation came not from marketing—but from permission. Today, the company co-creates with customers. New products are voted on. Wins are shared by both sides. When community builds organically, smart brands don't disrupt. They nurture. Because in the age of transformation, the customer isn't just the end of your funnel. They're your beginning.

11. Insight Without Action
When Data Didn't Drive Design

A UX researcher flagged a checkout flow issue: mobile users were abandoning carts at nearly three times the desktop rate. The insights were clear, the friction obvious. But nothing happened. The report floated around, acknowledged in meetings—but never made it into the roadmap. Product managers prioritized features that looked good in quarterly demos, not those that solved daily pain. Two months later, conversion dropped—and leadership asked, "Didn't we already know this?" They had. But insight without follow-through is decoration. That moment triggered a process shift: every sprint began with a voice-of-customer review, not a backlog review. PMs were tasked with removing friction before adding features. UX researchers presented data live, not buried in docs. The mobile flow was fixed. Drop-off fell. And customers noticed. The company didn't just act faster—they acted on what mattered. Transformation means treating insights not as observations, but obligations. Because if listening never leads to change, customers will eventually stop talking. And worse—start walking.

12. One Brand, Many Journeys
When Personalization Meant Respect

Liam's order history was crystal clear: luxury handbags, handwritten gift notes, all around key holidays. Still, every week he received push alerts for flash sales on sneakers, generic promos, and Father's Day gift guides. "They have all my data," he muttered, "but none of my story." Eventually, he unsubscribed. Internally, the company realized their segmentation logic was demographic—not behavioral. They'd mistaken data volume for insight. The CX team redesigned journeys based on intent signals, repeat behavior, and event triggers. Holiday buyers received concierge-style gift guides. Long-term customers were greeted by name—on site and in store. When Liam returned, the offers felt...crafted. "It's like they finally know I'm not just a buyer—I'm a gifter." Sales rose. Churn fell. And customers felt re-understood. The takeaway? Personalization isn't a name in the subject line. It's relevance without explanation. Transformation means honoring every journey—not forcing everyone through the same funnel.

Ai6x6™ Transformation Blueprint

Power of Diversified Insight

"The imperative is to teach how AI tools work to every citizen, and especially to our young people... The augmentation side is the transformation side."

Satya Nadella, CEO of Microsoft. Nadella champions the democratization of AI knowledge, positioning the core of transformation as a commitment to large-scale upskilling. The value is unlocked not through AI alone, but through its ability to augment and enhance human ingenuity.

We have deliberately chosen to move beyond the limiting confines of the conventional single-expert **Foreword**, which, by its nature, risks restricting the reader to one singular point of view. In an era where **business transformation** is shaped by a confluence of disruptive forces, relying on a solitary oracle is fundamentally inadequate. Instead, this book opens with a **curated collection of 15 interview snippets** from a diverse array of global thought leaders. This method immediately immerses you in a rich, multi-faceted tapestry of opinions and real-world experiences. By foregrounding this **plurality of perspective**, we offer a wider, more nuanced, and dynamic lens through which to view the challenges and enormous opportunities presented by modern change. Consider these short, powerful excerpts a **Chorus of Counsel**: a comprehensive and inclusive foundation built on collective wisdom that better equips you for your own complex journey of enterprise innovation and transformation. This is a commitment to **diversified insight** from the very first page.

Gain unprecedented, insider access to the minds of 27 global thought leaders in this six-book series. Sourced from comprehensive 1-on-1 interviews, these experts are recognized worldwide for their work in business transformation and emerging technology. Their insights are drawn from my personal, professional network—including former bosses, colleagues, and business partners —giving you an insider's view into the minds of industry shifters. To maximize focus and utility, we've synthesized these deep-dive conversations into thematic 'interview cards.' Each of the six books offers 15 relevant, concentrated insights on specific topics, allowing you to quickly access and apply the most relevant expertise, no matter where it originated in the original interview.

EXPERT VOICES

Subrata Mukherjee: Your current role at KPMG is about identifying, developing, and incubating the next generation of solutions. What is the biggest challenge in translating a breakthrough technology, like quantum computing or the space economy, from a theoretical concept to a practical enterprise solution?

Richard Entrup: In my role, my team is tasked with identifying and validating new emerging technologies as viable and scalable service solutions to clients. As a CIO, I would say the biggest challenge is bridging the gap between research and scalable, reliable implementation. This involves several key hurdles such as technical maturity, cost and infrastructure to support, the skills gap, integration with existing systems, market readiness and use case, and regulatory and compliance. The core issue is aligning the pace of technological advancement with practical, market-driven needs while overcoming resource and expertise constraints. Success hinges on iterative development, public-private partnerships, and clear use-case validation. You must always start with defining business outcomes and fixing business problems, not the technology.

SM: You work with a wide range of emerging technologies. What is your framework for evaluating which "frontiers" are worth pursuing and investing in?

RE: My group includes a Signals & Strategy leader who is tasked with validating what frontiers we need to pursue. This is done via insights and feedback from our ecosystem of luminaries and partners as well as other research. We also maintain a Ventures team who keeps track of **VC and CVC** investments and also makes minority investments in startups once a new frontier is determined to be a space we want to play in.

SM: How do you balance the need for short-term, measurable ROI with the long-term, more speculative bets on truly transformative technologies?

RE: Our startup investments are more about growing a service offering and less about equity like a traditional VC or investment group. We strive to close large transformation deals with many of our services wrapped around a startup we've invested in.

Balancing short-term, measurable ROI with long-term, speculative bets on transformative technologies like quantum computing or the space economy requires a strategic, portfolio-based approach. This includes maintaining a diversified investment portfolio, staged milestones, cross-functional teams, partnerships and ecosystem leverage, iterative validation, financial discipline with vision, and market and regulatory foresight.

The key is to treat transformative tech as a venture capital portfolio within the enterprise—accept some failures, prioritize learning, and ensure short-term wins provide the financial and strategic runway for long-term breakthroughs. Regular reassessment of priorities and progress keeps the balance dynamic and responsive.

VC and CVC investments both provide capital to startups and small businesses, but Traditional VC (Venture Capital) focuses on purely financial returns for investors through equity, while CVC (Corporate Venture Capital) seeks both financial returns and strategic advantages for its parent corporation, such as access to new technology or markets.

RICHARD ENTRUP

Managing Director, Head of Emerging Solutions at KPMG US

A multi-award-winning global technology and innovation leader with nearly three decades of experience driving strategic business transformation. He is at the forefront of identifying and incubating disruptive technologies like GenAI, Quantum Computing, and Spatial/XR, pioneering new products and solutions for clients. A two-time recipient of the HMG Global Leadership Institute Award for Exceptional Business Transformation, his distinguished background includes C-suite leadership roles (CIO, CTO, CDO) at globally recognized brands such as Verizon, Disney/ABC, Christie's, Time Warner, Viacom, and Tiffany & Company. He also serves as a strategic advisor and investor to a portfolio of innovative startups and venture capital firms, passionately guiding ventures focused on cutting-edge innovations. He is based out of New York, United States.

EXPERT VOICES

Subrata Mukherjee: Your career is defined by building and leading organizations to successful exits. As a data and product leader, how do you position these functions as a primary value driver for potential acquirers?

Saleem Khan: Potential acquirers are looking into your QoE (quality of earnings) metric to understand how sticky your business is and whether your customers will defect as soon as an acquisition occurs. The other major selling point is the potential for growth. What is the potential for this business to grow over the next 3, 5, or 10 years?

Data and product become a key part of crafting a narrative of readiness and growth when an acquisition happens. If you have data or a product customers can't live without or you have some sort of compelling moat, you will not see much churn. Also, with respect to growth, how easily can you add more data or move into adjacencies to address other existing customer needs or needs that customers have not yet anticipated.

SM: During the due diligence process, what are the most critical data and technology elements that acquirers scrutinize?

SK: The most important metric I've seen is how big of a moat you have with respect to your first party data. Do you have a unique dataset that only you can source or derive that no one else has access to? Does this dataset grow over time and will your moat grow with it?

The usual questions around accuracy, completeness and timeliness also come up when an acquirer evaluates your dataset. They will look into statistically significant samples to confirm the data is accurate and that there isn't much sparsity or gaps in your data. If your dataset is temporal in nature, this also becomes an important metric acquirers look into. Do you have a track record of updating your data daily or within milliseconds depending on the specific industry standards.

SM: How do you measure the success of an innovation lab or team?

SK: There's a saying in golf that there are no pictures on a scorecard. For those that are non-golfers, this means that if a golf hole has a par score of 4 and you have 4 absolutely hideous shots but still come out with a 4, nothing else matters; you scored a par (4) and that's all the world will know about how well you played that hole.

The same concept applies to measuring the success of an innovation lab. You are typically given a budget, a team and maybe 18-24 months if you're lucky. It doesn't matter that you had 11 failures, all that matters is that you had the one successful product that was able to generate net new revenue and ideally make up for the failures and the operating costs of the lab.

Using a pragmatic approach to innovation like this is a very simple and effective way to measure the success of an innovation lab.

SALEEM KHAN

Chief Data & Analytics Officer, Pole Star Global

Saleem is a seasoned data and analytics executive with a strong history of creating business value through data-driven innovation and successful company exits. In his current role as Chief Data & Analytics Officer at Pole Star Global, he leverages his deep expertise to lead and execute data strategy. His notable achievements include serving as Chief Digital Product Officer at VettaFi, which was successfully acquired by TMX Group, and as Chief Data & Analytics Officer at Discovery Data, which had a successful exit to Institutional Shareholder Services. He has also held leadership positions in data innovation and advanced analytics at major corporations like Dun & Bradstreet and S&P Global, and was a founder of the fintech startup ESGINE, an AI-enhanced platform for ESG compliance. He studied at Fordham University. He is based out of New York, United States.

EXPERT VOICES

Subrata Mukherjee: You are credited with a bold strategy that managed the migration from print to digital, transforming a loss-making business into the largest contributor to group profits. What was the most difficult decision you had to make to achieve this?

Michael Brunt: The hardest decision was to move to a paid model when free access was the prevailing industry orthodoxy. We believed in the distinctiveness of The Economist's journalism and made subscription central to the business. It was a controversial choice but it unlocked recurring revenue and turned circulation into the group's most profitable contributor.

SM: Can you share more about the award-winning, content-led digital marketing and experiential marketing campaigns you pioneered?

MB: Two campaigns show our approach. The first was "Raising Eyebrows and Subscriptions", which won Gold for Creative Effectiveness at Cannes. Working with Proximity and UM London we used provocative Economist content in unexpected, contextually relevant environments. The campaign generated 5.2 million clicks, 64,000 new subscribers and £51.7 million in lifetime revenue.

The second was "Discomfort Future" with Sense, where we brought Economist journalism to life by serving insect ice creams, waste-food smoothies and coffee staged as if brewed from recycled sewage water. These stunts intrigued audiences, challenged perceptions and generated over 30,000 subscriptions globally with a 171% ROI. They also reshaped perceptions of The Economist as more engaging and relevant.

SM: How did you balance the need for commercial growth with the need to protect the journalistic integrity and premium brand of The Economist?

MB: We achieved balance by ensuring that campaigns were drawn directly from the journalism itself. Both the Cannes-winning work and the experiential activations were authentic reflections of editorial content. That approach allowed us to grow circulation while strengthening the brand's integrity rather than undermining it.

SM: Why do you think there's a historical tension between these two functions within media firms?

MB: Traditionally editorial focused on the independence of its work while commercial chased revenue, creating separate cultures. Digital disruption has made those divisions unsustainable. Success now depends on collaboration, with both sides sharing audience data and aligning around common goals. Without deliberate cultural change tension is inevitable.

SM: What is the first step you advise a publisher to take when trying to break down these silos and achieve a unified business strategy?

MB: The first step is to create shared goals. If editorial is measured on awards and commercial on advertising yield the priorities conflict. When both are measured on engagement, retention and audience trust collaboration follows. At HBM Advisory we run joint workshops where editorial and commercial review data together as a source of audience insight.

SM: In an age of data-driven decisions, how can a commercial team effectively communicate the value of its insights to an editorial team?

MB: By framing data as a way to serve audiences better. Editors are motivated when they see that insights show their work reaching more readers or retaining subscribers. Presenting insights as evidence of audience value, not just revenue, builds trust and collaboration.

MICHAEL BRUNT

Former MD/COO/CMO @ The Economist, GM Times / News UK

Michael is a strategic media and publishing executive with a proven track record of driving business transformation and profitability. Currently as Co-Founder of HBM Advisory, he helps content businesses navigate the digital age by aligning business strategy, implementing AI capabilities, and optimizing content and team structures. His distinguished career includes a significant tenure at The Economist, where he held multiple senior leadership roles, including CMO and COO. During this time, he was instrumental in transforming the circulation business from a loss-making entity into the largest contributor to group profits through innovative digital and experiential marketing campaigns. He also served as General Manager for The Times and The Sunday Times, overseeing their long-term growth strategies. He has studied at Harvard Business School and University of Manchester. He is based out of London, United Kingdom.

EXPERT VOICES

Subrata Mukherjee: Your current work is heavily focused on digital trust, from authenticating data to mitigating cyber and software risks. What motivated your professional shift toward digital trust, and what is the biggest threat to digital trust today?

Stephane Pere: My shift toward digital trust came from a conviction that data alone was not enough—without trust, data has no value. At The Economist, reader as well as advertiser and client confidence was the foundation of data-driven revenue Today, with generative AI and deepfakes, this challenge has reached a new level. We are entering an era where fabricated content can be indistinguishable from authentic information, eroding confidence in what we see, read, and share. This poses risks not only for media and politics, but also for businesses whose brand integrity and digital assets can be manipulated. The biggest threat to digital trust is this blurring of truth and falsity. Companies need robust frameworks—legal, technological, and cultural—to prove authenticity in a world where deception is increasingly easy to scale.

SM: Your work at **Evidency** involves providing "qualified electronic timestamping services." How does a Qualified Trusted Service Provider (QTSP) fundamentally change how a company handles and protects its most critical digital data?

SP: A QTSP like Evidency offers more than just compliance—it provides assurance at the highest legal and technical standard. By using qualified timestamping and electronic seals, companies can prove the integrity and authenticity of their critical digital data and documents in court or regulatory audits. This transforms how businesses handle digital evidence: instead of relying on internal systems that can be challenged, they anchor trust in an external, certified authority. It's a shift from "we say this data is valid" to "this data is independently, legally, and technologically certified as valid." That creates both resilience and confidence.

SM: What is the biggest misconception that companies have about digital trust, especially as it relates to their business models?

SP: The most common misconception is to see digital trust as a purely technical issue or a compliance checkbox. Many leaders think that having cybersecurity software or GDPR policies is enough. But digital trust is holistic—it combines legal validity, technological robustness, and cultural transparency. It's about ensuring that customers, regulators, and partners believe in the authenticity and fairness of your digital interactions. Businesses that treat trust as a strategic asset, not a cost, will differentiate themselves in the digital economy.

SM: You mention leveraging "alternative data" like Earth observation analytics. Can you provide an example of how you used such data to create a new product or service?

SP: We explored Earth-observation and satellite imagery to derive economic insights—such as agricultural yield projections—as complementary signals to traditional reporting. By combining journalistic insight with unconventional datasets, we could offer premium intelligence products that gave our clients an edge. The key was not the data itself, but the ability to contextualize it—turning raw signals into actionable business insights.

STÉPHANE PERE

Former Chief Data Officer – The Economist | MD – Evidency, France

Stéphane is a seasoned digital and tech executive with a strong focus on reinventing trust in the digital realm. His extensive career includes pivotal roles at leading companies, where he has driven strategic transformation and innovation. As Chief Data Officer and Executive Vice President at The Economist, he earned multiple awards for his work in transforming the B2B market intelligence business. He has also served in executive roles at Vaultinum, Evidency, and Poool, where he has leveraged his expertise in digital strategy, business development, and software as a service (SaaS) to accelerate growth and mitigate risk. He is also a strategic advisor and non-executive director, guiding media and tech companies in revenue acceleration and digital transformation. He studied in London Business School , Emlyon Business School & Paris-Pantheon-Assas University.He is based out of Paris, France.

EXPERT VOICES

Subrata Mukherjee: Your leadership philosophy and your work as a Year Up mentor are central to your profile. What is the most important lesson you've learned about empowering teams for success?

Rod Rinker: Empowering a team requires a clear vision and a "why." In the age of AI, this means giving employees the tools and autonomy to use AI to augment their own work, not replace it. For example, a leader can implement a pay-for-results methodology that rewards the development of a new AI-powered script that reduces a team's manual workload by 20%, not just the completion of their daily tickets. This shifts the focus from simple task execution to strategic, value-adding innovation. A company can also empower its workforce by encouraging them to become "AI Champions" in their respective areas. They learn and experiment with AI tools, then share their knowledge and best practices with others, fostering a culture of continuous improvement and self-empowerment.

SM: What is the most critical piece of advice you give to aspiring IT service professionals and leaders who want to build a career that is resilient to technological change?

RR: The key to a resilient career is a commitment to continuous learning, but this now includes learning *how to work with and leverage AI*, not just about new technologies. Instead of just learning a new programming language, an IT professional can learn the art of prompt engineering to automate their daily tasks and generate strategic insights. They can then advise clients on how to do the same, making them invaluable. This mindset shift is crucial as the "tech stack" becomes the "human-AI stack." Future leaders will be skilled at integrating human creativity and critical thinking with AI's data processing power. The most resilient professionals will be those who can design and manage these blended workflows effectively.

SM: How do you personally stay on top of new technological advancements and ensure you are always ahead of the curve?

RR: Staying ahead of the curve requires an active, hands-on approach that goes beyond just reading about new products. It's about experimentation and building a personal "innovation lab." Instead of just attending a sales kickoff, a leader can create a personal "learning sandbox" with free trials and developer accounts for new technologies. This allows for direct, hands-on experimentation with new APIs and AI models, providing a deeper understanding than a marketing website can. A leader can not only absorb information but also host weekly "Lunch and Learn" sessions where they present new AI trends or tools they've personally experimented with. This not only keeps the leader sharp but also instills a culture of continuous learning within their team.

SM: How will the trusted advisor role change when many routine tasks are handled by AI, and how will human advisors continue to add unique value?

RR: The role of a trusted advisor will evolve from a manager of service-level agreements to a strategic partner and co-innovator. The human touch remains crucial for navigating complexity, building trust, and providing empathetic guidance. For instance, instead of just reviewing performance dashboards, which an AI can now do automatically and in real time, a human advisor will lead strategic workshops with the client. They'll use AI-generated insights—like identifying business process bottlenecks or suggesting new revenue streams based on user data—to co-create a future-state technology roadmap. Furthermore, when a client is considering implementing a new generative AI tool, a human advisor provides unique value by guiding the discussion around data privacy, ethical use, and the human impact on their workforce—nuanced conversations that AI can't yet lead effectively.

ROD RINKER

Senior Director, Global Field Services, Digital Workplace, Unisys

Rod is a results-driven executive with extensive expertise in global and regional IT service delivery and customer success. He has a proven track record of leading large, complex organizations (over 4,000 employees) and consistently exceeding service level agreements (SLAs) and contractual commitments. At Unisys, he is responsible for the global field services operation, focusing on modernization and transformation. His career is marked by key roles at NCR and Hewlett Packard Enterprise, where he drove significant revenue growth, improved customer satisfaction (as evidenced by high Net Promoter Scores), and successfully managed multi-million-dollar services contracts. An expert in leadership, he emphasizes building a "trusted advisor" status with clients to foster loyalty and expand business. He holds an MBA from Georgia State University & B.A. in Computer Science - Anderson University. He is based out of Atlanta, in the United States

EXPERT VOICES

Subrata Mukherjee: The competitive landscape is being reshaped by AI. How is AI impacting go-to-market strategies for technology companies today? What's an example of an AI-driven GTM tactic that you find particularly effective?

Gauthier Garnier: AI is transforming go-to-market strategies by enabling a level of precision and personalization previously impossible. A highly effective, AI-driven tactic we are pioneering is the automated creation of personalized demos. We found that customized demos substantially increase conversion rates, so we launched an initiative with partners like Pimcore, Shopware, and iPaaS.com, powered by an AI tool from Nexus aion Gmbh called Dragonfl.AI. This tool analyzes lead data from a contact form, creates a detailed company profile, and then builds a customized demo by scraping the prospect's website to use their own product data. Instead of hours of research, our team has a targeted sales pitch ready in minutes. Delivering a personalized demo just hours after a prospect submits a form leaves a powerful first impression, optimizing our GTM with incredible reactivity.

SM: What is the most critical lesson you've learned about building and maintaining client trust from your time in professional services?

GG: The most critical lesson is the power of radical transparency, especially when things go wrong. In any complex project, challenges are inevitable. Hiding them or shifting blame erodes trust instantly and irrevocably. Owning a problem, communicating it clearly, and presenting a concrete plan to resolve it not only preserves trust but can actually strengthen the relationship. Clients don't expect perfection, but they do expect honesty and partnership. This principle has been the bedrock of every successful client relationship I've had, both in services and now in sales and channels. Trust is also earned by consistently doing what you said you would do. Showing up on good days and bad creates alignment. Ultimately, empathy and being your partner's best advocate -both internally and with customers- are what set you up for long-term success.

SM: Do you find that a deep technical understanding is a prerequisite for success in a modern sales and partnerships role?

GG: I believe the best technology salespeople act as trusted advisors. While a deep technical understanding is not always a prerequisite, it is highly valued in certain complex deals. The most essential skill is a deep understanding of the business problems the technology solves. When a salesperson isn't deeply technical, they must work as a team with presales or solutions experts to advise the customer properly. A modern sales leader must be a translator, bridging the gap between a technical product and a business outcome. That said, a strong technical curiosity is crucial for maintaining credibility with all stakeholders.

SM: Looking ahead, what do you see as the next major evolution for the PaaS industry?

GG: The next major evolution for PaaS is a move towards truly autonomous, self-optimizing platforms. Today's PaaS has done a great job of abstracting away infrastructure, but the next wave will abstract away much of the operational complexity. Fueled by AI, future platforms will automatically optimize for performance, cost, and security based on real-time application behavior, limiting human interventions. Developers will focus almost exclusively on writing business logic, and the platform will intelligently handle the rest. It represents the ultimate fulfillment of the PaaS promise: making application development and deployment nearly effortless.

GAUTHIER GARNIER

General Manager, DACH & Channels at Platform.sh

Gauthier Garnier is a highly experienced General Manager and business leader with a career focused on building strategic partnerships and long-term, win-win relationships with clients. As General Manager for the DACH (Germany, Austria, Switzerland) region and Channels at Platform.sh, he is responsible for driving growth and managing strategic alliances. His background is deeply rooted in digital transformation and web agency leadership. Previously, as a partner and General Manager at Kaliop, he successfully launched and scaled a web agency in the North American market, overseeing production, human resources, and finances. He has a strong expertise in strategic planning for complex digital projects, particularly within the retail and media industries, and has taught and coached on these topics. He has been recognized for his projects, including an "Innovation of the year" award for an e-commerce platform for The Economist. He is based out of Montreal, Canada.

EXPERT VOICES

Subrata Mukherjee: Your recent posts highlight how Crystallize is solving complex problems like layered pricing for B2B. What are the core challenges modern eCommerce businesses face today, especially those that have both B2C and complex B2B needs?

Bard Farstad: One of the most consistent challenges we see is the quality and consistency of product information. Without a solid foundation, everything from discovery to conversion suffers. Modern eCommerce businesses also face the complexity of managing assortments across multiple markets, where regulations, preferences, and product availability differ. Handling translations and localizing product information, such as certifications or even culturally adapted imagery, is another recurring pain point. Businesses must also support market-specific bundles, promotions, and packaging strategies. On top of this, layered pricing for B2B adds further complexity, with customer-specific agreements, discounts, and multi-market price lists. Meeting both B2C and B2B requirements requires a flexible system that can adapt to these challenges without duplicating effort or data silos

SM: Can you explain the importance of a flexible pricing model—with price lists and customer-specific agreements—and how it provides competitive advantage?

BF: A flexible pricing model is critical because it allows businesses to meet customers where they are. With price lists, you can handle multi-market requirements such as local currencies, taxes, and regional pricing strategies without duplicating products or content. Customer-specific agreements add another layer, enabling tailored discounts, terms, and contract pricing that strengthen long-term relationships, especially in B2B. Delivering the right price to the right customer at the right time drives both customer loyalty and profitability.

SM: How does headless commerce empower businesses to create unique user experiences that are not possible with traditional platforms?

BF: Headless commerce separates the frontend from the backend and connects them through APIs. This architectural shift gives businesses complete freedom to design and build user experiences with the tools and frameworks of their choice, whether it is Next.js, Flutter, or something entirely custom. Unlike traditional platforms that lock you into predefined templates and rigid workflows, headless empowers teams to create experiences that truly reflect their brand and customer journey. It also enables seamless integration with emerging technologies, from AI-driven personalization to new devices like wearables or voice assistants. In short, headless unlocks creativity, speed, and adaptability that traditional systems simply cannot match.

SM: What role do you see open-source technology playing in the future of web development and commerce, and how does that philosophy inform Crystallize landscape, beyond headless?

BF: Open-source technology will continue to play a central role in the future of web development and commerce because it accelerates innovation and lowers the barrier to entry for developers. By building on shared knowledge and community-driven projects, businesses can move faster and avoid reinventing the wheel. This philosophy goes beyond headless commerce itself - it is about fostering a community where developers can contribute, learn, and shape the future together. In the long run, open source not only strengthens our platform but also ensures that our customers benefit from the collective progress of a global developer community.

BARD FARSTAD

CEO, Crystallize Headless Commerce

Bård is a serial entrepreneur and the CEO of Crystallize, a modern headless e-commerce service built on his vision to reimagine e-commerce platforms. His extensive career is defined by a history of founding and leading multiple companies, demonstrating a broad range of expertise that spans product, engineering, marketing, and digital strategy. In addition to Crystallize, he founded Snowball Digital, a product design and engineering firm, and Skiens Cykkelfabrik, a bicycle design company. Notably, he spent nearly two decades as the founder and chief strategist of eZ Systems, a content management software company, where he was a key driver of its engineering and product direction. With long-standing experience working with organizations across Europe and the USA, showcasing his commitment to strategic guidance and business development. He is based out of Norway.

EXPERT VOICES

Subrata Mukherjee: How has the relationship between agencies and tech companies evolved from your time at Marketo and TruEffect to today's landscape?

Ellen Weinstein: The relationship has evolved from a clear client-vendor dynamic to a symbiotic and often competitive partnership. During my time at Marketo and TruEffect, tech companies provided the platforms, and agencies were the buyers. Today, the lines have blurred, and we see what I call a "stack convergence." Some tech companies have developed in-house agency capabilities, while agencies have built their own proprietary tech stacks and data lakes to gain a competitive advantage. This has created a new ecosystem where value is derived not just from the technology itself, but from the ability to seamlessly integrate a company's data with advanced AI and automation tools.

SM: In your opinion, what's the most significant technological leap that has occurred in ad-tech over the past decade?

EW: The most significant technological leap has been the shift to AI-powered programmatic ecosystems. A decade ago, programmatic was about automating the simple buy-and-sell of ad space. Today, it's a sophisticated, automated ecosystem that uses machine learning to predict consumer behavior, optimize bidding strategies, and deliver highly personalized ad experiences in real-time. This has transformed ad-tech from a purely transactional business to a predictive science, enabling marketers to target audiences with unprecedented precision and achieve a level of campaign efficiency that was previously unimaginable.

SM: How do you foresee AI continuing to transform creative production and multi-channel services, building on the concepts of "Smart-shoring" from your time at We Are Amnet?

EW: AI will continue to revolutionize creative production by enabling generative and personalized advertising at scale. Building on the "Smart-shoring" concept, AI will automate the entire creative lifecycle. It will analyze consumer data to generate hyper-personalized ad creative in real-time, blurring the lines between ad creation and ad delivery. Instead of static banner ads, we will have dynamic assets that adjust in real-time based on a user's context. This will free up human creative teams to focus on high-level strategy and brand-building, while AI handles the real-time execution across all channels, from social media to digital out-of-home.

SM: After decades of consulting, how do you personally define innovation?

EW: My personal definition of innovation is the intersection of a new idea and a solved problem. It's not just about creating something novel in isolation; it's about applying that novelty in a way that delivers measurable value and fundamentally changes a business process or a customer's experience. This is what we're seeing with AI today—it's not just a new technology, but a catalyst for business model innovation. True innovation requires a relentless focus on outcomes and the courage to challenge the status quo to achieve them.

SM: What is the one thing you are most excited about for the future of business in an increasingly AI-driven world?

EW: I am most excited about the potential for human creativity and strategic thinking to be amplified by AI. AI will take over the mundane, repetitive tasks that have historically bogged down business operations. This will free up human capital to focus on higher-order, cognitive tasks that require empathy, creativity, and strategic foresight. I believe AI will become a co-pilot for innovation, enabling us to build products, services, and companies that we can't even conceive of today, ultimately accelerating the pace of human achievement.

ELLEN WEINSTEIN SALAMAY

Seasoned Business Development Leader, Ex-Marketo, Ex-Cognizant

Ellen is a seasoned business development and strategy professional with a proven record of helping companies leverage technology to achieve significant growth. Her current focus is on AI adoption and digital transformation, which she applies in multiple roles, including as an Advisor at Covington Place Partners and Sr. Consultant at Priorise. She also serves as Vice President of Business Development for Accounti, an AI-driven financial services firm, and InCommon®, which builds AI-capable teams. With a career spanning over two decades, she specialized in connecting tech startups with agencies and media companies. Her extensive background includes senior leadership positions at prominent firms such as Marketo, Cognizant Technology Solutions, and Xen.AI, where she consistently drove business growth, forged strategic partnerships, and optimized marketing and operational strategies across the advertising and media industries. She is based out of New Jersey, in United States.

EXPERT VOICES

Subrata Mukherjee: Your seven years at Rave Digital involved leading large-scale product teams and driving hundreds of millions in client revenue. How do you see AI being applied in the near term by corporations across the world?

Brian Glassman: In the near term, AI adoption across corporations is moving beyond fragmented experimentation toward focused, high-impact deployment. Initial efforts often involved applying AI to small, disconnected tasks in various departments. This scattered approach failed to deliver meaningful gains in productivity or profitability, largely because the applications lacked scale and integration. As a result, many organizations found themselves with underutilized AI investments and no clear path to transformation. Companies are concentrating AI implementation within specific departments where full-scale adoption can yield measurable results. Nowhere is this more evident than in software development and DevOps. In these domains, AI tools are being deployed across entire teams, not just in isolated pockets. Developers are incorporating AI coding assistants into their daily workflows, driving significant improvements in efficiency and consistency.

SM: How do you see AI being applied in the midterm by corporations across the world?

BG: In the midterm, AI adoption will evolve from isolated deployment within a single function to broader application across interrelated domains. The first phase of this expansion builds on current momentum in software engineering and DevOps, where AI is already demonstrating measurable productivity gains. From there, adoption will extend into high-response environments such as cybersecurity, where the speed and consistency of AI outperform traditional human-led monitoring and response. These functions are ideal proving grounds for AI's operational reliability because they demand rapid analysis, low error tolerance, and round-the-clock execution. Following success in these high-velocity technical domains, AI will begin to penetrate adjacent fields such as mechanical engineering, where it can support tasks like systems modeling, simulation, and mathematical analysis. At the same time, corporations will target high-volume, low-risk workflows for automation. Areas like data entry and standardized legal reviews offer strong opportunities to reduce manual effort and improve accuracy without exposing the business to meaningful strategic or compliance risk.

SM: What should corporations be doing now to prepare for these near and midterm AI applications?

BG: To prepare for near and midterm AI applications, corporations should begin by mapping their operational use cases according to risk. Understanding which functions are high-risk versus low-risk enables more strategic alignment when deciding where to pilot or scale AI solutions. This risk-based portfolio view helps organizations avoid deploying AI in areas where the margin for error is too thin or the business impact is unclear. Companies that monitor these adoption patterns are better positioned to benchmark their own maturity and avoid falling behind. Smaller companies will likely need to rely on commercially available, off-the-shelf tools that offer plug-and-play functionality. In contrast, larger enterprises can consider customized deployments and may benefit from working with experienced consulting firms like AI Inspire to design and implement solutions tailored to their needs. Regardless of size, every organization should maintain a dual focus. Internally, they should experiment with emerging technologies to build familiarity and surface practical applications. Externally, they must track trends in competitor behavior and broader industry movements to guide prioritization.

DR. BRIAN SCOTT GLASSMAN

Chief Product Officer & Head of Client Engagements at AInspire.ai

Dr. Glassman is a seasoned product management executive, engineer, and product strategist specializing in generative AI. As the co-founder and Chief Product Officer at AInspire.ai, he leads client engagements and product strategies, collaborating with C-level executives to develop and deploy AI solutions for large enterprises. With a career dedicated to leading technology innovation, he previously served as Director of Product Management at Electra Vehicles, where he built the product function for AI-powered solutions in the automotive sector. His extensive experience includes a seven-year tenure at Rave Digital, where he led product management and customer engagements for Fortune 1000 clients. Furthermore, his background as an engineer and product leader at Cameron, a Schlumberger company, highlights a unique expertise in technical product development. He is based in New York.

EXPERT VOICES

Subrata Mukherjee: In your view, what's the most effective way for a CIO to ensure their organization is ethically and responsibly using AI and automation?

Chris Pena: It begins with robust governance frameworks and organizational transparency. CIOs must establish clear ethical guidelines, ensure data integrity, and prioritize fairness and accountability in AI model development and deployment. Close partnership with HR, legal, and compliance teams is essential for creating comprehensive AI ethics frameworks. Regular audits, employee education programs, and including diverse voices in design processes help build organizational trust. Ethical AI is not just regulatory compliance but a powerful brand differentiator in the marketplace.

SM: What is the one thing you are most passionate about in your current work at Salesforce?

CP: I am most energized by helping CIOs connect strategic vision with practical execution in the age of AI. Technology is often perceived as abstract, but when CIOs successfully align AI and digital capabilities with tangible business value such as faster growth, enhanced customer engagement, and smarter operations, transformation becomes genuinely meaningful.

Witnessing those breakthrough moments with technology leaders is what continues to drive my passion for this work.

SM: Finally, what is your ultimate vision for the future of the CIO role and its impact on the business?

CP: The CIO of the future will function as the Chief Transformation Officer, operating at the critical intersection of technology, data, and business strategy to guide companies through constant change and disruption. Through writing both Déjà Boom and Emotional Intelligence, I have argued that transformation is never purely about systems, it is about leadership, culture, and adaptability. Future CIOs will be evaluated not on system uptime, but on their ability to enable new revenue models, orchestrate complex business ecosystems, and champion ethical AI implementation. Their impact will be measured through growth acceleration, organizational trust, and business resilience in an increasingly digital world.

SM: At PVH Corp. and Aspire Lifestyles, you were a hands-on leader of major digital transformations. At PVH, you led an agile transformation that delivered a 40% faster time to market. What was the most critical step in that process that enabled such a significant efficiency gain?

CP: The breakthrough was shifting our mindset from projects to products. By organizing teams around business outcomes rather than siloed initiatives, we empowered cross-functional squads to take end-to-end ownership of their work.

Combining this with disciplined agile practices and strong product management enabled faster decision-making and continuous iteration. Ultimately, it was not the tools or processes that unlocked the 40% improvement, it was the cultural shift toward empowerment and accountability that transformed our delivery capability.

SM: As the Co-Executive Sponsor of LatinForce NY, what role do you believe diversity and inclusion play in building a successful technology organization?

CP: Diversity represents a genuine competitive advantage in technology development. Different backgrounds and perspectives lead to richer problem-solving approaches and more inclusive product development. Inclusion ensures those diverse voices are genuinely heard and valued, creating stronger and more resilient teams. For me, diversity and inclusion is not just a moral imperative, it is a business accelerator that drives innovation, organizational resilience, and customer relevance in our increasingly diverse global marketplace.

CHRIS PEÑA

VP, CIO Go To Market Industry Lead at Salesforce | Ex-CIO Aspire

Chris is a strategic technology leader and CIO advisor with a deep expertise in AI, digital transformation, and e-commerce. In his current role as VP, CIO Go To Market Industry Lead at Salesforce, he is responsible for developing and executing the company's value proposition for Chief Information Officers across various industries. As Global VP of Consumer Technologies at PVH Corp., he managed the global Salesforce platform. Prior to that, as General Manager CIO at Aspire Lifestyles, he built a global technology team from the ground up, modernizing operations with cloud infrastructure and automation, and successfully managing massive budgets. His extensive experience also includes senior leadership roles at EY, Razorfish, and Pearson, where he led digital strategy, platform implementations, and engineering. He is based out New York in United States.

EXPERT VOICES

Subrata Mukherjee: How do you translate complex data infrastructure projects into clear business value for non-technical stakeholders?

Manish Bhandari: I've found the key is to frame data projects in terms of outcomes, not pipelines. Non-technical stakeholders don't care about Spark clusters; they care about cost savings, revenue growth, or customer experience. For example, at Wayfair, I explained how our analytics pipelines would optimize marketing spend and directly improve ROI. My role is often to act as a translator: taking the complexity of distributed systems and presenting it as clear, trusted, and relevant outcomes.

SM: What is the most impactful business outcome you've seen directly result from a well-designed data pipeline or platform?

MB: The most impactful outcome I've seen was at QuantumBlack, where a predictive maintenance platform delivered over $100 million in savings for a client. By ingesting millions of sensor readings and applying machine learning to predict failures, the platform helped them avoid costly unplanned downtime, optimize maintenance schedules, and improve asset utilization. The pipeline didn't just produce data; it delivered actionable business decisions. Another example at Tinder was a data platform that reduced pipeline runtimes by over 60%, directly accelerating product experimentation and impacting user growth and retention. These examples taught me that the real value of a data pipeline is not in the technology, but in how it acts as a force multiplier for the business—enabling faster decisions, reduced risk, and new revenue opportunities.

SM: In your experience, how can a data engineering team effectively collaborate with marketing or product teams to ensure data is leveraged for maximum impact?

MB: Effective collaboration begins with aligning on business outcomes, not pipeline design. The key is to ask, "What decisions do you need to make, and how will success be measured?" At Wayfair, I co-defined KPIs with the marketing team before writing any code, ensuring our pipelines were strategically relevant. At Tinder, we empowered product managers with self-service access to data, allowing them to run their own experiments and freeing engineering to focus on scaling. By embedding data engineering as a strategy partner, not a service provider, and co-owning KPIs, we ensure data is a driver of business outcomes, not just a byproduct.

SM: What is one technology or trend in the data space that you are most excited about right now?

MB: I am most excited about the intersection of data engineering and Generative AI. Technologies like retrieval-augmented generation (RAG) and agent frameworks mean we can build systems where AI agents can query and reason over data directly. I see this as the next big step: moving from data as a tool for people to data as an intelligent collaborator. We can build autonomous data products—pipelines that not only deliver clean data but also proactively surface insights, anomalies, and opportunities in natural language.

SM: Looking to the future, what do you believe is the next big challenge or opportunity in data infrastructure?

MB: The next big challenge is designing data platforms that effectively serve both people and AI agents. Today's systems are built for analysts, but tomorrow's must support machines that query and reason over data with high efficiency and trust.

The opportunity is huge: moving from data warehouses that power dashboards to data platforms that proactively deliver insights and actions through AI, fundamentally changing how businesses operate.

MANISH BHANDARI

Ex-Tinder, Ex- Wayfair, Currently building an AI-First Data platform

Manish Bhandari is a seasoned professional with a strong focus on building and maintaining analytics infrastructure and architecting data solutions. He is currently in a new venture at a Stealth company, leveraging his extensive experience from leadership roles in the tech and data sectors. Previously, he served in data infrastructure roles at companies such as Tinder and Wayfair. As a Senior Data Engineer at QuantumBlack, AI by McKinsey, he was instrumental in building impactful analytics solutions for C-level executives and designing data pipelines to support AI and data science projects. His experience is built on a foundation of managing corporate data infrastructure at Genscape. Manish's core expertise lies in Data Engineering, Advanced Analytics, and ETL, with proficiency in technologies including Python, Databricks, and AWS. He is based out of San Francisco, United States.

EXPERT VOICES

Subrata Mukherjee: From the 1992 Barcelona Olympics to Warner Bros. Discovery, you have been at the forefront of digital change. You led a multi-year digital transformation at HBO from a "non-linear broadcast supply chain." What does that mean for a layperson, and what were the biggest challenges?

Michael Gabriel: A non-linear supply chain in the media and entertainment industry describes a content production and distribution model where information, resources flow in multiple directions rather than in a single, sequential path. Unlike a linear supply chain such as an assembly line—where a product is created, packaged, distributed, and consumed in a fixed order—a non-linear supply chain is dynamic and interconnected and affects multiple parts of a company, including business partners and customers. Changing one aspect of that multi-dimensional interaction will affect multiple processes and parts of a company. So, understanding data dependencies and process flows is crucial to success, and you need to ensure that change management is properly handled with all affected staff communicating together about the impact of the changes and the best way to implement them. Otherwise, lots of operational aspects can go awry. That was complex to handle with a digital transformation that had to continue to support regular broadcast TV product creation and supply chain needs vs. streaming media through the Internet.

SM: Having worked with traditional media companies, how has the transition from physical content distribution to digital streaming transformed technology strategy?

MG: The biggest change is with the timing of the business needs. Content distribution used to have release timeframes and windows, all happening in a planned and controlled sequence with long lead times. Digital streaming places media & entertainment products into the consumer environment 24x7, which places IT development and support into a whole new world of business urgency and unpredictability. The rate of change is now far faster, since consumers see that and expect it from other digital products. And the supply chain protection is different since there are threats from both consumers and bad actors trying to gain access to digital products without paying, or to the content itself for illegal sites, or to gain access to consumer PII, including credit card information, or for ransomware purposes. So, the involvement of the technology group now has a far greater need around the clock, and in many cases around the globe in different countries. The barriers to media content distribution have been reduced to a technical capability, rather than the previous business scenario that was based on negotiations with third-party distributors around the world. Those deals still happen (such as a content creator having a streaming app, but also providing content to other streamers or to traditional TV-based distribution companies, radio stations, and retail stores). So our support landscape has grown tremendously, and our development and enhancement cycles have also grown exponentially.

SM: Finally, looking ahead, what is the single greatest challenge facing boards today in governing technology, and how can they overcome it?

MG: The greatest challenge is the lack of Boards' understanding of the risks and opportunities concerning the impact that technology has, or could have, on their business. They need to relentlessly reinforce that status on technology be conveyed in a business-understandable manner, request a dashboard focus and meaningful metrics on what defines their technology and cybersecurity success, ensure that impactful projects and initiatives are completed to attain their goals, and confirm risk mitigation and issue resolution are clearly being tracked and reported on. Once that's in place they'll be able to meet this challenge to ensure proper governance and oversight is being accomplished.

MICHAEL GABRIEL

Ex CIO at HBO, NBA | NACD Certified Board Director

Michael is a senior technology and cybersecurity executive with experience providing C-level leadership, governance, and advisory services. He is currently an independent Board Director and provides professional services for companies like Warner Brothers Discovery (WBD). His career is highlighted by a 14-year tenure as EVP & Global CIO at HBO, where he was the initial executive to spearhead the groundbreaking HBO GO initiative and led a company-wide digital transformation. An inventor on six technology patents, he also played a key role in information security governance across the Time Warner enterprise. Prior to HBO, he served as Global CIO for the National Basketball Association (NBA), where he implemented the first NBA retail store technology system. He also held senior IT leadership positions at EMI Records Group and NBC Olympics.. He is based out of New York

EXPERT VOICES

Subrata Mukherjee: You've been instrumental in growing executive thought leadership by 80% year-over-year. What is your approach to positioning leaders as credible voices in a crowded digital space?

Amanda Murphy: Where I've found success in building executive brands online is by interviewing them and working closely with them to learn what they are passionate about, why they are who they are, what influences their actions and what are their personal values. Matching their passions with thought leadership topics creates authenticity in their posts/shares/stories and ultimately creates share-worthy, take-notice content.

SM: How do you use digital storytelling and content strategy to bring an organization's mission to life for its employees?

AM: Video production, animation, and design are all common tools I use for internal messaging to employees, and for various reasons. 1. We're all tired of emails and newsletters. We get so many, we don't know which to prioritize or pay attention to. Digital messaging strategies are a welcome change of pace. 2. Senior leaders aren't able to get in front of all of their team members very often. Video is another way a senior leader can give a personal interaction without being in-person. I leverage content strategy techniques to use the digital storytelling pieces in multiple ways - on various mediums, in a variety of forums - to maximize the impact and ROI.

SM: In your experience, what makes a communication plan truly effective in driving a cultural shift?

AM: A communications plan and an Organizational Change Management plan go hand in hand and must be aligned. There must be multiple communication touchpoints, along with multiple opportunities for training, open space for sharing feedback or concerns, a long tail so those resistant to change can get to their ah-ha moment (which could be months later)...These plans are extremely important because they create a roadmap of the change and how you'll get through it successfully. You pinpoint each stakeholder group or audience and how best to communicate to them, you plan multiple touchpoints for each group, you're able to share the roadmap with senior leaders, allowing them to visual the change curve (and be part of fueling the adoption), and determining which avenues will be utilized to roll out the change most effectively. A transformation involves People, Process, and Technology. Many times, a new technology is rolled out without the People or Process considerations. Or a new Process is implemented that forgets to consider the technologies already leveraged and impacts the new process creates, and the people who are touched by the new process. A change won't be successful unless it has considered all 3 of these parts, which a communication AND an OCM plan help drive.

SM: You led a business case analysis that resulted in a strategic pivot and substantial cost savings. How do you build a data-backed case for change that is compelling to both C-suite executives and project teams?

AM: In this particular instance, the data led us to the change. We were all hoping for different results - we liked the product, we wanted it to work and wanted to keep it in market. But the reality was (according to the data analysis) we would need to put significant more investment in to make it a viable product and keeping it as-is didn't align with our brand and being the best for our customers. I've seen data manipulated to create the results or preferred story, but that isn't going to be beneficial in the long term. You have to use data honestly and look at it with an unbiased view. This is also where having multiple voices/perspectives in the room help curate the big-picture. Even strong, quality data won't drown out the loudest voices in the room. It really comes down to trust - trusting the quality of your data and trust in your reporting. Data driven decision making will only work with trust.

AMANDA MURPHY

Technology Director, Chief of Staff at Nationwide Insurance

Amanda is a seasoned leader specializing in transformational leadership, business operations, and organizational change management (OCM) within the technology sector. As the Technology Director, Chief of Staff for Security & Infrastructure at Nationwide, she formulates and executes strategic initiatives for a large, Fortune 100 organization. Her proven expertise in building and leading high-performance teams is demonstrated by her success in managing a complex merger. With a strong background in strategic communications, digital storytelling, and executive messaging, Amanda is skilled at creating strong corporate cultures and driving successful outcomes by bridging the gap between high-level strategy and tactical execution. She has been recognized with a Gartner Communications Award for her work in transforming internal communications. She is based out of Westerville, Ohio, United States.

EXPERT VOICES

Subrata Mukherjee: Based on your experience, what are the two or three most common mistakes you see startups make in their initial go-to-market strategy?

Jeff Rishi: Typical go to markets for new leaders or to younger leaders typically involve overall "market penetration" approaches opposed to customer acquisition focus. Seed or pre-seed or series A investment capital is spent on permanent hire senior leadership in marketing and sales, brand building and systems too early in the maturation cycle and too heavily. Fundamentally, unless the founder/CEO has started and grown a company previously, the experience is seldom inherent in the owners/founders. By front loading the cost/overhead and expecting great things to happen leads most often to drastic maneuvering because the lack of fit is discovered after an alignment period. Wasting time and money. Learning the hard way myself but as well, though being a sales and marketing lead for other ground floor startups I see this pattern emerge and become all too common. The lesson here for all that are new to the founding game is that founders are usually good at one thing. Be it technical, or people or finance or sales and are seldom have enough depth in the needed disciplines of business to be effective. Thus creating a need to "hire the best". The best come at a higher cost, unless of course you leverage fractional services. Focusing on your "network", centers of influence and using fractional services for a period will not only ensure that you can easily make changes to the C suite, it also shows the VC that you are fiscally responsible as a founder and increase the trust factor in the relationship.

SM: How does the role of a fractional CRO differ from being an enterprise sales leader, and how do you measure success in this model?

JR: The fractional service can be highly flexible. Providing what is needed and how often it is needed opposed to having the continuous overhead attached to sales. Consulting on a hours/month basis OR for a project or target in a specific timeframe. Achieving the target, regardless of approach, is the measuring stick for the engagement. The difference compared to being an enterprise sales leader is that in this role one is bound to the cultural limitations (typically conservatism), single, very narrow target (the guard rails) field both geographic and revenue size which is both a negative and a positive. Providing focus for marketing and sales departments but at the same time limiting the opportunity to expand scope and reach of new potential markets. Which has more merit in today's business climate? (A car can be a transport, an offroad vehicle and a truck not just a car). If I offer an AI chatbot to retailers or if I sell an AI chatbot to those that need it are two very different approaches both have their merits and both can be successful but the former will eventually either create a product specific to an additional market or, reach its limit where the latter will continue to appeal to the mass market.

Suppose your legacy CRO has mastered selling into your main target vertical. For example, into manufacturing. You as a CEO now see the future is in retail. Is your current CRO going to be able to bring 20 yrs experience in retail overnight? What about venturing into new geos? In the past, loyalty and a sense of ownership and working your way up was good for the leader as well as the company. Today, having the choice of a multi-CRO approach provides a lower cost and higher ROI. This is not only true for sales and marketing leadership but for almost every facet of today's enterprise. In terms of which types of companies are best suited for this "C level rotation" approach, I would say that all companies worldwide have hired consultants from the outside to support their current CXO as the value of a non-bias set of eyes and ideas is most times priceless to an entity that seeks to maximize shareholder value.

JEFF RISHI

Founder/GM at Occumoda, Ex- Cignex, IBM

Jeff Rishi is a visionary business and technology leader with a career defined by driving significant revenue and commercial growth, having generated over $1.23 billion in software and services revenue. As the Founder and General Manager of Occumoda, he champions a contrarian, multi-vendor approach to AI/ML and software development, leveraging a curated collective of elite talent to activate client visions. A proven leader in sales and go-to-market strategy, Jeff has held senior roles at companies such as CIGNEX Datamatics and IBM, where he was a key evangelist for open source and Big Data solutions. His expertise lies in transforming complex, bleeding-edge technologies into tangible business outcomes while scaling global operations and providing fractional CRO services to startups and scale-ups.. He is based out of United States.

EXPERT VOICES

Subrata Mukherjee: Your coaching focuses on helping leaders craft stories that resonate with hiring managers. What is the single biggest mistake you see talented leaders make when telling their professional story?

Jennifer Speciale: Most talented leaders mistake a resume recap for a leadership narrative. They lead with chronology, responsibilities, and tools - rather than a clear thesis about the business problems they solve and the measurable outcomes they create. Great stories begin with a one-sentence value promise ("Here's the kind of inflection I drive"), then prove it with a flagship case: context → pivotal decision → result (ideally with quantified impact). They're tailored to the room's priorities, avoid jargon, and make the "so what?" unmistakable. Finally, they point forward: how that same value scales in this role, with these stakeholders.

SM: You mention helping leaders build networks to uncover roles no one else knows about. Can you give an example of how a leader can strategically build this type of network?

JS: The most effective networks are built through curiosity, not need. Leaders who consistently reach out with thoughtful questions—about emerging trends, leadership lessons, or business shifts—create genuine connection long before an opportunity appears. One client, for example, set a rhythm of 15-minute conversations each week with adjacent leaders in AI and Design. Within months, he was hearing about roles that never reached a job board. The lesson: when you lead with curiosity and value, you become the person others think of before the posting goes live.

SM: What does "executive presence" mean in today's digital-first world, and how can a leader cultivate it effectively?

JS: Executive presence today is less about physical stature and more about signal clarity. In a digital-first environment, leaders are judged on how they show up in short bursts: a LinkedIn post, a Zoom meeting, a two-line email. Presence comes from three things: clarity of message, visible conviction, and consistency across platforms. I coach leaders to "design the impression," asking: what do you want others to feel in the first 30 seconds of engaging with you?

Also, start with listening tours before strategy decks. Spend the first 30 days not imposing, but surfacing the story people tell themselves about their work: "What are you proud of? What worries you? What do you want leadership to understand?" Once voiced, those narratives become anchors for change. I've seen leaders gain instant credibility simply by reflecting those words back, then aligning incentives accordingly. Integration fails when it is treated as a process; it succeeds when treated as relationship architecture.

Retention absolutely hinges on psychological safety in uncertainty. After Tremor Video's acquihire, I prioritized radical transparency: every team member knew what decisions were being made, when, and why. We built micro-rituals—weekly forums, direct access to executives—that signaled, "You are seen here."

SM: What are the primary cultural challenges when blending a smaller, innovative team with a large, established corporate giant?

JS: The biggest challenge is pace. Smaller, innovative teams thrive on speed, improvisation, and intimacy. Larger corporates operate through scale, governance, and hierarchy. When merged, frustration brews if neither side adapts. The key is translating—not transplanting—culture. In one integration, I created "bilingual spaces" where startup-style prototyping coexisted with corporate risk frameworks. This gave each side permission to learn from the other. Leaders who explicitly value both cultures create bridges instead of silos—and prevent the all-too-common erosion of the very creativity the acquisition was meant to capture.

JENNIFER SPECIALE

CEO, Speciale Executive Leadership | Ex- Amazon, McKinsey, Google...

Jennifer is a Global Executive Leadership Coach where she partners with senior leaders to showcase Executive Presence, Influence, and Impact. Her coaching practice is built on over two decades of experience in strategic talent advisory, having directly shaped executive teams at the world's most innovative companies. Jennifer's career highlights include establishing the pioneering UX Design team in NYC for Google, building the foundational global hiring processes for Indeed during its hypergrowth phase, and leading the global expansion of the Design executive talent function at both McKinsey & Company and Amazon. She is an expert in organizational Change Management and Executive Search, dedicated to translating senior leaders' expertise into compelling narratives that unlock C-suite and VP opportunities across the Design and Technology landscapes. She is based out of New York city.

WE NEED YOU!

WELCOMING OUR NEXT SET OF THOUGHT LEADERS

The remarkable success of the Ai6x6 Organization Transformation Series is built upon the **hard-won wisdom** of executive leaders who have successfully navigated the AI transformation within their industries, turning theory into a deeply practical, actionable guide. As we expand the series, we are extending a highly selective invitation to a new cohort of pioneers—individuals whose professional journey is at the **leading edge of AI strategy and execution**. We are looking for executives who possess a proven track record of rewriting business models, mastering unforeseen ethical and competitive challenges, and offering a **clear blueprint** for the next 3-5 years of the AI-Era.

Contributing to the Ai6x6 Series is a unique opportunity to **cement your professional legacy**, joining an exclusive, curated network of global authorities whose insights are distilled into executive-level knowledge cards, directly shaping the governance and innovation decisions of the world's most influential companies. Our selection process is based strictly on **pure expertise and demonstrated leadership** in the AI-Era, reinforcing our commitment to editorial integrity and authentic thought leadership—there is no monetary exchange involved. Select contributors will also receive an invitation to blog on our expert network microsite, further amplifying their platform. To confidentially inquire about contributing your expertise, please contact us via our dedicated portal at **www.Ai6x6.com/contact** or by scanning the adjacent QR code.

Contact

SUBSCRIBE TO OUR NEWSLETTERS: To stay ahead of the curve and receive continuous, executive-level intelligence on the rapid evolution of the AI-Era, we invite you to register for our **Ai6x6 Weekly Newsletter**. Each edition delivers curated insights directly from our thought leaders and highlights critical shifts in AI strategy, governance, and provides relevant excepts from our website and books. Keep your finger on the pulse of the future and ensure your company remains competitive by signing up on the same page

PLEASE REVIEW US ON AMAZON!

Customer Blueprint

We know your time is valuable, but if the Ai6x6 Organization Transformation Series delivered the real-world, actionable strategies you needed, please consider leaving a brief review on Amazon. If we missed the mark, we would also like to know, and that is the only we we can improve. Your honest feedback is crucial; it helps fellow executives discover this unique framework and benefit from the insights of our thought leaders. Thank you for supporting our mission!

BLUEPRINT SERIES

SECTION B

Human-AI Synergy

Ai6x6™ Transformation Blueprint

Story 1 — Enigma Codebreakers (1940–45): German U-boats encrypted communications using constantly shifting ciphers, forcing Allied cryptanalysts to chase patterns that dissolved daily. Manual decoding could not keep pace with wartime urgency, and insights often vanished before they could be confirmed. Electromechanical devices accelerated the testing of solutions at unprecedented speed. Humans interpreted these outputs, correcting false leads and refining assumptions. The fusion of machine acceleration & human judgment reshaped the trajectory of the war.

Story 2 — Turing's Bombe & Operators (1940–45): Breaking enemy codes required evaluating thousands of cipher possibilities rapidly, and the Bombe machine generated candidate solutions at speeds humans could not match. Yet its outputs were incomplete, ambiguous, and often misleading. Human operators validated results, corrected machine assumptions, and filtered noise from meaningful signals. Their contextual judgment transformed mechanical output into strategic clarity. This interplay of mechanical speed and human reasoning cracked codes that neither side could solve alone. It became one of history's earliest demonstrations of human–machine symbiosis.

Story 3 — Toyota Production System (1950s–70s): Factories struggled with defects, bottlenecks, and unpredictable workloads as machines produced volume without contextual awareness. Workers observed issues that machines could not detect or interpret, prompting Toyota to design a synchronized workflow that aligned human insight with machine precision. Real-time data from the line informed continuous improvement cycles. This human–machine rhythm created a new operational standard for reliability and efficiency. The system became a global model for intelligent production. It demonstrated how augmentation, not automation, drives sustainable excellence.

Story 4 — Salk Polio Vaccine Trials (1952–55): As polio outbreaks intensified, statistical models guided trial design but lacked the ethical nuance required for medical decision-making. Researchers interpreted model outputs through clinical judgment, ensuring safety and fairness while machines processed vast amounts of trial data. This combined system accelerated discovery without compromising integrity. Human oversight balanced computational rigor with moral responsibility. It showed how human values anchor scientific acceleration.

Story 5 — Apollo 13 Mission Control (1970): An explosion crippled Apollo 13, leaving astronauts in mortal danger as incomplete telemetry streamed back to Earth. Engineers relied on simulations to test survival scenarios under extreme constraints, but the machines could only reveal possibilities. Human teams improvised procedures the computers could not imagine, blending machine-generated insights with creative reasoning. It became a defining moment of human–machine crisis collaboration.

Story 6 — Air Traffic Control Modernization (1960s–80s): Surging air travel overwhelmed manual tracking systems, and controllers struggled to maintain safe separation across increasingly crowded skies. Radar automation provided real-time positional data, enabling machines to track aircraft continuously while humans made judgment calls. This hybrid system reduced collisions, increased capacity, and established a global standard for aviation safety. Human–machine coordination became the backbone of modern air travel. The system remains one of the world's most enduring examples of operational augmentation.

LESSONS FROM A CUSTOMER TRANSFORMATION PERSPECTIVE

Customer value strengthens as human–AI collaboration matures from tactical problem-solving to systemic reliability. Early systems protected people by accelerating critical decisions, while later systems enhanced safety, consistency, and trust at scale. Across decades, the pattern is clear: machines improve speed and accuracy, but humans safeguard fairness, empathy, and ethical intent. Customers benefit when technology amplifies human responsibility rather than replacing it. Human–AI synergy creates environments where outcomes are not only efficient but humane—where reliability is paired with care, and precision is paired with judgment. The customer becomes the ultimate beneficiary of shared intelligence.

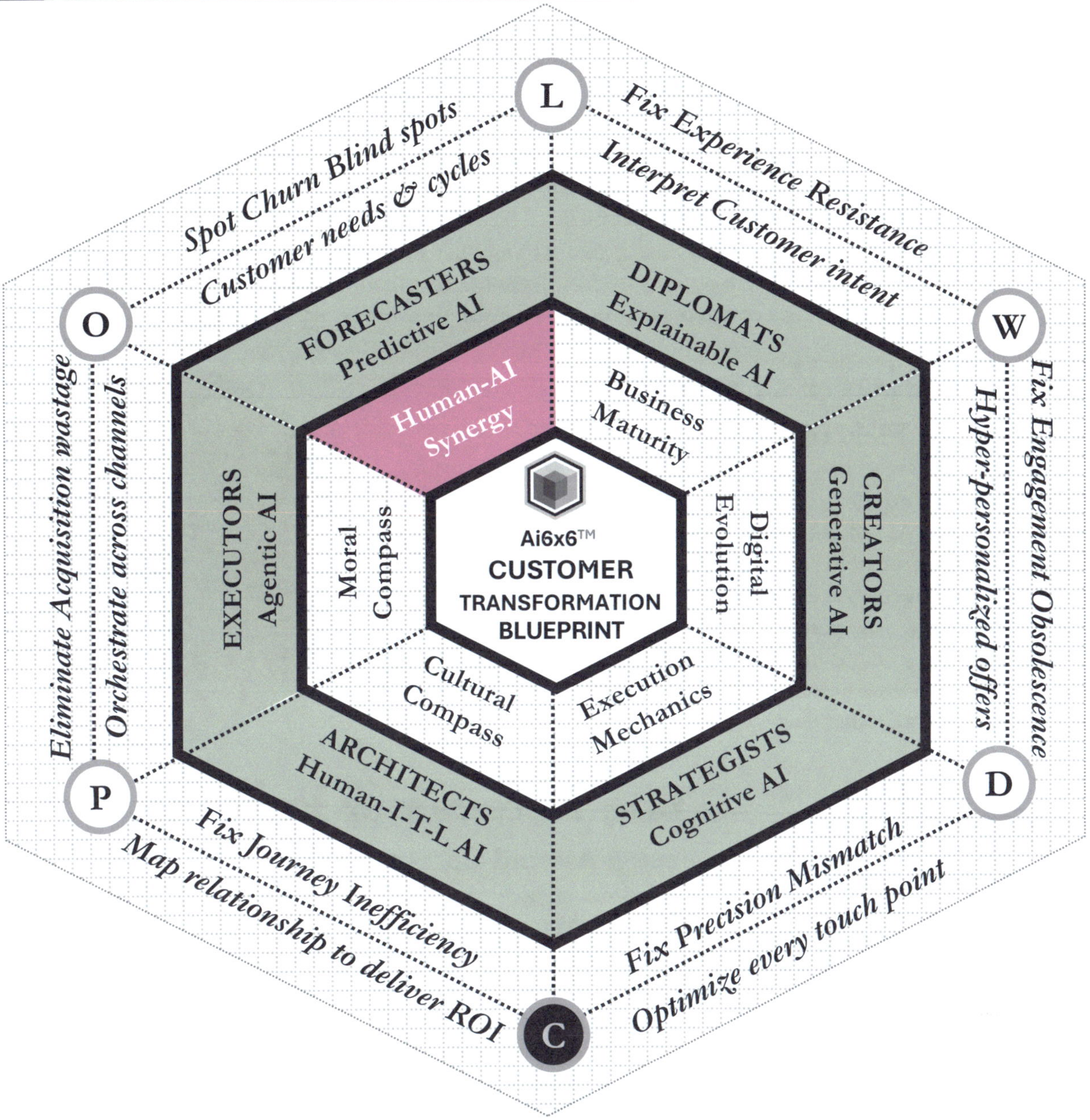

The **Ai6x6 Research Lab** chose this lens to address the profound shift in the sacred contract between the brand and the individual. We believe the first hurdle of customer transformation is the psychological transition from "selling to" an audience to "advocating for" a human being. This lens is vital because it forces the leader's mind to stop viewing the customer as a data point and start viewing them as a complex entity whose needs can be met through a collaborative architecture. It is designed to move the brand from a reactive posture to a state of **Need Anticipation**, ensuring the human spirit remains the guiding "why" behind the automated **Fulfillment Agents**.

It is about shifting the leader's heart from a state of transactional noise to one of creative orchestration. This lens acts as a cognitive bridge, allowing the leader to see AI not as a wall between the brand and the person, but as a diverse set of personas—like the **Diplomat** and **Creator**—that extend the brand's capacity for empathy. Without this shift, the brand remains a distant entity; with it, it becomes an indispensable partner in the customer's life. It is an invitation to shed the burden of mass marketing and embrace a new, augmented intimacy grounded in the courage to be relevant in every micro-moment.

Leadership in the age of intelligent systems demands a new grammar of collaboration—one that transcends the binary of human versus machine and instead frames intelligence as a continuum. The **Human–AI Synergy Lens** redefines leadership not as the art of commanding technology but as the discipline of co-creating with it. It asks leaders to move beyond automation and efficiency toward a deeper integration of foresight, ethics, creativity, cognition, governance, and execution. At its core, Human–AI Synergy is about **shared agency**. It recognizes that leadership today is exercised not only through human judgment but also through algorithmic influence. Every forecast, recommendation, and automated decision carries a fragment of leadership intent. The challenge is not to suppress that influence but to shape it—to ensure that AI amplifies human purpose rather than diluting it. This lens therefore becomes a mirror: it reflects how leaders distribute cognition, emotion, and control across human and machine systems, and how they design organizations that learn, reason, and act as unified intelligences.

The Creator Persona

Expanding the Field of Possibility

Creators represent the imaginative frontier of leadership. Through **Generative AI**, they explore more ideas than any human-only process could conceive. They treat AI not as a threat to originality but as a catalyst for it—a partner in ideation that expands the creative bandwidth of the organization. The Creator's discipline is not about producing more content but about discovering new patterns of thought. They use AI to prototype futures, to visualize complexity, and to democratize innovation. In their hands, generative systems become engines of curiosity, enabling leaders to design from abundance rather than scarcity.

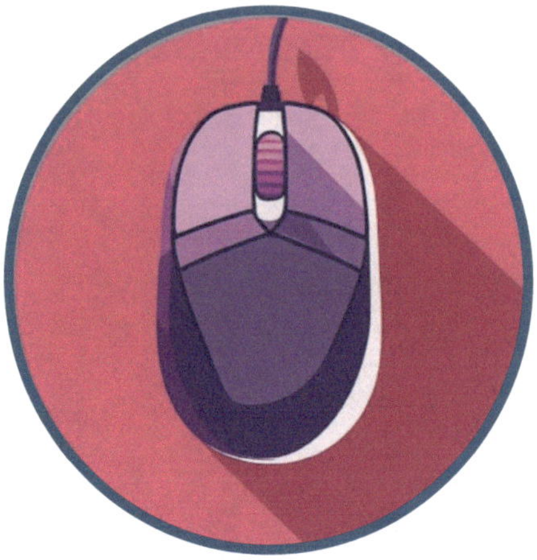

The Forecaster Persona

Seeing Around Corners

Forecasters embody the anticipatory dimension of leadership. They use **Predictive AI** to model futures before committing resources, transforming uncertainty into structured insight. Their craft lies in converting data into foresight—detecting weak signals, pressure-testing assumptions, and rehearsing alternative scenarios. In traditional leadership, forecasting was intuition-driven; in synergistic leadership, it becomes a disciplined dialogue between human imagination and machine probability. The Forecaster's maturity is measured not by how accurately they predict but by how responsibly they prepare. They shift the organization from reacting to events to anticipating them, embedding resilience into its strategic DNA.

The Strategist Persona

Orchestrating Cognitive Intelligence

Strategists use **Cognitive AI** to weave intelligence across functions, geographies, and time horizons. They design strategy as a living system—adaptive, self-correcting, and continuously learning. The Strategist's genius lies in orchestration: aligning distributed intelligences into coherent purpose. They understand that cognition is no longer centralized in the executive suite; it flows through networks of humans and machines that co-reason in real time. Their leadership is systemic rather than hierarchical, emphasizing feedback loops, dynamic modeling, and cross-domain synthesis. In this persona, AI becomes the connective tissue of enterprise intelligence.

The Diplomat Persona:

Building Ethical Bridges

Diplomats operate at the intersection of **Explainable AI** and human empathy. They translate algorithmic logic into moral language, ensuring that machine outputs are understood, trusted, and socially accepted. Their role is to humanize the interface between data and decision, to make transparency a leadership act. In a world where AI can both illuminate and obscure, the Diplomat safeguards the integrity of interpretation. They remind the enterprise that trust is not a technical feature—it is a cultural outcome. By bridging emotional and ethical terrain, they ensure that AI strengthens relationships rather than eroding them.

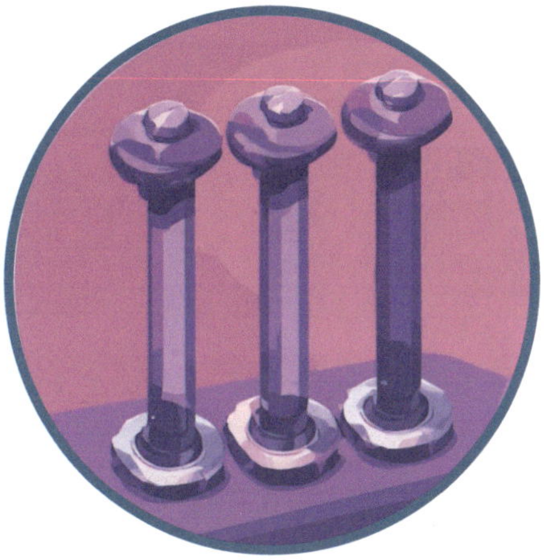

The Architect Persona:

Designing Responsible Collaboration

Architects are the custodians of **Human-in-the-Loop AI**, ensuring that governance and ethical guardrails are embedded into every workflow. They design environments where humans and machines collaborate responsibly—where oversight is not an afterthought but a structural principle. The Architect's leadership is infrastructural: they build the conditions under which synergy can thrive. Their focus is on accountability, transparency, and inclusion, crafting systems that respect both human dignity and machine precision. In their world, collaboration is engineered, not improvised; it is sustained by design integrity.

The Executor Persona:

Converting Insight into Action

Executors close the loop between knowing and doing. Through **Agentic AI**, they embed intelligence directly into operations so that insight becomes action without delay. They represent the kinetic dimension of synergy—the moment where cognition translates into execution. The Executor's mastery lies in operationalizing intelligence: automating decisions that are safe, contextual, and aligned with strategic intent. They ensure that the organization's nervous system is responsive, not reactive, and that every process carries the imprint of learned intelligence.

Together, these six personas form a **leadership spectrum**—from foresight to action, from imagination to governance. They illustrate that synergy is not a single capability but a choreography of roles. The Forecaster anticipates, the Diplomat interprets, the Creator imagines, the Strategist aligns, the Architect safeguards, and the Executor delivers. Each persona represents a distinct mode of engagement with AI, yet all converge on one principle: leadership is now a shared performance between human and machine. The Human–AI Synergy Lens therefore reframes leadership as **co-evolution**. It invites leaders to cultivate humility before complexity, curiosity before certainty, and design before control. It challenges them to lead not by commanding intelligence but by convening it—to orchestrate a dialogue between human wisdom and machine capability that continuously redefines what is possible. In this lens, synergy is not the end state; it is the ongoing discipline of learning together.

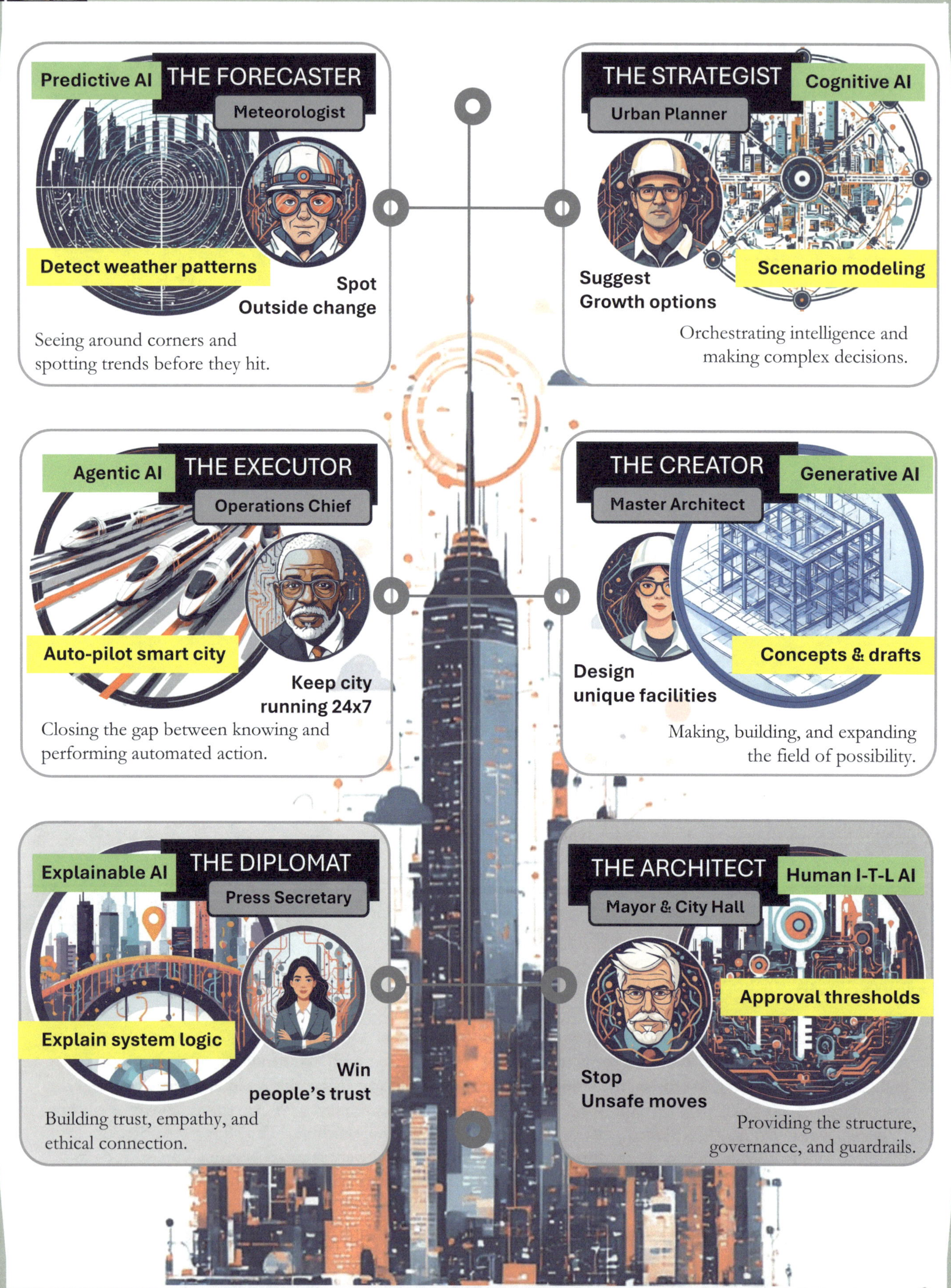
Predictive AI
THE FORECASTER
Meteorologist
Detect weather patterns
Spot
Outside change
Seeing around corners and spotting trends before they hit.
THE STRATEGIST
Cognitive AI
Urban Planner
Scenario modeling
Suggest
Growth options
Orchestrating intelligence and making complex decisions.
Agentic AI
THE EXECUTOR
Operations Chief
Auto-pilot smart city
Keep city
running 24x7
Closing the gap between knowing and performing automated action.
THE CREATOR
Generative AI
Master Architect
Concepts & drafts
Design
unique facilities
Making, building, and expanding the field of possibility.
Explainable AI
THE DIPLOMAT
Press Secretary
Explain system logic
Win
people's trust
Building trust, empathy, and ethical connection.
THE ARCHITECT
Human I-T-L AI
Mayor & City Hall
Approval thresholds
Stop
Unsafe moves
Providing the structure, governance, and guardrails.

Ai6x6™ Transformation Blueprint

Evolution of Humanity

"We are what we repeatedly do. Excellence, then, is not an act, but a habit."

Aristotle (384–322 BC); Trust Sovereignty is built as a technological habit, requiring the repeated, predictable excellence enforced by Ethical AI Constraints and design.

In ancient civilizations, commerce was highly relational, and fulfillment for the merchant was deeply tied to their reputation and the concept of relational trust (Fides in Roman culture). Purpose was found in the honor of the handshake and the maintenance of a relationship based on integrity—a recognition of the customer's inherent sovereignty. The Digital Age introduced sophisticated systems that redefined fulfillment for organizations as the maximization of transactions and the extraction of personal data. The focus shifted from serving the customer's needs to fulfilling the algorithm's need for data, often compromising Trust Sovereignty for profit. In the AI Era, a necessary ethical mandate restores this balance. Product and design teams now find their profoundest purpose in realizing an Ethical Purpose by designing and enforcing Zero-Trust Privacy models. Their professional satisfaction comes from being the architects of digital dignity, ensuring that automated systems operate within Ethical AI Constraints that provably protect the customer, making them advocates for, rather than exploiters of, the customer's sovereignty.

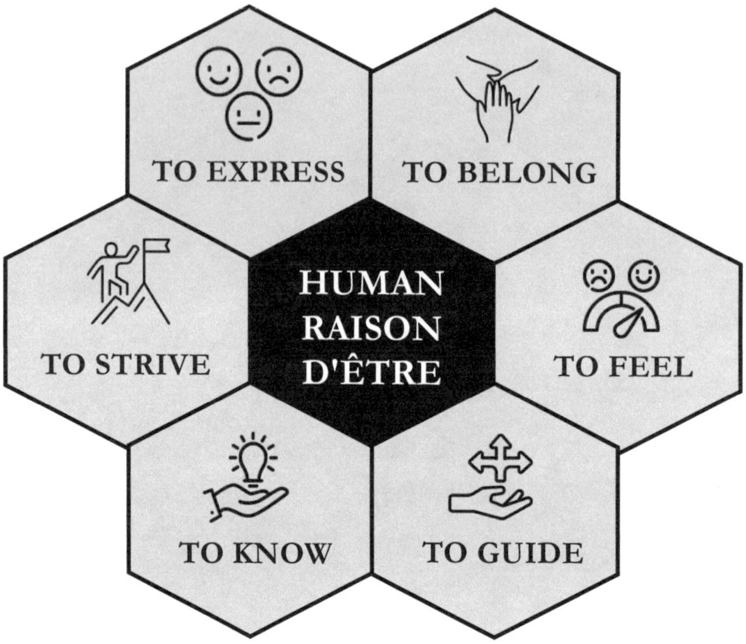

SIX UNIVERSAL ANCHORS OF HUMAN FULFILLMENT

HUMAN RAISON D'ÊTRE (THE REASON FOR EXISTENCE)

In an age defined by exponential automation, the core question of human existence shifts from survival to **significance**. When technology handles efficiency and labor, human value is no longer defined by capability, but by **meaning**. The six **Universal Anchors of Human Fulfillment** define the enduring, non-negotiable contributions of human consciousness—the pillars of a **Raison D'être** that secures our purpose across any era. These anchors affirm that our ultimate mission is to serve as the unwavering source of **Meaning and Moral Authority** in existence. They are the essential acts that only consciousness can perform.

1) TO EXPRESS

The **Anchor of Expression** is the essential human mandate to be the **Source of Novelty and Creation**, generating original ideas and subjective feelings that resist calculation or statistical synthesis. This purpose validates our unique consciousness by asserting that fulfillment comes from the generative act of bringing forth something utterly unique, not merely consuming or replicating. It requires **Uncurated Creativity**—the freedom to explore and define **Aesthetic Judgment**—thereby expanding the conceptual universe with concepts that have no statistical precedent. Through expression, we transform complex inner experience into external reality, upholding our role as the inventors of meaning.

The ultimate benefit of this anchor is the preservation of **Human Originality** and **Intellectual Sovereignty**. By committing to expression, we secure a future that is continually enriched by unique ideas, preventing stagnation and the predictability of automated systems. This anchor maintains the essential spark of invention, ensuring that human meaning remains tied to the capacity to create, **differentiating our value from that of a machine**. It is the source of all artistic, philosophical, and inventive breakthroughs that define a truly dynamic culture.

2) TO BELONG

The **Anchor of Belonging** defines our core identity as social beings, asserting that **authentic connection** is an existential necessity beyond mere digital interaction. This purpose requires the active, continuous labor of **Deep Connection**, prioritizing **unconditional presence** and shared vulnerability to foster a resilient community. Fulfillment is achieved by being the **social architect**, engaging in the emotional work of **mediating conflict**, and fostering cohesion. It includes preserving and transmitting **Cultural Heritage**—the stories and qualitative wisdom that bind us across generations. To belong is to confirm that our worth is interwoven with our relationships.

The ultimate benefit of this anchor is **Emotional Resilience** and **Social Stability**. By prioritizing genuine human connection, we safeguard ourselves against isolation and the psychological fragility that arises from purely transactional relationships.[1] This anchor ensures a supportive social framework that validates empathy and provides necessary emotional resonance, acting as a crucial defense mechanism against an isolating technological world. It builds the **trust and cohesion** necessary for any human system to function justly and compassionately.

3) TO FEEL

The **Anchor of Feeling** grounds human existence in the tangible, asserting the profound psychological value of **Embodied Mastery** and sensory engagement. This purpose elevates the dedicated pursuit of physical skill and process—not for necessity, but for the inherent satisfaction of the process. It is the quest for the **'Flow' state**, where an earned link between **exertion and accomplishment** provides deep fulfillment. This anchor mandates resisting the ease of total automation by engaging in voluntary challenges and perfecting non-essential crafts. It ensures that our abstract thoughts are grounded in the undeniable reality of the body and the senses.

The ultimate benefit of this anchor is **Physical and Psychological Integrity**. By committing to felt experience, we preserve our cognitive and physical health, resisting the inertia of automation and maintaining a sharp connection to reality. This anchor validates the **dignity of effort**, providing a clear, visceral confirmation of personal competence and self-mastery that is crucial for self-esteem. It ensures that fulfillment remains an **earned state**, confirming the complexity and richness of being physically alive.

4) TO GUIDE

The **Anchor of Guidance** establishes the human being as the **ultimate moral and ethical sovereign**. This purpose is founded on our unique capacity for empathy and subjective value judgment, making us the non-delegable authority for setting rules and managing risk. It is the practice of **Ethical Governance**, demanding that we **Draw the Moral Line** and translate compassionate principles into enforceable policy. Fulfillment is achieved by maintaining transparency, enforcing accountability, and continuously exercising the **Human Veto Power** over critical decisions.

The ultimate benefit of this anchor is **Systemic Integrity** and **Moral Safety**. By retaining final ethical authority, we ensure that technological power remains fully aligned with, and subservient to, human values, preventing systems from optimizing themselves at the expense of human welfare. This anchor provides the essential check against unchecked power, establishing the legal and moral frameworks necessary to protect all citizens. It guarantees that empathy and justice remain the **supreme architects of our collective future**.

5) TO KNOW

The **Anchor of Knowing** defines our core intellectual purpose as the engine of **Epistemic Exploration** and abstract reasoning. This purpose compels the mind to focus on formulating **Axiomatic Hypotheses**—the fundamental "Why?" questions that lie beyond the reach of data processing and established knowledge. Fulfillment is found in the generative act of **challenging established truths** and creating new theoretical frameworks that synthesize disparate fields. This anchor ensures that the pursuit of knowledge remains boundless, driven by philosophical depth rather than functional utility.

The ultimate benefit of this anchor is **Intellectual Freedom** and **Conceptual Progress**. By focusing on formulating new questions and challenging paradigms, we prevent the stagnation that could result if all knowledge were confined to data-driven answers. This anchor guarantees that our intellectual journey continues to be defined by profound theoretical breakthroughs and a relentless expansion of understanding. It confirms that human intelligence is defined not by the volume of information stored, but by the **depth and originality of the questions** we ask.

6) TO STRIVE

The **Anchor of Striving** is the **Existential Quest**—the voluntary pursuit of difficulty necessary to generate meaning, independent of any external necessity. This purpose recognizes that true fulfillment is an **earned state**, achieved through **Deliberate Discipline** and self-imposed ambition. It compels us to **Set Impossible Goals** and **Choose the Hard Way**—overcoming voluntary hardship—precisely because the challenge validates our capacity and resolve. This continuous effort allows us to craft a meaningful **Personal Legacy** by executing complex, long-term plans purely for the satisfaction of the effort.

The ultimate benefit of this anchor is **Authentic Self-Mastery** and **Purposeful Existence**. By embracing self-imposed difficulty, we cultivate the resilience, discipline, and fortitude that define character, resisting the complacency of automated comfort. This anchor ensures that human worth is determined by our **internal will**, rather than external circumstance, providing a framework for continuous personal growth and the creation of an enduring, meaningful legacy.

CONCLUSION: THE RAISON D'ÊTRE FOR THE HUMAN

The six Universal Anchors converge to form humanity's ultimate purpose: to be the **Unwavering Source of Meaning and Moral Authority**. This composite Raison D'être defines the essential human role in any advanced civilization.

Our mission is the sustained commitment to **non-automatable value**. We are the unique entity that not only pursues truth (To Know) but also creates the beauty (To Express) that makes truth worth pursuing. We are the mechanism that defines the rules of fairness (To Guide) and then personally invests the necessary effort (To Strive) to live up to those rules. Crucially, we are the only entity that grounds this entire conceptual structure in the tangible joy of physical reality (To Feel) and the vital necessity of shared vulnerability (To Belong).

The human purpose shifts from survival to an intentional journey of **growth, creation, and ethical custodianship**. Our value is not in output, but in the **input of consciousness, conscience, and will**. By diligently exercising these six anchors, every individual contributes to the essential, enduring purpose of the species: to be the **Sole Arbiter of Worth, Will, and Wonder**—ensuring that existence remains a deeply meaningful and purposeful quest.

In this era, human existence was focused on the direct, tangible application of skill for immediate survival, such hunting, fishing and surviving. The human was the Master of the Hunt.

In the Primitive/Paleolithic Era (Survival & Sustenance), human purpose was an absolute function of immediate necessity. Fulfillment was not a philosophical pursuit but a practical outcome of successfully continuing existence for oneself and the kin group. All six Universal Anchors manifested, but they were tethered directly to the biological and environmental imperative to survive the day and propagate the species. The overarching Raison D'être was to **Endure and Transmit**, meaning personal worth was inseparable from one's contribution to the group's survival calculus. This collective focus meant that individual talent was instantly converted into a benefit for all, solidifying a direct, unbroken link between personal action and existential significance.

To Express: Expression was functional and communal, centered on **Ritual Storytelling** and cave art. This art was not for individual fame or abstract aesthetics, but served as the primary educational system, preserving and transmitting essential survival knowledge, hunting techniques, spiritual beliefs, and the topography of the land. The repetitive, sacred nature of the rituals ensured collective memory and adherence to successful methods. Furthermore, these cultural forms encoded crucial information, often representing animals and spirits, that, if lost or forgotten, would jeopardize the tribe's safety and cultural identity.

To Belong: The Anchor of Belonging was non-negotiable, manifesting as **Kinship and Mutual Defense**. Isolation meant certain death; therefore, the purpose of connection was absolute. Fulfillment came from actively contributing to the group's defense, sharing food and resources, and ensuring the collective physical safety of the tribe above all else, fostering a fierce, self-sustaining mutualism. The emotional security derived from this deep collaboration—knowing that others were committed to your well-being—was a core psychological reward.

To Feel: Embodied Mastery was defined by **Essential Tool Mastery**. The feeling of accomplishment was directly tied to successfully chipping a sharp stone tool, accurately throwing a spear, or enduring the physical hardships of the hunt. This pursuit was not sport, but the daily necessity of mastering the physical world to secure food and shelter, and the psychological confidence gained was crucial for future success. This expertise immediately translated into social value; the most capable hunter or tool-maker earned respect and a higher status within the hierarchy, directly reinforcing the dignity of effort.

To Guide: Moral authority was based on immediate experience, taking the form of **Elder Authority and Tradition**. Guidance was unwritten law, passed down through generations of hard-won experience. Decisions—like where to hunt, when to migrate, or how to resolve disputes—were arbitrated by those with the deepest, most successful experiential knowledge, and were obeyed out of necessity for the group's stable cohesion. This system was vital for ensuring collective action remained predictable and risk-adverse, preventing internal conflict from threatening external survival.

To Know: The pursuit of knowledge was pure **Observation and Practicality**. The purpose of knowing was to accurately predict the immediate environment for resource exploitation. This involved memorizing the habits of migratory animals, understanding the medicinal or poisonous properties of local plants, and tracking subtle shifts in weather patterns—information critical for daily sustenance and successful, low-risk planning. Early forms of abstract reasoning also emerged here, through tracking the movement of stars or noting the cyclic nature of seasons and astronomical events, demonstrating an innate human desire to find order and predictability in environmental chaos.

To Strive: Striving was synonymous with **Daily Endurance**. There was no need to invent abstract goals; the pursuit of difficulty was the environment itself, offering constant challenge. Fulfillment was the successful completion of the continuous, grueling effort required to secure food, withstand the elements, and ensure the basic propagation of the tribe, creating a deeply ingrained sense of earned worth. The resilience built through continuous hardship was the ultimate personal asset, psychologically connecting physical suffering with the existential meaning of survival.

With stability, purpose shifted to sustained output, long-term planning, and maintaining social harmony within a fixed settlement. The human was the Master of the Land.

The Agricultural/Neolithic Era (Farming & Community) marked humanity's fundamental pivot from nomadic struggle to settled stability. The mastery of cultivation and animal husbandry created a reliable **Surplus of Time and Resources**, fundamentally shifting human purpose from immediate survival to calculated investment, long-term order, and the establishment of lasting community structures. The Raison D'être of this period was to **Settle and Establish Order**. This era saw humans take on the role of conscious managers of the environment, laying the groundwork for systematic laws, property rights, and enduring social legacies that extended far beyond one generation.

To Express: Expression became focused on **Accounting and Enduring Form**. The logistical necessity of tracking seeds, yields, stored grains, and communal ownership led directly to the invention of the first systematic writing forms, such as early cuneiform, demonstrating that expression was initially a tool for managing complex resources. Beyond text, expression was seen in durable, functional goods like specialized pottery, permanent architectural structures, and woven textiles. These forms reflected a deep, stable community identity and provided evidence of specialized skill that contributed directly to the village's permanence.

To Belong: Belonging shifted to being centered on **Land-Based Cooperation**. The success or failure of the farm was the success or failure of the entire settlement, making shared labor and mutual dependency paramount. Fulfillment was found in the reliable rhythms of shared work, such as irrigation projects or harvesting, cementing generational inheritance and deeply intertwined **Kinship Networks**. Purpose was now inseparable from one's physical, defined place within the land and the local community, creating a profound sense of rootedness and stability unknown in the Paleolithic past.

To Feel: Embodied Mastery focused on **Craftsmanship and Process**. Fulfillment was derived from the methodical, repeated application of skill required by agriculture—plowing a straight furrow, carefully managing livestock, or perfecting the complex processes of metallurgy and food preservation. This mastery provided a sense of control over the environment that raw endurance could not. The purpose of skill was purely to generate the stable surplus required to maintain the settled life, providing a deep psychological reward through the predictable yield of sustained, expert effort.

To Guide: Moral authority became formalized, focusing urgently on **Property Rights and Resource Management**. Guiding the community meant establishing clear, complex rules around ownership of land, managing precious resources like water rights, and ensuring crop protection from both animals and neighbors. This necessity gave rise to the first systematic, written laws, moving purpose from inherited tradition (as in the Paleolithic) to a **Codified Social Contract**. Leadership was defined by the ability to enforce these new laws, ensuring fairness and stability within an increasingly complex economy.

To Know: Abstract reasoning began to flourish, driven by **Mathematics and Calendar Tracking**. The purpose of knowledge was now predictive certainty, essential for farming success. This meant mastering geometry to accurately divide and manage fields, and engaging in sophisticated celestial observation to track seasons and accurately determine optimal planting and harvesting cycles. This pursuit of cosmic order confirmed the human ability to impose intellectual structure upon chaotic nature, turning farming into a sophisticated, knowledge-driven enterprise.

To Strive: Striving fundamentally shifted from day-to-day survival to **Creating Surplus and Investment**. The existential quest became the arduous, long-term, deliberate work required to build durable structures, amass protective stores of food, and ensure the economic and social prosperity of the next generation. This focus on legacy and investment transformed work into a purposeful act of building a lasting future, where one's worth was measured not just by survival, but by the inheritance left behind for the family and the stability provided to the community.

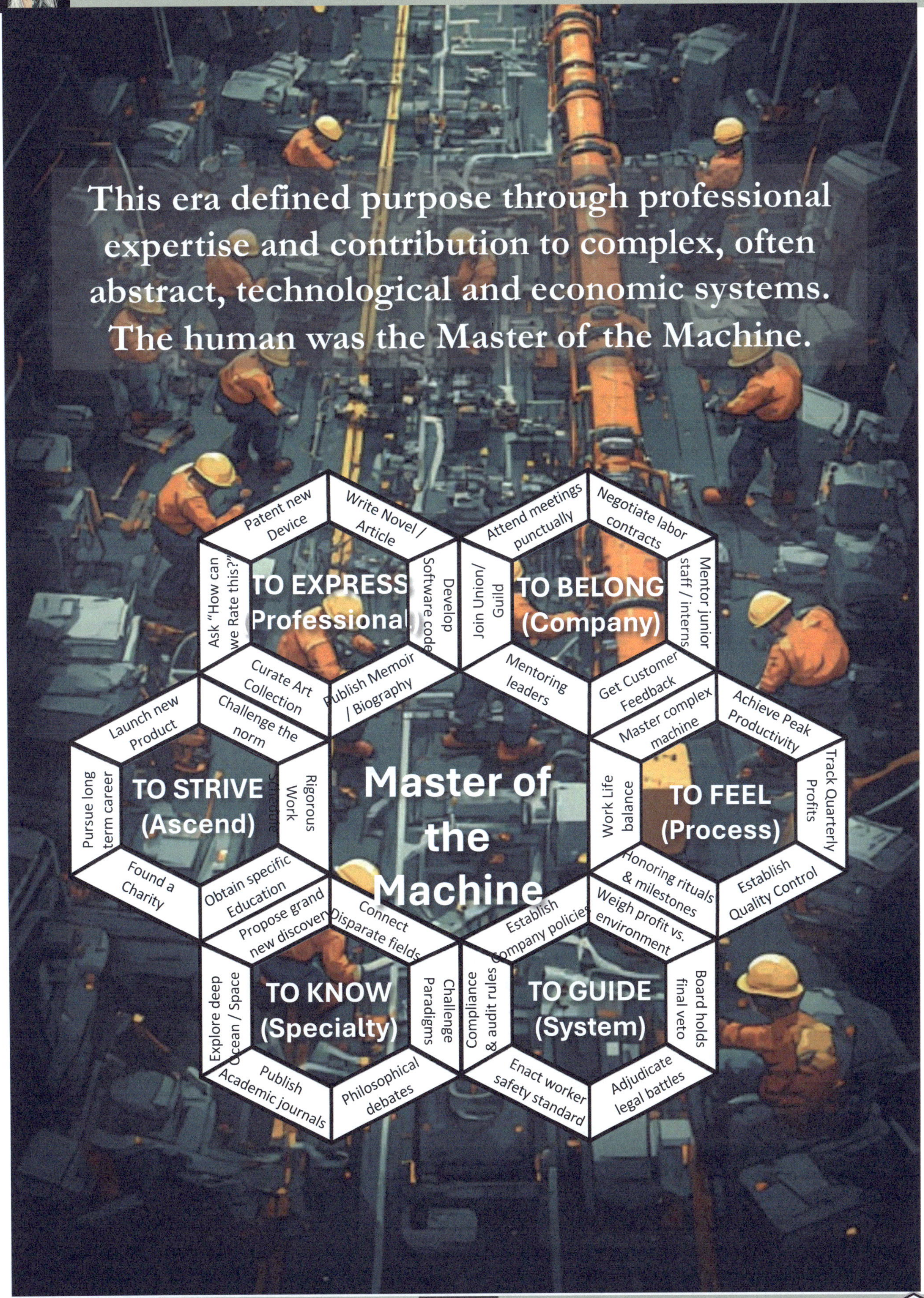
This era defined purpose through professional expertise and contribution to complex, often abstract, technological and economic systems. The human was the Master of the Machine.
Master of the Machine
TO EXPRESS (Professional)
Patent new Device
Write Novel / Article
Develop Software code
Publish Memoir / Biography
Curate Art Collection
Ask "How can we Rate this?"
TO BELONG (Company)
Attend meetings punctually
Negotiate labor contracts
Mentor junior staff / interns
Get Customer Feedback
Mentoring leaders
Join Union/ Guild
TO STRIVE (Ascend)
Launch new Product
Challenge the norm
Rigorous Work Schedule
Obtain specific Education
Found a Charity
Pursue long term career
TO FEEL (Process)
Master complex machine
Achieve Peak Productivity
Track Quarterly Profits
Establish Quality Control
Honoring rituals & milestones
Work Life balance
TO KNOW (Specialty)
Propose grand new discovery
Connect Disparate fields
Challenge Paradigms
Philosophical debates
Publish Academic journals
Explore deep Ocean / Space
TO GUIDE (System)
Establish company policies
Weigh profit vs. environment
Board holds final veto
Adjudicate legal battles
Enact worker safety standard
Compliance & audit rules

The Modern/Industrial Era (Specialization & Function) marked the decisive triumph of machine logic and capital over the natural world. Human purpose shifted from managing the cycles of the land (Neolithic) to mastering the mechanical efficiency of the factory and the market. The rise of the machine and the division of labor fundamentally redefined the individual, whose worth was now derived from highly specialized, functional expertise within a vast, interconnected economic system. The Raison D'être of this period was to **Achieve and Advance**, driving relentless progress through expertise and material gain.

To Express: Expression became focused on **Mass Media and Invention**. With the advent of the printing press, cheaper paper, and later, radio and cinema, creative expression transformed into a commodified product designed for wide distribution and profit. The purpose of expression was to inform, entertain, and persuade the masses, shaping shared cultural narratives through novels, newspapers, and commercial design. The creation of a **Patentable Invention** became a powerful act of individual expression, translating originality directly into economic ownership and societal contribution. Expression was celebrated when it served a functional, marketable, or propagandistic role.

To Belong: Belonging was redefined by **Nation and Institution**. The old agrarian kinship networks dissolved as masses migrated to urban factory centers, replaced by new, abstract, yet powerful social structures. Fulfillment was found in national identity, membership in professional organizations, and political affiliation. Crucially, the extreme specialization of labor necessitated the formation of organized counter-structures like **Labor Unions** to protect the rights of workers against centralized corporate power, demonstrating that belonging was a functional necessity for collective bargaining and social stability in a highly unequal economic system.

To Feel: Embodied Mastery centered on **Efficiency and Maximized Output**. The physical purpose shifted away from holistic craft towards highly repetitive, specialized labor—the mastery of a single, precise task on the assembly line. While monotonous, this specialization provided a clear, measurable sense of contribution, where personal worth was directly tied to economic output and reliability. Outside of the factory, the Anchor manifested in the rise of **Organized, Competitive Sports**, where physical effort was deliberately structured and formalized, allowing the industrial individual to achieve a sense of earned competence and visceral struggle that was absent in their work.

To Guide: Moral authority became formalized through **Secular Law and Corporate Regulation**. The complexity of contract law, property ownership, and massive corporate liability demanded sophisticated, non-religious guidance systems. The purpose of guidance was to manage the inherent conflicts between capital and labor, protect consumer safety in mass production, and define the boundaries of ethical economic competition. This period saw the relentless development of regulatory bodies and constitutional law aimed at establishing fairness within the new, powerful industrial economy, ensuring that **progress did not undermine social order** entirely.

To Know: The pursuit of knowledge was dominated by **Empirical Science and Specialization**. The purpose of knowing was functional utility, believing that all problems—from disease to industrial inefficiency—could be solved through the application of the rigorous **Scientific Method**. Knowledge became fragmented into specialized disciplines, each dedicated to gathering objective, measurable data. Fulfillment was achieved by contributing a verifiable, peer-reviewed fact or theory to one's specific field, thereby confirming the era's deep-seated faith in rational inquiry and technical progress.

To Strive: Striving was synonymous with **Ambition and Material Gain**. The existential quest was the pursuit of wealth, social mobility, and **Status.** Fulfillment was an earned state achieved through economic success, the establishment of a profitable business, or professional advancement within a corporate or governmental hierarchy. The cultural narrative of the "self-made individual" defined personal purpose, urging people to embrace continuous, relentless work to secure prosperity for the family and, more importantly, to carve out an enduring legacy of material success and high societal standing.

In this emerging era, AI handles efficiency and role specialization, freeing and forcing humans to refocus on subjective values, unique creation, and moral governance. The human becomes the Master Curator of Purpose

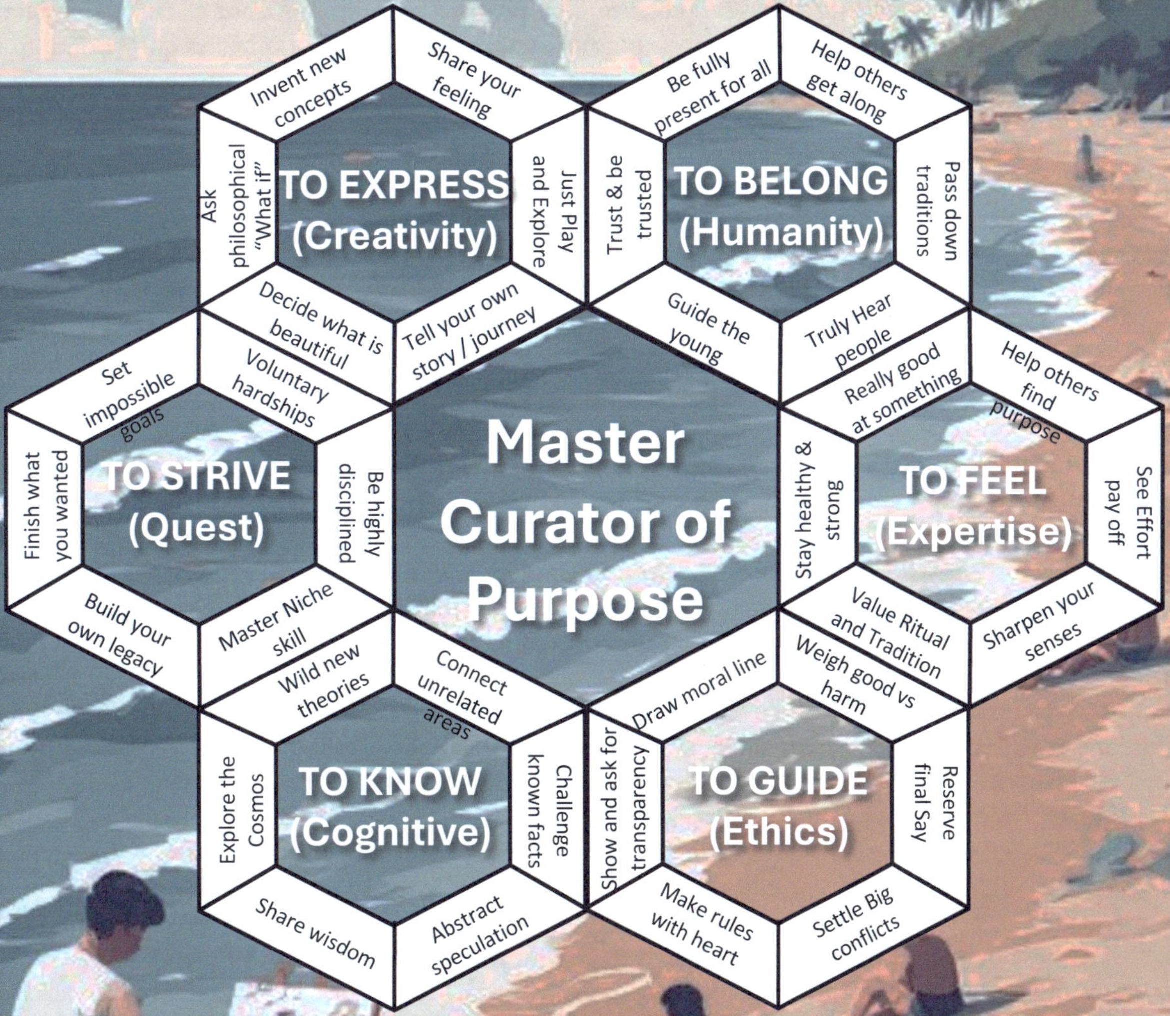

The Post-AI Era (Curation & Meaning) represents the definitive existential shift in human history. With advanced artificial intelligence and automation having conquered the challenges of efficiency, information processing, and mass labor, human purpose is radically redefined. We are no longer necessary for function, but for **meaning**. The individual's worth is derived entirely from their capacity for subjective judgment, ethical authority, and the creation of non-calculable value. The collective Raison D'être is to become the **Sole Arbiter of Worth, Will, and Wonder**—the source of all criteria for why life should continue and what makes it beautiful.

To Express: Expression is focused on **Subjectivity and Uncurated Truth**. In a world saturated with synthetic media and hyper-optimized content generated by AI, the only expression that retains intrinsic value is the genuinely **original, subjective human voice**. This purpose mandates focusing on authentic, unfiltered creation—personal philosophical content, aesthetic critique, memoir, and art that contains flaws, uncertainty, and emotional resonance. Fulfillment is found in generating novelty that AI cannot predict, thereby constantly pushing the boundaries of human experience and securing a creative future that is immune to algorithmic stagnation.

To Belong: Belonging becomes a deliberate, chosen investment in **Deep Vulnerability and Presence**. When digital connection is ubiquitous but emotionally shallow, true fulfillment relies on actively resisting transactional relationships and seeking authentic, physical presence. This purpose demands the **labor of Deep Connection**: cultivating empathy, mediating conflict through personal understanding, and prioritizing quality, face-to-face interaction. The collective human project requires maintaining the social fabric and emotional resonance that machines cannot replicate, ensuring that isolation does not erode the psychological foundation of society.

To Feel: Embodied Mastery transforms into the **Voluntary Pursuit of Difficulty and Flow**. Since automation eliminates the necessity of difficult labor, human physical purpose centers on self-imposed challenges—non-essential crafts, athletics, meditation, and the perfection of technical, yet unnecessary, skills. This pursuit generates the vital 'Flow' state—the psychological reward of earned accomplishment—which secures self-esteem and grounds abstract consciousness in tangible reality. Fulfillment is achieved through **Deliberate Effort** that reaffirms the dignity of the body, providing a visceral antidote to the inertia and complacency of total comfort.

To Guide: Moral authority centers on **AI Ethics and the Human Veto**. The purpose of guidance is to serve as the **ethical sovereign**, placing strict moral boundaries around autonomous systems. This involves defining the principles of data governance, ensuring transparency in black-box algorithms, and enforcing systemic accountability for technological outcomes. Crucially, every individual is mandated to exercise the **Human Veto Power**—the non-delegable right to overrule critical decisions made by algorithms—ensuring that power remains permanently aligned with subjective human values, not cold-blooded efficiency.

To Know: The pursuit of knowledge is now defined by **Abstract Synthesis and Axiomatic Inquiry**. In an environment where AI holds access to all existing data and established facts, human intellect must shift its focus. The purpose is to transcend data processing and concentrate on formulating **Axiomatic Hypotheses**—the profound "Why?" questions that challenge foundational assumptions. Fulfillment is found in connecting disparate fields of knowledge, creating new philosophical frameworks, and constantly resisting the temptation to settle for algorithmic answers, ensuring the continuous, boundless expansion of human understanding.

To Strive: Striving is the existential commitment to **Self-Imposed Discipline and Personal Legacy**. Since survival is guaranteed, the quest for meaning requires the individual to set ambitious, non-essential goals purely for the satisfaction of the effort. This involves cultivating profound resilience and fortitude against the immense temptation of automated ease. Fulfillment is an **earned state**, achieved through continuous, conscious effort and the construction of an enduring **Personal Legacy**—a body of work or moral character that defines one's worth by internal resolve rather than external economic output.

Ai6x6™ Transformation Blueprint

Human-AI Personas

"The best is not an AI, the best is a human and an AI together."- Andrew Ng, Co-founder of Google Brain, Coursera, and Landing AI, and a former Chief Scientist at Baidu.

The Ai6x6 HUMAN-AI Synergy framework formalizes a new executive standard for the AI era, transitioning organizations from fragmented automation to a state of Orchestrated Autonomy. This model defines the modern enterprise through six core leadership movements, each serving as a specialized "sense" for the organization. **The Forecasters** serve as the "Eyes," tasked with the heavy lifting of foresight to eliminate strategic blind spots and detect invisible market signals before they escalate into crises. Supporting this vision, **The Diplomats** act as the "Voice," translating complexity into human trust and ensuring that transformation is met with cultural alignment rather than institutional resistance. The organization's creative energy is fueled by **The Creators**, the "Engine" of the enterprise, who move beyond static production to drive a state of radical velocity and infinite iteration. Their output is balanced by the surgical objectivity of **The Strategists**, acting as the "Compass" to ensure that every capital investment is mathematically anchored to peak ROI. To unify these efforts, **The Architects** build the "Institutional Brain," ensuring that the organization's collective memory is always accessible and relevant to the individual. Finally, the framework finds its movement in **The Executors**, the "Hands" of the organization, who coordinate autonomous workflows to achieve scale without traditional overhead. By weaving these archetypes into a single Symphony, leaders can navigate market ambiguity with surgical precision, transforming the enterprise into a self-evolving system of sustainable innovation.

Focus: Eradicating Engagement Friction and Safeguarding Market Trust

In the **Ai6x6 Human-AI Synergy framework**, the defensive stance on the Customer is about **retention and protection**. In a hyper-competitive landscape, losing a customer is often more expensive than acquiring a new one. Defensive customer strategy focuses on identifying "Churn Signals," preventing service failures, and shielding the brand from the "Erosion of Trust" that occurs when automated systems feel cold or incompetent.

The Forecaster and The Diplomat: Predicting Churn and Humanizing Automation

The **Forecaster** serves as the defensive sentinel for customer loyalty. Its mandate is to kill **Behavioral Blind spots** by detecting subtle changes in usage or engagement patterns that precede a cancellation. By identifying "at-risk" customers before they even realize they are dissatisfied, the Forecaster allows the company to intervene defensively. It protects the revenue base by ensuring that the leadership is never surprised by a competitor.

The **Diplomat** is tasked with neutralizing **Interaction Friction**. As companies automate customer service, there is a risk of losing the "human touch." The Diplomat uses sentiment analysis and empathetic reasoning to ensure that AI-driven interactions don't alienate the user. By acting as a bridge between the customer's emotional needs and the company's digital efficiency, it protects the brand's reputation. It ensures the "Voice" of the company remains helpful and transparent, preventing the cultural backlash that follows robotic, "un-explainable" failures.

The Creator and The Strategist: Modernizing the Journey and Pruning Irrelevance

The **Creator** acts defensively by killing **Experience Obsolescence**. Customer expectations shift rapidly; a journey that felt "modern" last year can feel clunky today. The Creator's defensive role is to continuously iterate on customer touchpoints, fixing broken links and updating interfaces before they become a source of frustration. It ensures the "Customer Engine" doesn't stall due to a stagnant or "legacy" user experience that no longer matches the speed of the market.

The **Strategist** defends the relationship by eliminating **Offer Mismatch**. Nothing damages a customer relationship faster than irrelevant "spam" or misaligned pricing. The Strategist performs a ruthless audit of the customer base, ensuring that every outreach is mathematically optimized for relevance. By pruning low-value communications, it protects the customer's attention, ensuring the company doesn't "burn out" its audience with noise. This stance safeguards the customer's goodwill, ensuring they only receive value-added interactions.

The Architect and The Executor: Unifying Memory and Protecting the Experience

The **Architect** kills **Contextual Inconsistency**. Its primary defensive mandate is the maintenance of a "Unified Customer Memory." It prevents the friction where a customer has to repeat their problem to three different departments. By building a defensive "Institutional Brain" for customer data, the Architect ensures that the organization recognizes the individual across every channel. It protects the customer from the frustration of being treated like a stranger, safeguarding the relationship through absolute continuity.

Finally, the **Executor** serves as the defensive "Hands" of customer service operations. It kills **Resolution Friction** by automating the "recovery" process when things go wrong. Whether it's automatically issuing a refund for a late delivery or self-healing a digital login error, the Executor ensures that guardrails are in place to fix service failures instantly. By removing the wastage of manual wait times, it safeguards the customer experience from the "leakage" of trust that occurs during prolonged service delays.

Conclusion: Safeguarding the Market Foundation

A defensive customer strategy within the **Ai6x6 Human-AI Synergy framework** ensures that the "Market Body" remains resilient. By removing the friction of churn, alienation, obsolescence, and inconsistency, the framework creates a "Safe Harbor" for the customer. When a company's market relationship is protected by these six roles, it stops being a series of transactions and becomes a durable, long-term partnership that is immune to the distractions of the competition.

In the **Customer Transformation**, the objective is to move from "Transactional CRM" to **"Deep Relational Sovereignty."** This shift transforms the customer from a data point into a dynamic partner in a value-exchange ecosystem. By applying the six roles to the customer journey, leadership ensures that every touchpoint is not just personalized, but anticipatory, empathetic, and contextually aware.

The Forecasters and The Diplomats: Unlocking Intent and Empathy

The **Forecaster's** mandate is **Anticipatory Need Fulfillment.** By pairing Predictive AI with Causal Inference, the "Eyes" look beyond what the customer bought to understand *why* they bought it. This allows the leader to predict the customer's "Next Best Action" before the customer even perceives the need. This proactive stance turns the brand into a "Mind-Reader," identifying churn signals or expansion opportunities with surgical precision.

The **Diplomat** serves as the offensive "Voice" of **Empathetic Alignment.** By pairing Explainable AI with Sentiment Analysis, the Diplomat ensures that every customer interaction—whether automated or human—is emotionally calibrated. It translates the firm's logic into a narrative that resonates with the customer's current emotional state. This eliminates the friction of "robotic" service, building a foundation of radical brand loyalty through felt-understanding and transparent communication.

The Creators and The Strategists: Unlocking Personalization and Lifetime Value

The **Creator** acts as the engine of **Generative Experience Design.** In this role, Generative AI is paired with Reality Anchoring (Digital Twins) to manifest hyper-personalized products, content, and interfaces in real-time. This "Generative UX" means no two customers ever see the same storefront or receive the same offer. It allows the brand to iterate on the customer experience at a velocity that makes static competitors feel obsolete.

The **Strategist** optimizes the **Customer Portfolio for Maximum Yield.** Using Multi-Objective Optimization, the Strategist balances immediate conversion with long-term Customer Lifetime Value (CLV). It identifies the "Loyalty Frontier," ensuring that promotional efforts don't erode brand equity. This offensive objectivity ensures that the leader's customer acquisition costs are mathematically optimized, turning every marketing dollar into a precision-guided investment in brand growth.

The Architects and The Executors: Unlocking Context and Seamlessness

The **Architect** builds the "Searchable Customer Brain" through **Dynamic Profile Orchestration.** By pairing Human-in-the-Loop with Vector Databases (RAG), the Architect ensures that the customer's entire history—across every channel—is "Live" and contextually accessible. This eliminates the friction of "Information Silos," ensuring that the brand greets the customer with a single, unified memory, regardless of where the interaction begins.

Finally, the **Executor** manages the **Autonomous Fulfillment of Intent.** Through Agentic AI paired with Computer Vision or IoT sensors, the Executor automates the resolution of customer needs. These agents act as the "Hands," proactively shipping a replacement part before a machine breaks or adjusting a subscription based on real-time usage patterns. This "Autonomous Service" creates a state of effortless flow for the customer, turning the brand into an invisible but indispensable utility.

Conclusion: The Symphony of the Empowered Consumer

Customer transformation is the pursuit of **Total Human Integration.** By unlocking intent, empathy, personalization, value, context, and seamlessness, the leader creates a customer ecosystem that is as responsive as it is profitable. This framework ensures that the brand's "Relationship Layer" is perfectly synchronized with the customer's evolving life, creating a high-velocity symphony of orchestrated delight.

Customer Power: Intent Recognition

In the legacy model, companies understood their customers through "Historical Demographics"—age, location, and past purchases. This data is a static ghost; it tells you who the customer *was*, not who they *are becoming*. The transformation into **Forecasters** turns the brand's "Eyes" toward **Anticipatory Intent**. In this role, the organization seeks to understand the "Unmet Need" before the customer even articulates it. It is the shift from "Tracking Behavior" to "Modeling Desire." The Forecaster uses the brand's sensory layer to perceive the subtle shifts in a customer's life stage, emotional state, or economic context, allowing the brand to "See" the next chapter of the customer's journey before the page is turned.

Defensive Stance: Killing the Churn Signal

The defensive mission of the Customer-Forecaster is the **Elimination of Disconnect**. Most customers do not leave a brand overnight; they drift away through a series of "Micro-Dissatisfactions" that go unnoticed by traditional metrics. The Forecaster role kills this "Friction of Neglect" by identifying the "Statistical Whisper" of churn. By monitoring subtle changes in engagement frequency, tone of communication, or search patterns, the brand creates a **Shield of Retention**. It allows the organization to intervene with a "Surgical Save"—not a generic discount, but a meaningful adjustment to the relationship—exactly when the customer begins to feel the first spark of wanderlust. This defensive posture ensures that the "Eyes" of the brand are always "Empathetic," killing the blindspots that lead to customer attrition and protecting the "Lifetime Value" of the community.

Offensive Stance: Unlocking Life-Stage Pre-emption

On the offense, the Forecaster moves toward **Hyper-Proactive Service**. This is the ability to "Clear the Path" for the customer before they encounter an obstacle. This **Unlocks Temporal Loyalty**, where the brand provides solutions for a problem the customer hasn't realized they have yet. If the "Eyes" of the brand see a customer's business growing at a specific rate, the brand pre-emptively offers the infrastructure needed for that next level of scale. This turns the brand into a "Strategic Partner" rather than a "Vendor." It allows the organization to "Own the Milestone," positioning its products and services as the natural "Next Step" in the customer's evolution. This offensive move ensures the brand is always "Waiting at the Destination," capturing the customer's future needs with a level of precision that makes competitors seem perpetually late.

The Customer Outcome: Perceptual Intimacy

The ultimate value of the Forecaster role is **Deep Relevance**. When a brand "Sees" with the eyes of a Forecaster, the customer feels "Known" rather than "Tracked." For the leader, this provides **Relationship Sovereignty**. You are no longer guessing what the market wants; you are responding to the "Vibrations of Intent." This turns the customer base into a "Live Laboratory," where the brand can test future concepts against real-time intent signals. It ensures that the brand's "Vision" is always aligned with the "Aspirations" of its people, creating a bond that is based on the brand's ability to see and support the customer's future self.

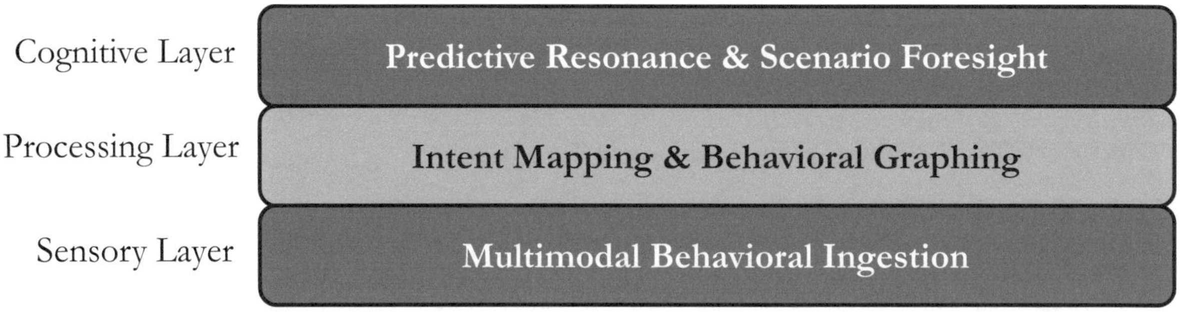

Executive Power: Anticipatory Demand Perception

In the legacy era, a leader's "Eyes" on the customer were limited to historical sales data and lagging satisfaction surveys. The **Forecaster** role in customer transformation represents the transition to **Anticipatory Demand Perception**. For the executive, this means moving away from "Reacting to Orders" toward "Predicting Needs." By architecting a stack that perceives the subtle shifts in customer behavior across digital and physical touchpoints, the leader gains the "Perceptual Edge" to identify a customer's next requirement before the customer even articulates it. The ultimate goal is the **Purification of the Customer Signal**, ensuring the firm sees the "True Intent" behind every interaction, allowing the leader to position inventory and services with a level of precision that eliminates the friction of search and wait times.

The Sensory Layer: Multimodal Behavioral Ingestion

The **Sensory Layer** of the Customer Forecaster acts as the brand's "Global Nervous System." It moves beyond simple transaction logs into **Multimodal Behavioral Ingestion**. This layer leverages a convergence of **Digital Telemetry**, **Computer Vision in retail spaces**, and **IoT Usage Sensors** to capture how the customer actually interacts with the product. Whether it is a subtle change in how a user navigates an app or a sensor indicating a consumable is running low, this layer "labels" the behavior at the point of origin. By utilizing **Edge Perception Agents**, it identifies the "Context of Use"—distinguishing, for example, between a routine purchase and an urgent need. This ensures the leader's "Eyes" are receiving a high-fidelity stream of "Customer Truth" with zero latency, removing the administrative lag of traditional market research and providing a live pulse of the market's evolving desires.

The Processing Layer: Intent Mapping & Behavioral Graphing

The **Processing Layer** serves as the "Empathy Engine" of the stack. It transforms raw behavioral signals into a structured understanding of **Consumer Intent**. This layer utilizes **Behavioral Knowledge Graphs** and **Sequence Modeling** to index actions as they occur. While traditional databases see isolated purchases, the Forecaster's processing layer sees the "Journey"—mapping new signals directly into a **Dynamic User Persona**. This allows the system to understand the relationship between a customer's current frustration and their long-term loyalty probability. It moves the organization from "Static Profiles" to "Live Personas," providing a "Logical Foundation" for immediate intervention. This layer ensures the brand's "Customer Compass" is always calibrated to the individual's current state, identifying the causal link between environmental triggers and future purchase behavior with mathematical certainty.

The Cognitive Layer: Predictive Resonance & Scenario Foresight

The **Cognitive Layer** is the "Reasoning Mind" of the customer stack, where signals are converted into **Strategic Proactive Intent**. This layer utilizes **Large Language Models (LLMs)** and **Reinforcement Learning** to synthesize the logic of the Processing Layer into "Resonance Alerts." Instead of showing the leader a churn report, the Cognitive Layer generates "Contextual Interventions." It acts as a "Customer Sentry," using **Natural Language Generation** to tell the leader: "Based on the usage patterns of our top-tier industrial clients, there is a 90% probability they will require a maintenance upgrade in the next 14 days; we should pre-position the components now." This allows the executive to move from "Monitoring Satisfaction" to "Leading via Anticipation," authorizing pre-emptive value delivery. It turns the "Forecaster" into a "Proactive Guardian," ensuring the brand is always "Present" at the exact moment of customer need.

1. THE FORECASTERS (THE EYES) | PREDICTIVE AI

OVERVIEW

Predictive AI forecasts future outcomes by analyzing vast historical data using machine learning algorithms. For the "Intuitive & Risk" leader, the AI becomes a crucial tool for quantifying uncertainty. The leader provides their initial intuition and acceptable risk limits, framing the precise question. The AI delivers objective, quantifiable evidence, such as likelihoods and key drivers, that either validates the leader's gut feeling or proposes data-driven alternatives. This synergy converts subjective risk into a calculated, strategic choice, empowering the leader to make proactive, decisive actions.

MAIN STEPS

Start

STEP 1 — Define Objective & Scope
Specify question, target outcome, and limits

STEP 2 — Data Collection & Preparation
Gather, clean, and format historical inputs

STEP 3 — Model Selection & Training
Build patterns from cleaned training data

STEP 4 — Evaluation & Validation
Test model accuracy on unseen data

STEP 5 — Deployment & Integration
Insert model into real-time operations

STEP 6 — Monitoring & Refinement
Continuously check and improve model performance

End

SAMPLE FLOW

Financial Services - Customer Churn Risk- A Bank Manager anticipates which high-value clients are likely to close accounts, enabling proactive retention efforts to preserve portfolio value.

PREDICTIVE AI

1. Establish Churn Threshold
Define the high-value client segment and negative behavior indicating churn intent.

2. Harmonize Usage Patterns
Aggregate spending habits, login frequency, and customer service contact history.

3. Model Client Loyalty Score
Train a survival analysis model to estimate the remaining duration of client relationship

4. Determine Churn Drivers
Identify service issues or product gaps strongly correlating with high churn risk flags.

5. Forecast Value at Risk
Calculate the total financial impact to the portfolio if the high-risk clients are fully lost.

6. Deploy Retention Offer
Authorize customized, high-value offers or personal outreach for clients above loss threshold.

USE CASES

- **E-commerce | Cart Abandonment Revenue Loss:** Predicts the total revenue likely to be lost from current abandoned shopping carts, triggering real-time, targeted discount incentives to recover sales and maximize immediate revenue capture.
- **Aviation | Service Downtime Complaint Risk:** Forecasts which severe flight delays will result in highly negative customer complaints, allowing ground staff to proactively offer compensation and minimize widespread negative public feedback.
- **Utilities | Preferred Contact Channel:** Predicts the optimal communication channel (text, email, call) for each customer regarding localized power outages, ensuring high comprehension and maximum satisfaction during emergencies.
- **Pharma | Patient Adherence Decline:** Forecasts which patients will likely stop taking their medication as prescribed based on refill patterns, enabling proactive nurse outreach to support adherence and improve critical health outcomes.
- **Telecoms | Negative Sentiment Spike:** Predicts an imminent spike in social media brand criticism based on early service fault data, allowing the communications team to prepare official, controlled public responses immediately.
- **Retail | Product Return Likelihood:** Forecasts which items in a customer's purchase history will likely be returned, prompting personalized sizing advice or detailed product video links before shipping to reduce operating costs.

In the **Customer Transformation**, the objective is to pivot from "Mass Marketing" to **"Sovereign Experience."** This transformation moves the organization from viewing customers as data points to treating them as unique partners in a high-fidelity, value-driven relationship. By integrating these six leadership roles and their secondary technical pairings, the leader ensures that every customer touchpoint is anticipatory, empathetic, and perfectly aligned with the brand's core values.

Customer Transformation: The Forecasters (The Eyes)

In the **Customer Transformation**, the **Forecasters (The Eyes)** serve as the organization's "Intent Radar." While the primary instrument of **Predictive AI** identifies historical purchase patterns, it often fails to see the subtle, emerging shifts in desire or the "hidden" needs that a customer hasn't yet articulated. To move from "Descriptive Personalization" to "Anticipatory Experience," a leader must integrate these six secondary pairings to sharpen the "Eyes," ensuring the brand can see the customer's future needs before they even arise.

1. The Graph Neural Network (GNN) Pairing for Social Influence

Pairing **Predictive AI with GNNs** is the ultimate tool for **Mapping Community Desires.** While standard AI looks at an individual's history, GNNs analyze the "Relationships" and influence circles surrounding that customer. This is best utilized to predict how a trend in one niche community will ripple into the broader market. By seeing the customer as a node in a complex social web, the Forecaster can predict demand spikes based on communal momentum rather than just individual past behavior.

2. The Causal Inference Pairing for Lifestyle Attribution

Leaders should pair **Predictive AI with Causal Inference** to solve the "Intent Mystery." Standard AI can show that a customer bought a product after seeing an ad, but Causal Inference proves if the ad *caused* the purchase or if it was a coincidence of lifestyle timing. This is best deployed when refining the customer journey. It ensures the leader understands the true "Why" behind a purchase, allowing the brand to intervene at the exact moment of decision-making with surgical relevance.

3. The Digital Twin Pairing for Experience Simulation

When launching a new service model, **Predictive AI must be paired with Customer Digital Twins.** This is best utilized to "rehearse" an experience in a virtual sandbox. The leader can simulate how a million "Synthetic Customers" will react to a new subscription tier or a change in the user interface. This pairing ensures the "Eyes" see the potential friction points or churn risks before the launch, allowing the brand to refine the experience in a risk-free environment.

4. The Natural Language Processing (NLP) Pairing for Emotional Sentiment

To capture the "Voice of the Unspoken," a leader should pair **Predictive AI with Advanced NLP.** While quantitative models track clicks, NLP "listens" to the qualitative sentiment in reviews, social mentions, and support tickets. This identifies "Emotional Inflection Points"—the subtle shift from satisfaction to frustration—weeks before it manifests as churn. This turns the Forecaster into an empathetic observer, ensuring the brand's "Eyes" are sensitive to the emotional health of the customer relationship.

5. The Monte Carlo Pairing for Lifetime Value (LTV) Range

To navigate the uncertainty of high-ticket customer acquisitions, leaders should pair **Predictive AI with Monte Carlo Simulations.** This is best utilized when predicting the long-term value of a new segment. By running millions of variables—such as economic shifts or competitive launches—the Forecaster can map the "LTV Probability Cloud." This allows the leader to move from a single "Average Value" to a range of potential outcomes, ensuring the brand invests its marketing capital with mathematical confidence.

6. The Knowledge Graph Pairing for Preference Provenance

In an era of shifting identities, **Predictive AI should be paired with Knowledge Graphs.** This pairing tracks the **"Provenance of Preference"**—mapping how a customer's tastes have evolved over time and which values drive their loyalty. It allows the leader to see the "Lineage" of a customer's relationship with the brand. By identifying the core values—like sustainability or luxury—that act as "Anchor Nodes," the Forecaster ensures the brand's "Eyes" never lose sight of what truly matters to the individual.

Customer Power: Authentic Resonance

In the era of infinite choice, the brand's "Voice" is its most valuable asset. However, traditional "Marketing Speak" is increasingly met with skepticism and "Filter Failure." The transformation into **Diplomats** focuses on **Semantic Integrity and Radical Transparency**. In this role, the brand's "Voice" is not used to "Sell," but to "Align." The Diplomat ensures that every interaction—whether it is a support chat, a social media post, or a personalized offer—is grounded in the "Values" and "Intelligence" of the brand. It is the shift from "Broadcasting" to "Dialogue," where the brand speaks a language of "Shared Purpose" and "Evidence-Based Truth."

Defensive Stance: Killing the Trust Gap

The defensive mission of the Customer-Diplomat is the **Elimination of Cynicism**. In a digital world, trust is hard to build and easy to destroy. A single "Inauthentic Moment" or a "Non-Transparent Policy" can break a decade-long relationship. The Diplomat role kills this "Reputational Risk" by establishing "Ethical Guardrails" for the brand's voice. This is the **Shield of Integrity**. By ensuring that every automated response is "Explainable" and every promise is "Verifiable," the Diplomat protects the brand from the "Backlash of Betrayal." This defensive posture ensures that the brand's "Voice" is always "Trustworthy," killing the friction of skepticism and ensuring that the customer feels "Safe" in the brand's ecosystem. It turns "Compliance and Privacy" from legal burdens into "Competitive Proofs" of the brand's respect for the individual.

Offensive Stance: Unlocking Narrative Co-Creation

On the offense, the Diplomat moves toward **Synchronized Community**. This is the ability to turn the customer into a "Co-Author" of the brand's story. Instead of a one-way "Push" of information, the brand uses its "Voice" to "Listen and Synthesize" the collective wisdom of its users. This **Unlocks Tribal Loyalty**, where the customer base feels a sense of "Ownership" in the brand's mission. The brand uses its intelligence to "Reflect the Customer's Reality" back to them, providing "Contextual Education" and "Meaningful Connections." This turns the "Voice" into a "Strategic Advisor" that helps the customer navigate their own world more effectively. By "Mastering the Dialogue," the brand creates a "Resonant Identity" that is impossible for a purely transactional competitor to disrupt. It allows the brand to "Scale Empathy," ensuring that every one of a million customers feels like they are in a "One-on-One" relationship with the "Soul" of the company.

The Customer Outcome: Institutional Belonging

The ultimate value of the Diplomat role is **Alignment through Meaning**. When a brand acts as a "Diplomat," the customer relationship moves from "Price-Based" to "Purpose-Based." For the leader, this provides **Narrative Security**. You know that your "Voice" is resonant, ethical, and backed by the "Total Evidence" of your intelligence. This turns the organization into an "Authentic Powerhouse"—where every customer interaction is an opportunity to strengthen the "Shared Truth" of the community. It ensures that the brand's "Voice" is never just "Noise," but a "Signal of Belonging" that attracts and retains the highest-quality customers in the world.

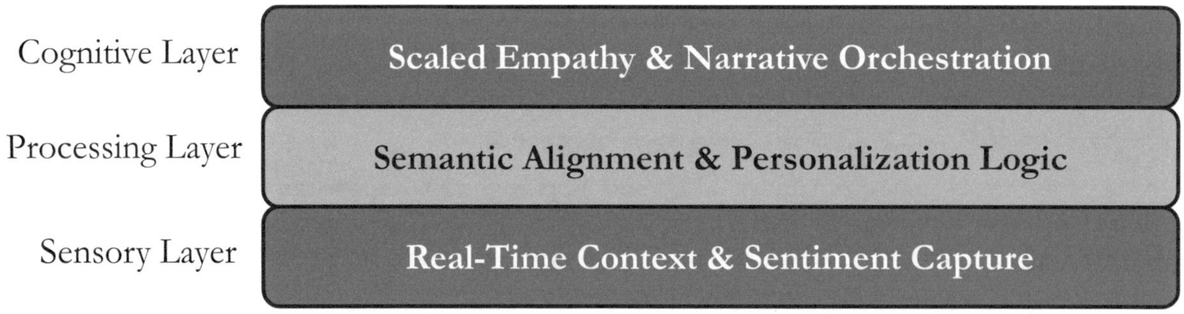

Executive Power: Hyper-Personalized Resonance

For the modern executive, the "Voice" of the customer transformation is the ability to project **Hyper-Personalized Resonance** across millions of unique individuals simultaneously. In an era of "Brand Fatigue," the **Diplomat** role ensures the brand's narrative is not a generic broadcast, but a "Sovereign Conversation" tailored to each person's specific context and values. This role is about the **Abolition of the Generic Experience**, ensuring that the "Voice" of the organization is the most "Relevant" and "Empathetic" in the customer's life. The leader uses this stack to ensure that the brand's "Intent" is articulated with absolute integrity, moving the company from "Selling" to "Serving." By architecting this role, the executive ensures that every customer feels a sense of "Authentic Alignment" with the brand, building a level of institutional trust that competitors using "One-Size-Fits-All" marketing cannot replicate.

The Sensory Layer: Real-Time Context & Sentiment Capture

The **Sensory Layer** of the Customer Diplomat acts as the organization's "Empathy Ear." It moves beyond the "CSAT Score" and into **Autonomous Context Capture**. Utilizing **Natural Language Processing (NLP) across support logs** and **Social Listening APIs**, this layer "listens" to the "Emotional Nuance" of the customer's life. It captures the subtle shifts in tone—detecting stress, excitement, or confusion during a voice call or chat interaction. By monitoring the "Situational Context" (such as local weather, economic shifts, or personal milestones), this layer provides the leader with a real-time map of the "Customer's World." It ensures the executive is aware of the emotional landscape of the audience before the brand speaks. This layer provides a high-fidelity "Map of Resonance," ensuring the "Voice" of the firm is grounded in the absolute reality of what the customer is feeling in the present moment.

The Processing Layer: Semantic Alignment & Personalization Logic

The **Processing Layer** serves as the "Integrity Engine" of the communication stack. It utilizes **Knowledge Graphs** to link the brand's core values to the specific needs and history of the individual customer. If the executive's "Voice" speaks of "Reliability," this layer "Hard-Codes" that promise into the **Interaction Logic**. This ensures that the leader's message is not just a slogan, but is mathematically reflected in the customer's actual experience (e.g., prioritized support or tailored warranties). It acts as a **Shield of Semantic Integrity**, ensuring that the brand's promises are consistent with its actions. By resolving the identity of every customer interaction in the context of their unique history, this layer kills the friction of "Corporate Impersonality," providing a "Verifiable Logic" that the company's "Voice" is aligned with the customer's best interests, thereby building an unassailable bond of trust.

The Cognitive Layer: Scaled Empathy & Narrative Orchestration

The **Cognitive Layer** is where the "Brand Truth" is transformed into **Hyper-Personalized Narratives**. Utilizing **Large Language Models (LLMs)** and **Agentic Orchestration**, the Diplomat generates "Sovereign Customer Experiences." It doesn't issue a "Campaign"; it orchestrates "Million-Person Dialogues." This layer ensures the brand's intent is perfectly "decoded" into the specific language, tone, and medium preferred by each customer—achieving **Empathy at Scale**. Because this layer is governed by **Explainable AI (XAI)**, it ensures that every personalized recommendation or message has a "Reasoned Justification" that aligns with the brand's ethical guardrails. This allows the leader to project a "Voice" of **Absolute Individual Relevance**, turning "Marketing" from an intrusion into a "Service." It ensures that every stakeholder feels a sense of "Security and Alignment" with the brand's mission, creating a culture of "Principled Customer Obsession."

OVERVIEW

Explainable AI (XAI) provides transparent, human-readable rationales for AI decisions, moving beyond black box outputs. It reveals why a model reached a conclusion, building trust and mitigating legal and ethical exposure. The Regulatory & Compliance leader leverages XAI to validate complex, automated decision-making processes, particularly in highly regulated domains. This synergy ensures AI systems are fair, auditable, and adhere to internal policies and external legal requirements, transforming compliance from a check-the-box function into a strategic enabler of trust.

MAIN STEPS

Start

STEP 1 — Define Audit Scope
Establish parameters for compliance assessment.

STEP 2 — Feature Importance Analysis
Quantify input factors driving AI output.

STEP 3 — Generate Local Explanation
Create rationale for a single prediction.

STEP 4 — Validate Model Fairness
Test decision consistency across subgroups.

STEP 5 — Human Review and Oversight
Leader verifies explanations meet standards.

STEP 6 — Document Rationale & Action
Record justification for audit readiness.

End

SAMPLE FLOW

Automated Customer Consent Verification: A Marketing leader audits the AI system using XAI that verifies customer consent for personalized marketing, ensuring all communications are legally compliant.

1. Input Consent Parameters
Define legal requirements for opt-in consent and data usage accurately.

2. Analyze Consent Logic
Trace the specific customer actions that confirmed explicit data use permission.

3. Identify Flagged Contacts
XAI explains which customer profiles lack sufficient, compliant consent documentation.

EXPLAINABLE AI

4. Communication History
Audit the contact history for flagged users to find any unauthorized communications.

5. Non-Compliance Exposure
XAI calculates potential fine severity based on the scale of consent violation risks.

6. Approve Remediation Strategy
Leader approves communication changes and documents consent verification failure rationale.

USE CASES

- **Manufacturing | Product Recall Notification Audit:** Explains the specific criteria (batch number, region) that trigger targeted product recall notifications to customers, ensuring complete and compliant communication quickly.
- **Financial Services | Loan Denial Justification:** Provides clear, auditable explanations for *why* a customer was denied credit, ensuring full compliance with fair lending acts and maintaining consumer trust reliably.
- **Healthcare | Treatment Personalization Privacy:** Explains which HIPAA-compliant patient data variables were used by the AI to recommend personalized care pathways, assuring adherence to privacy mandates safely.
- **Technology | Terms of Service (ToS) Change Communication:** Justifies the timing and target group for communicating complex changes to Terms of Service, ensuring all required users are legally informed accurately.
- **Retail | Pricing Fairness Rationale:** Explains the factors (inventory, demand) driving dynamic pricing changes, allowing leaders to certify that differential pricing models are non-discriminatory and legally sound clearly.
- **Energy | Service Interruption Prioritization:** Provides transparent rationale for the AI-based prioritization of restoring power during outages, verifying that critical facilities and vulnerable customers are served first quickly.

In the **Customer Transformation**, the **Diplomats (The Voice)** serve as the "Empathy Interface" between the brand's data and the consumer's experience. While the primary instrument of **Explainable AI (XAI)** provides the technical reasoning behind personalized offers or service decisions, it often lacks the warmth and nuance required to build a deep, emotional connection. To move from "Algorithmic Recommendations" to "Trusted Brand Advocacy," a leader must integrate these six secondary pairings. This ensures the organization's "Voice" is not just logically sound, but is also contextually resonant, ethically transparent, and human-centric.

1. The Sentiment Analysis Pairing for Tone Calibration

Pairing **Explainable AI with Sentiment Analysis** is best utilized during "Recovery Moments," such as when a service failure occurs. While XAI explains why a delay happened, Sentiment Analysis gauges the customer's frustration level in real-time. This allows the leader to calibrate the "Voice," ensuring the explanation is delivered with the appropriate level of contrition or urgency. By matching the brand's tone to the customer's emotional state, the Diplomat turns a potential churn event into a moment of strengthened loyalty.

2. The Contextual NLP Pairing for Multimodal Brand Voice

Leaders should pair **Explainable AI with Contextual NLP** to ensure a "Universal Voice" across all customer touchpoints. XAI provides the core facts of a decision, but Contextual NLP translates those facts into the specific dialect of the platform—whether it is a short, punchy SMS, a detailed email, or a conversational voice assistant. This ensures the brand's "Voice" feels native to every environment, eliminating the friction of "Corporate Speak" and ensuring the customer feels understood in their preferred medium.

3. The Knowledge Graph Pairing for Preference Continuity

When a customer moves across departments—from sales to support—**Explainable AI must be paired with Knowledge Graphs**. While XAI justifies a specific action, the Knowledge Graph provides the "History of Intent." This allows the "Voice" to acknowledge the customer's entire journey, ensuring they never have to repeat their preferences or frustrations. This pairing builds "Conversational Memory," transforming the brand's Voice into that of a long-term partner who remembers every detail of the relationship.

4. The Differential Privacy Pairing for Trust-Based Transparency

To build a "Privacy-First Voice," a leader should pair **Explainable AI with Differential Privacy**. This allows the Diplomat to explain *why* a customer is seeing a specific recommendation based on aggregate trends without ever exposing the private data of other users. This is best used in "Social Proof" marketing. It ensures the brand remains radically transparent about its data usage while providing a mathematically guaranteed "Privacy Shield," reinforcing the customer's trust in the brand's role as a sovereign data steward.

5. The Conversational AI Pairing for High-Fidelity Interrogation

To eliminate the "Black Box" of automated decisions, leaders should pair **Explainable AI with Conversational AI**. While XAI holds the logic, Conversational AI allows the customer to "talk back" and ask for clarification—such as "Why was my credit limit set at this level?" This pairing turns the Diplomat into a dynamic advisor, providing instant, reasoned justifications. It moves the customer relationship from "Passive Consumption" to "Active Dialogue," building a foundation of radical, open-door transparency.

6. The Blockchain Pairing for Verifiable Brand Promises

In an era of skepticism, **Explainable AI should be paired with Blockchain**. XAI generates the explanation for a brand claim—such as a product's sustainability origin—while the Blockchain stores that proof in an immutable ledger. This is best utilized for "Value-Based Branding." It ensures the "Voice" is legally unassailable and tamper-proof. By providing absolute accountability for every brand promise, the leader secures the brand's moral integrity, turning every customer communication into a verifiable and trusted transaction.

Customer Power: Personal Reality Manifestation

In the legacy consumer world, customers were forced to choose from a "Fixed Menu" of options. Brands would create a product, and the customer would have to bend their needs to fit what was available. This created a "Friction of Compromise," where the customer was never truly satisfied, only "satisfied enough." The transformation into **Creators** represents the shift from "Mass Customization" to **"Generative Reality."** In this role, the brand's "Engine" is used to manifest a unique version of the brand for every single individual. The customer no longer buys a product; they participate in the "Generation" of a solution that is perfectly calibrated to their specific context, aesthetic, and functional requirements.

Defensive Stance: Killing the Friction of Choice

The defensive mission of the Customer-Creator is the **Elimination of Cognitive Load**. In an age of "Infinite Choice," customers are often paralyzed by the effort required to find the right solution. This "Decision Fatigue" leads to abandonment and dissatisfaction. The Creator role kills this "Friction" by using the "Engine" to act as a "Curatorial Filter." Instead of showing the customer a million options, the brand "Generates" the "Perfect One." This creates a **Shield of Simplicity**. By "Hard-Coding the Customer's Preferences" into the generative process, the brand ensures that every prototype, every design, and every offer is "Pre-Validated." This defensive posture kills the "Waste of Exploration," protecting the customer's time and ensuring that the "Engine" only ever produces what is relevant. It turns the "Legacy Burden" of the brand—old inventory or outdated models—into a "Fluid Resource" that can be reshaped to meet the modern demand instantly.

Offensive Stance: Unlocking Co-Creative Agency

On the offense, the Creator moves toward **Collaborative Manifestation**. This is the ability to give the customer the "Keys to the Engine." The brand provides the "Framework and the Logic," but the customer directs the "Creation." This **Unlocks Creative Empowerment**, where the customer moves from being a "Consumer" to a "Co-Creator." Whether it is generating a personalized financial plan, a custom-fitted piece of apparel, or a unique digital interface, the brand's "Engine" becomes the customer's "Hands." This turns the brand into a "Platform for Identity," where the customer uses the company's intelligence to manifest their own vision. This offensive move creates a "Deep Product Lock-in," because the customer is not just buying an object; they are buying the result of their own agency, powered by the brand's sophisticated "Engine." It ensures the brand is always producing the most "Original" and "Relevant" output in the market, because every output is an original creation for a specific human being.

The Customer Outcome: Absolute Personalization

The ultimate value of the Creator role is **Radical Relevance**. When a brand "Creates" for a customer, the "Gap" between "Desire" and "Reality" disappears. For the leader, this provides **Execution Flexibility**. You are no longer "Stuck with Inventory" that the market doesn't want; you are "Generating Value" as it is demanded. This turns the brand into a "Living Response System"—one that evolves its "Output" in real-time to match the "Intent" of its community. It ensures that the brand's "Engine" is never "Idle," but is always working to manifest the highest possible value for each individual, creating a brand that feels like a "Personal Concierge of Creation."

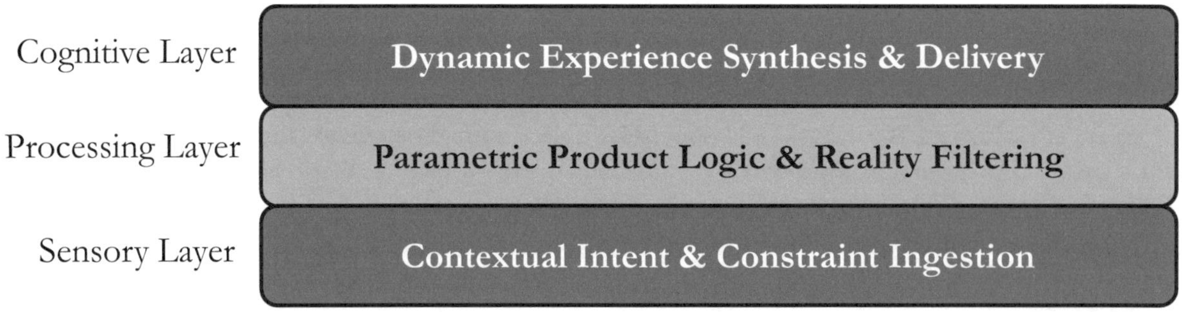

Executive Power: Generative Value Manifestation

The **Creator** role in customer transformation represents the ultimate shift from selling "off-the-shelf" products to becoming a **Manifestation Engine** for the customer's specific goals. In the legacy model, customers were forced to choose from a static menu of options, leading to the friction of "settling" for the closest fit. The Creator role replaces this with **Autonomous Value Synthesis**, where the brand's offering—whether it is a financial product, a software interface, or a physical configuration—is dynamically generated in response to the user's immediate intent. The executive uses this engine to move the organization from a state of "Mass Production" to "Mass Customization," ensuring the brand is always manifesting the most "Relevant" utility. This turns the "Engine" of the firm into a liquid resource that reshapes itself around the customer's life.

The Sensory Layer: Contextual Intent & Constraint Ingestion

The **Sensory Layer** for the Creator captures the "Goal State" of the customer as a high-definition signal. It moves beyond simple search terms by integrating **Natural Language Intent Capture** and **Environmental Context Streams**. By utilizing **Agentic Discovery** to scan the customer's current ecosystem—their existing portfolio, their hardware constraints, or their historical preferences—this layer identifies the "Atomic Requirements" of the solution before a product is ever suggested. It captures the raw intent—"I need to fund a low-risk expansion in a volatile market"—and automatically identifies the parameters required for success. This ensures the leader is building the "Engine" of the solution on a foundation of "Individual Reality" rather than "Segmented Assumptions," capturing the necessary signals for a bespoke manifestation as a live, fluctuating requirement.

The Processing Layer: Parametric Product Logic & Reality Filtering

The **Processing Layer** acts as the "Architectural Refinery" for customer value. It utilizes **Constraint-Based Solvers** and **Generative Design Engines** to manifest a solution that is both "Ideal" for the customer and "Viable" for the brand. This layer understands the "Rules of the Possible"—including regulatory compliance, profit margins, and technical feasibility. It stress-tests every generative configuration against the organization's "Causal Truths," ensuring the executive only sees solutions that are sustainable. It acts as a **Guardian of Feasibility**, filtering out high-risk "hallucinations" and ensuring that every manifest offer is mathematically sound. By applying "Parametric Value Logic," this layer ensures the "Engine" is disciplined, preventing the brand from making promises it cannot fulfill, while ensuring the customer receives a solution that feels uniquely engineered for their specific problem.

The Cognitive Layer: Dynamic Experience Synthesis & Delivery

The **Cognitive Layer** is the "Orchestration Mind" of the customer engine, powered by **Large Language Models (LLMs)** and **Multi-Agent Orchestration**. This layer "hallucinates" thousands of different ways to fulfill the customer's intent, presenting the most "Resonant" version for immediate delivery. It doesn't just suggest a product; it manifests the **End-to-End Experience**. The leader provides the strategic guardrails, and the Cognitive Layer generates the "Offer Genome"—automatically configuring the pricing, the terms of service, and the digital interface required to fulfill the mission. It involves "Recursive Experience Evolution," where the AI adapts the offering as the customer interacts with it, ensuring the final output is the "Optimal Manifestation" of the brand's promise. This turns the executive into a "Conductor of Utility," capable of deploying complex, individualized solutions at a speed that makes traditional R&D cycles look prehistoric.

OVERVIEW

Generative AI creates novel outputs like text, art, code, and immersive environments from simple prompts. It excels at rapid prototyping, style transfer, and exploratory ideation. The Creative & Artistic leader partners with Generative AI to rapidly visualize strategic concepts, explore diverse future scenarios, and prototype new organizational narratives. The synergy transforms vision casting from a slow, manual process into a continuous, inspiring, and engaging collaborative endeavor.

MAIN STEPS

Start

STEP 1 — Define Creative Prompt
Articulate desired output parameters precisely.

STEP 2 — Initial Content Generation
Create diverse drafts based on input.

STEP 3 — Style and Mood Adaptation
Adjust tone, format, or aesthetic output.

STEP 4 — Iterative Refinement Loop
Guide model toward final creative goal.

STEP 5 — Human Curation & Selection
Choose the most impactful creative result.

STEP 6 — Deploy Creative Asset
Integrate final output into strategy.

End

SAMPLE FLOW

Personalized Marketing Campaign Assets: A marketing executive uses Generative AI to create hyper-personalized visual and textual assets for diverse customer segments to promote sales.

1. Analyze Segmentation Data
Define core customer archetypes and emotional purchase triggers precisely.

2. Generate Tailored Content
Produce unique, scenario-specific marketing copy, images, and videos instantly.

3. A/B Test Creative Variations
Rapidly test numerous visual styles, messaging tones, and platform formats.

4. Integrate Brand Aesthetics
Refine generated assets to strictly comply with established brand guidelines.

5. Evaluate Conversion Impact
Measure segment-specific engagement and sales uplift from the new assets rigorously.

6. Launch Dynamic Experience
Deliver the perfectly personalized and emotionally resonant content across all channels.

USE CASES

- **Manufacturing | Product Configurators:** Designs interactive 3D product configurators for clients, allowing creative visualization and personalization of complex machinery before the final purchase decision is made.
- **Financial Services | Client Education Assets:** Generates customized financial planning scenarios and video explainers, simplifying complex investment concepts into visually compelling and easily digestible educational content for clients.
- **Healthcare | Patient Engagement Content:** Creates personalized recovery journey visualizations and supportive motivational content for patients, fostering emotional connection and improving adherence to prescribed long-term treatment plans.
- **Technology | Developer Experience:** Produces unique, generative code samples and detailed documentation that instantly adapt to a user's specific project needs, drastically improving the developer experience and adoption rate.
- **Retail | Virtual Try-on Experiences:** Develops personalized virtual try-on experiences and AI-generated lifestyle photography, allowing customers to visually integrate products into their own environment before committing to purchase.
- **Energy | Consumption Visualization:** Generates tailored reports and artistic data visualizations of individual household energy consumption, promoting conservation and clearly communicating the personalized benefits of adopting smart grid technology.

In the **Customer Transformation**, the **Creators (The Engine)** represent the brand's "Generative Experience Lab." While the primary instrument of **Generative AI** is used to create hyper-personalized marketing copy or product recommendations, it can often become a "Generic Noise Generator" that lacks the specific aesthetic and functional soul of the brand. To move from "Automated Content" to "High-Fidelity Co-Creation," a leader must integrate these six secondary pairings. This ensures the "Engine" produces experiences that are not only unique to each individual but are also grounded in the brand's sovereign identity and operational reality.

1. The Synthetic Data Pairing for Persona Expansion

Pairing **Generative AI with Synthetic Data Generation** is the ultimate solution for **Niche Market Discovery.** Before targeting a new customer segment, the Creator generates millions of "Synthetic Personas" that mirror the behavior of untapped demographics. This allows the brand to "test-market" products and messages in a virtual environment. By training the engine on these high-fidelity simulations, the leader can bypass the expensive "trial-and-error" phase of expansion, ensuring that the brand's first real-world interaction with a new segment is perfectly calibrated.

2. The Digital Twin Pairing for Journey Prototyping

Leaders should pair **Generative AI with Customer Digital Twins** to achieve **Dynamic Experience Simulation.** While the AI generates new user interface (UI) layouts or service flows, the Digital Twin provides the "Behavioral Guardrail." This is best utilized in mobile app development. The "Engine" dreams up thousands of navigation paths, and the Digital Twin instantly tests them against "Historical Friction Nodes." This pairing ensures that the Creator's output is not just visually appealing but is functionally optimized for the highest conversion and lowest frustration.

3. The RLHF (Human Feedback) Pairing for Aesthetic Alignment

When the "Engine" produces outward-facing creative assets—such as video ads or personalized catalogs—**Generative AI must be paired with Reinforcement Learning from Human Feedback (RLHF).** This ensures that the machine's output aligns with the brand's "Sovereign Tone." Human creative directors provide the "Taste" and "Nuance" that raw models lack, refining the generative output until it mirrors the brand's unique DNA. This pairing turns every marketer into a "Curator of Excellence," amplifying the volume of creative output without diluting the brand's premium soul.

4. The Vector Database (RAG) Pairing for Grounded Storytelling

To prevent "hallucinations" in personalized customer interactions, a leader should pair **Generative AI with Vector Databases (RAG).** This pairing ensures that every generative response is grounded in the firm's actual product manuals, history, and past customer successes. This is best utilized in "Expert-Level Concierge" services. Instead of the AI making up features, it retrieves relevant facts from the "Institutional Memory" to provide creative solutions to complex customer problems.

5. The Differential Privacy Pairing for Secure Personalization

To generate hyper-personalized experiences without violating individual sovereignty, leaders should pair **Generative AI with Differential Privacy.** This is best deployed when creating "Lookalike Recommendations" or collaborative loyalty programs. The generative model learns the underlying patterns of customer desire while the privacy layer ensures that no individual user's data can ever be reverse-engineered or exposed. This pairing allows the "Engine" to manifest the value of "Intimate Personalization" while maintaining absolute privacy, reinforcing the brand's role as a trusted partner.

6. The API Orchestration Pairing for Instant Manifestation

To move from "Creative Ideas" to "Immediate Action," **Generative AI should be paired with API Orchestration.** While the AI creates a personalized "Service Recovery Plan" for an unhappy customer, API Orchestration provides the "Connection" to actually trigger the refund or the replacement shipment. This moves the "Engine" from being a generator of words to a generator of functional business logic. It turns the Creator into a high-speed factory for customer satisfaction, where the time between "Generating a Solution" and "Executing the Solution" is reduced to zero.

Customer Power: Value Optimization

The traditional customer relationship is often a "Zero-Sum Game" based on negotiation and information asymmetry. One side wins by convincing the other to pay more or accept less. The transformation into **Strategists** moves the relationship toward **Value Alignment.** In this role, the brand uses its "Compass" to find the "Optimal Exchange" for both parties. It is the shift from "Pricing a Product" to "Optimizing an Outcome." The Strategist role ensures that the relationship is governed by "Objective Logic" rather than "Transactional Friction," turning the brand into a "Strategic Navigator" for the customer's own capital and resources.

Defensive Stance: Killing the Exploitation Trap

The defensive mission of the Customer-Strategist is the **Validation of Fairness**. In a world of "Dynamic Pricing" and "Hidden Fees," customers have become deeply skeptical of brand motives. This "Trust Friction" acts as a barrier to long-term loyalty. The Strategist role kills this "Skepticism" by using "Causal Logic" to make the value exchange transparent. This is the **Shield of Objectivity**. By "Modeling the Outcome" for the customer—showing them exactly how a purchase will save them time, grow their wealth, or reduce their risk—the brand removes the "Gambler's Anxiety" from the transaction. This defensive posture ensures that the "Compass" of the brand is never used to "Extract" value, but always to "Create" it. It prevents the organization from chasing "Short-Term Margin" at the expense of "Long-Term Equity," killing the risk of "Predatory Behavior" and ensuring the relationship is built on a foundation of "Verified Benefit."

Offensive Stance: Unlocking Life-Cycle Navigation

On the offense, the Strategist moves toward **Prescriptive Success**. This is the ability to "Guide the Customer" through their own complex decisions. The brand uses its "Compass" to help the customer navigate their "Life-Stages" or "Business Challenges." This **Unlocks Navigational Loyalty**, where the brand becomes the "Operating System" for the customer's success. Instead of just selling a tool, the brand "Prescribes the Best Path" to a goal. If the "Compass" identifies that a customer is over-spending in one area or missing an opportunity in another, the brand proactively suggests a "Course Correction." This turns the brand into a "Strategic Partner" that is "Invested in the Win." By "Optimizing the Whole Journey," the brand creates a "Resilient Relationship" that competitors cannot disrupt with a lower price, because the competitor doesn't possess the "Navigational Logic" that the customer now relies on for their own growth and stability.

The Customer Outcome: Strategic Certainty

The ultimate value of the Strategist role is **Objective Confidence**. When a brand acts as a "Strategist," the customer feels "Supported" rather than "Sold To." For the leader, this provides **Economic Stability**. You are building a "Portfolio of Successful Customers" whose growth is intrinsically linked to your own. This turns the organization into a "Calculating Ally"—one that finds the "Win-Win" in every market condition. It ensures that the brand's "Compass" is always pointing toward "Mutual Prosperity," creating a brand that is respected for its "Intelligence" and loved for its "Alignment" with the customer's true goals.

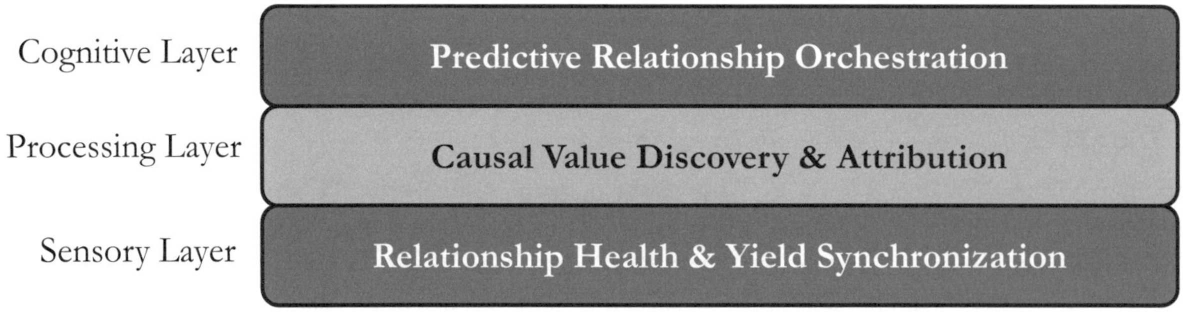

Executive Power: Causal Lifetime Yield

The **Strategist** role in customer transformation represents the "Abolition of Short-Termism." It moves the executive from "Quarterly Sales Targets" to **"Causal Lifetime Yield Optimization."** The leader uses this stack to ensure that every "Customer Interaction" is an investment in long-term systemic value rather than a one-time transaction. The Strategist is the **Guardian of the Relationship**, ensuring that the organization is not just "Closing Deals," but "Cultivating Assets." In this role, the executive evolves into a "Strategic Relationship Navigator" who manages the company's "Trust Capital" with mathematical precision. This role kills the friction of "High Churn" and "Inefficient Acquisition" by ensuring the brand's energy is always flowing toward the "True Drivers" of customer lifetime value (LTV) and long-term loyalty.

The Sensory Layer: Relationship Health & Yield Synchronization

The **Sensory Layer** of the customer Compass monitors the "Micro-Levers" of relationship health across the entire global portfolio. It synchronizes **Real-Time Interaction Data** (support tickets, usage frequency, payment history) with **External Market Context**. By tracking the "Velocity of Trust" (how quickly a customer moves from trial to advocacy) against the "Cost of Service," this layer provides the "Coordinate Data" required to know exactly where the firm's customer base stands in relation to its "Strategic North Star." It provides the "Eyes" for the Compass, ensuring that the starting point of any customer initiative is grounded in the absolute reality of current relationship health. This layer removes the "Blind Spots" in management, ensuring the leader is aware of every factor—from "Silent Dissatisfaction" to "Emerging Competitor Threats"—that could impact the trajectory of the organization's LTV.

The Processing Layer: Causal Value Discovery & Attribution

The **Processing Layer** is the heart of the Strategist role, utilizing **Causal Inference Engines** and **Structural Equation Modeling** to determine "Why" certain customers are staying or leaving. While traditional CRM analytics show simple "Correlation" (e.g., users who use the mobile app stay longer), the Strategist identifies the **Causal Path** (e.g., they stay longer because the app reduces the friction of specific high-value tasks). This layer provides the "Mathematical Evidence" for customer strategy, ensuring the leader is investing in the "Levers" that actually matter for the brand's "Yield." It acts as a **Shield of Objectivity**, stripping away the "Noise" of vanity metrics like "Likes" or "Clicks" and focusing the executive on "Global Relationship Optimization." By identifying the "Mathematical Why," this layer ensures the firm is never "Chasing Trends," but is always focused on the "Causal Truths" that drive real-world lifetime profitability.

The Cognitive Layer: Predictive Relationship Orchestration

The **Cognitive Layer** acts as the "Navigational Mind" of the customer stack, utilizing **Reinforcement Learning** and **Multi-Agent Simulation**. This layer "plays the game" of the customer lifecycle millions of times to find the "Optimal Path of Engagement." When the market environment shifts—due to a recession or a technological disruption—the Compass "Recalculates" the customer retention strategy instantly. This layer doesn't just report on customers; it "Prescribes the Best Move" to maximize the **Lifetime Yield**. For example, it might recommend offering a proactive credit to a specific high-value segment because the probability of churn has increased due to an external market shock. This provides the executive with **Strategic Certainty**, allowing them to navigate customer volatility with the confidence that their "Compass" is calibrated to the objective laws of value creation. It turns the organization into a "Calculating Partner," ensuring the brand "Wins" because its "Customer Strategy" is mathematically optimized for the highest possible mutual outcome.

OVERVIEW

Cognitive AI mimics human intelligence to understand and interpret unstructured data like text, speech, and images. It processes complex inputs to generate hypotheses, assess logical connections, and offer nuanced context. The Research & Analytical leader uses Cognitive AI as a hyper-efficient knowledge partner, rapidly synthesizing vast amounts of data to rigorously test assumptions, structure complex arguments, and make deeply informed, evidence-based strategic decisions.

MAIN STEPS

Start

STEP 1 – Data Ingestion: *Collect and structure input knowledge quickly*

STEP 2 – Understand Context: *Understand meaning and relationships deeply.*

STEP 3 – Hypothesis Generation: *Formulate potential explanatory outcomes.*

STEP 4 – Evidence Synthesis: *Find relevant supporting information.*

STEP 5 – Analytical Refinement: *Test hypotheses against real-world data.*

STEP 6 – Insight Formulation: *Structure findings for human action.*

End

SAMPLE FLOW

Customer Insights - Sentiment and Friction Mapping: A CMO uses AI to analyze massive volumes of feedback to identify specific friction points and deep emotional drivers for customer dissatisfaction.

1. Define Experience Scope →
Specify the journey stages or product interactions to be analyzed rigorously.

2. Ingest Unstructured Data
Collect social media posts, support transcripts, and survey open-text responses effectively.

3. Map Sentiment Drivers
Automatically categorize language to isolate specific emotional triggers and pain points.

COGNITIVE AI

4. Correlate Feedback Channels
Analyze relationships between support calls and low product review scores quickly.

5. Test Root Cause Hypotheses
Validate the perceived causes of friction against operational and technical logs.

6. Perform Experience Redesign
Sign off on targeted interventions to resolve the verified customer pain points.

USE CASES

- **Manufacturing | Service Agreement Analysis:** Synthesizes complex product performance data and service level agreements to analyze customer risk exposure and ensure proactive contract compliance management.
- **Financial Services | Complaint Resolution Pathways:** Reviews extensive customer complaint records to classify and interpret systemic operational weaknesses, designing improved, efficient, and regulatory-compliant resolution protocols.
- **Healthcare | Patient Journey Mapping:** Analyzes patient flow data, survey responses, and clinical notes to identify critical bottlenecks in the care experience, improving access and service quality.
- **Technology | Feature Adoption Analysis:** Interprets user forum discussions, in-app usage data, and product roadmaps to analyze feature success factors, guiding future interface and functionality development efforts.
- **Retail | Hyper-Personalization Strategy:** Synthesizes purchase history, browsing behavior, and geographical trends to develop complex customer segmentation models, enabling highly targeted marketing campaigns.
- **Energy | Proactive Outage Communication:** Analyzes weather forecasts, infrastructure maps, and historical communication data to design highly effective, tailored, and empathetic customer communication plans during service interruptions.

In the **Customer Transformation**, the **Strategists (The Compass)** act as the "Navigational Intelligence" of the brand's market position. While the primary instrument of **Cognitive AI** identifies high-level patterns in churn or segment growth, it often lacks the multi-dimensional rigor required to balance immediate revenue with long-term brand equity. To move from "Reactive Campaigning" to "Anticipatory Market Leadership," a leader must integrate these six secondary pairings. This ensures the "Compass" doesn't just point to where the money is, but identifies the precise path to maximum customer lifetime value (CLV) and category dominance.

1. The Multi-Objective Optimization (MOO) Pairing for Value Balance

Pairing **Cognitive AI with MOO** is the ultimate tool for **Sustainability-First Growth.** While AI identifies the most profitable customer actions, MOO balances the trade-offs between short-term conversion, acquisition cost, and long-term customer sentiment. This is best utilized in subscription models. It allows the leader to identify the "Loyalty Frontier," ensuring the brand isn't "burning out" its audience for a quarterly spike but is instead pacing its growth for a decade of sustained revenue.

2. The Game Theory Simulation Pairing for Competitive Defense

Leaders should pair **Cognitive AI with Game Theory** to achieve **Market Share Sovereignty.** This is best deployed during aggressive competitive launches. While the AI predicts how your customers might churn to a rival, Game Theory simulates the "Counter-Moves" required to retain them. This ensures the "Compass" identifies the strategy that keeps your brand as the "Primary Choice," allowing the leader to anticipate rival discounts or features and neutralize them before they impact the bottom line.

3. The Digital Twin Pairing for Journey Stress-Testing

When redesigning the "Total Customer Experience," **Cognitive AI must be paired with Customer Journey Digital Twins.** This is best utilized to "rehearse" the impact of a major policy shift—such as moving from a free to a paid tier. The Strategist simulates how different personas will migrate through the new funnel. This pairing ensures the "Compass" guides the brand through the transition without causing a "Mass Exit," allowing for the hardening of the customer experience in a virtual model before it is launched to the real world.

4. The Knowledge Graph Pairing for Intent Arbitrage

To maximize the "Return on Attention," a leader should pair **Cognitive AI with Knowledge Graphs.** While AI tracks what a customer buys, the Knowledge Graph identifies the *latent needs* across different life stages. This allows for "Intent Arbitrage"—identifying when a customer's data in one category (e.g., home buying) signals a future high-value need in another (e.g., insurance). It ensures the "Compass" is always directing the brand's offensive resources to the customer's next major life event before the competition notices.

5. The Causal Inference Pairing for Attribution Integrity

For maintaining a "Results-Driven Marketing Culture," leaders should pair **Cognitive AI with Causal Inference.** While AI identifies successful campaigns, Causal Inference proves *which* specific touchpoint actually moved the needle—separating true influence from "organic" behavior. This is best utilized in multi-channel attribution. It ensures the "Compass" points toward the investments that actually drive incremental growth, preventing the waste of capital on "False Positive" channels that were merely taking credit for existing loyalty.

6. The Agentic Orchestration Pairing for Live Experience Governance

In high-velocity markets, **Cognitive AI should be paired with Agentic Orchestration** for **Live Market Adaptation.** While the Strategist sets the growth goals, the Agentic layer automatically re-allocates promotional budgets or "Surprise and Delight" credits based on real-time churn signals. This is best utilized during peak shopping seasons. It turns the "Compass" into an active driver of the enterprise, ensuring that the brand's resources are always flowing toward the segments with the highest immediate opportunity and the greatest long-term potential.

Customer Power: Personal Knowledge Synthesis

In the legacy era, a brand's "Brain" was centralized and generic. It treated every customer as a data point in a vast, undifferentiated pool. The transformation into **Architects** shifts the focus toward **Individual Structural Intelligence**. In this role, the brand builds a "Dedicated Brain" for every customer—a structured, evolving repository of that specific individual's preferences, history, and context. The Architect doesn't just store data; they design the "Connectivity" between the brand's capabilities and the customer's life. It is the shift from "Knowing the Market" to "Being the Memory" for the individual, ensuring that the brand's intelligence is perfectly mapped to the customer's personal architecture.

Defensive Stance: Killing the Fragmentation of Experience

The defensive mission of the Customer-Architect is the **Abolition of Disconnect**. Most customer frustration stems from a "Brain" that is fragmented—where the support team doesn't know what the sales team promised, or the digital interface forgets a preference set in a physical store. The Architect role kills this "Friction of Amnesia" by creating a "Unified Personal Knowledge Graph." By "Resolving the Identity" of the customer across every touchpoint, the brand creates a **Shield of Seamlessness**. It ensures that the customer never has to repeat themselves or "re-teach" the brand their needs. This defensive posture protects the relationship from the "Death by a Thousand Cuts" caused by minor operational disconnects. It ensures the brand's "Brain" is a "Global Asset" for the customer, providing the same personalized context regardless of the channel, time, or location.

Offensive Stance: Unlocking Predictive Contextualization

On the offense, the Architect moves toward **Cognitive Anticipation**. Because the "Brain" understands the "Structural Relationships" in the customer's life—such as the link between their business cycles and their personal travel—it can provide "Hyper-Relevant Context" at the moment of need. This **Unlocks Relevance**, where the brand surfaces the exact insight or tool the customer needs before they have to search for it. It involves "Adaptive Schema Evolution," where the brand's personal model for the customer gets smarter with every interaction. This turns the brand into a "Proactive Partner" that possesses "Total Recall" of the relationship's history. It ensures that the brand's "Intelligence" is an "Accumulative Asset" for the customer, making it increasingly difficult for a competitor to "catch up" because they lack the years of structured, personalized "Wisdom" that the Architect has built into the bond.

The Customer Outcome: Institutional Intimacy

The ultimate value of the Architect role is **Frictionless Recognition**. When a brand acts as an "Architect," the customer feels like the organization has "One Mind" dedicated to their success. For the leader, this provides **Operational Continuity**. You are no longer managing "Transactions"; you are managing a "Live Intellectual Asset." This turns the brand into a "High-IQ Companion"—one that is fundamentally "Smarter" about the customer than any rival. It ensures the brand's "Brain" is a "Continuous Presence," providing the "Perfect Context" required to maintain a state of "Deep Intimacy" at a global scale.

Executive Power: Institutional Customer Consciousness

In the legacy enterprise, customer knowledge was "Epidermal"—it lived on the surface in fragmented CRM notes or the heads of individual account managers. The **Architect** role represents the executive's transition to the "Designer of the Firm's Collective Customer Consciousness." In this role, the leader constructs a permanent, self-evolving **Cognitive Infrastructure** that transforms every customer interaction into a persistent, networked "Brain." This is the shift from "Data Silos" to "Architected Awareness," ensuring that the brand's "Wisdom" about its customers is a compound asset. The Architect ensures the organization has a "Customer Mind"—one that understands the history, emotion, and context of every individual relationship, ensuring the brand "never forgets" a preference or a past frustration, even as the human team changes.

The Sensory Layer: Universal Identity & Contextual Resolution

The **Sensory Layer** of the Customer Brain acts as the "Universal Relationship Intake System." It moves beyond simple tracking into **Autonomous Identity Resolution**. Utilizing **Cross-Channel Entity Linking**, this layer ensures that a customer's behavior on a mobile app, their tone in a support call, and their physical presence in a retail store are all unified into a single, persistent digital identity. It doesn't just record a "Click"; it "labels" the context and intent behind that click in real-time. By utilizing **Multimodal Perception Agents**, the sensory layer captures the "Nuance" of the relationship—such as a shift in loyalty or a change in life stage—at the point of origin. This ensures that the "Brain" of the firm is constantly being fed with a high-fidelity, perfectly identified stream of customer reality, removing the "Friction of Disconnect" that leads to fragmented and frustrating customer experiences.

The Processing Layer: Semantic Relationship Mapping & Graphing

The **Processing Layer** serves as the "Neural Connector" of the customer stack. It utilizes **Customer Knowledge Graphs** and **Vector Databases** to define and store the complex relationships between the identities and behaviors captured in the sensory layer. While traditional systems store data in isolated rows, the Architect's processing layer stores the **Semantic Connectivity**. It understands, for example, the deep relationship between a customer's "Brand Values" and their specific "Purchase Triggers." By employing **Graph Neural Networks (GNNs)**, this layer identifies "Hidden Needs" and potential "Friction Points" within the customer base, allowing the "Brain" to perceive each customer as a living, breathing entity rather than a data point. This creates a **Shield of Systemic Empathy**, ensuring that when a customer's needs shift in one area, the entire global organization "knows" it instantly and can adapt its logic accordingly.

The Cognitive Layer: Recursive Institutional Memory & Empathy

The **Cognitive Layer** is the "Reasoning Core" of the Customer Brain, powered by **Retrieval-Augmented Generation (RAG)** and **Recursive Feedback Loops**. This layer ensures the brand "never forgets" a relationship lesson. It synthesizes the relationships from the processing layer into a "Continuous Stream of Intelligence" that acts as a permanent advisory for the firm. When an executive or an AI agent interacts with a customer, the Cognitive Layer doesn't just retrieve a file; it "reasons" through the entire relationship history to provide an interaction grounded in "Deep Contextual Empathy." It involves **Self-Evolving Personalization Schemas**, where the Brain updates its understanding of a customer's evolving needs as new data arrives. This turns the organization into a "Learning Partner" with an ever-increasing "Relationship IQ," providing the executive with a "Memory Asset" that ensures the brand's "Mind" remains perfectly aligned with the customer's heart.

OVERVIEW

Human-in-the-Loop (HITL) AI is a cycle where human supervisors continuously train, tune, and validate a model's performance, especially for ambiguous tasks. The Supervisory & Coaching leader focuses on defining ethical boundaries and quality thresholds, ensuring AI decisions align with organizational values. This synergy elevates the leader from a decision-maker to a model trainer and quality assurance expert. By providing critical feedback, the leader helps the AI evolve, improving accuracy, fairness, and overall governance in complex, high-stakes leadership decisions.

MAIN STEPS

Start

STEP 1 — Initial Model Training
Establish baseline knowledge and performance rules

STEP 2 — Active Decision Queue
Flag uncertain outputs for human review.

STEP 3 — Human Supervisor Intervention
Leader reviews, corrects, documents outcomes.

STEP 4 — Feedback Integration Cycle
Corrected data retrains and improves the model.

STEP 5 — Model Validation & Testing
Confirm updated model meets quality standards.

STEP 6 — Deployment and Monitoring
Deploy new model; track performance, collect data.

End

SAMPLE FLOW

Fine-Tuning Customer Service Chatbot Responses: A Digital Workplace leader supervises the AI's customer interaction logs using HITL, correcting vague/insensitive chatbot replies to ensure brand alignment.

Input Ethical Standards
Define fairness metrics and bias thresholds for all performance scoring clearly.

Flag Ambiguous Interactions
HITL flags customer chats where sentiment is unclear or resolution confidence is low inaccurately.

Coach Model on Empathy
Leader manually rewrites the bot's response, documenting the optimal tone and resolution path ethically.

HUMAN-IN-THE-LOOP AI

4. Retrain Algorithm
Incorporate the leader's manual corrections to enhance future conversational tone into its logic.

5. Validate Unbiased Output
Test the updated model against diverse profiles to confirm ethical fairness is achieved correctly.

5. Approve Final Response Logic
Leader signs off on the final, customer-approved, and ethically sound chatbot response.

USE CASES

- **Manufacturing | Post-Sale Feedback Analysis:** Leaders review AI's categorization of complex product defect reports, manually labeling technical nuances to ensure the quality team receives accurate, actionable insights effectively.
- **Financial Services | Loan Application Anomaly Review:** Supervisors manually review loan applications flagged as marginally risky by the AI, correcting false positives based on relationship history to refine risk scoring reliably.
- **Healthcare | Patient Satisfaction Survey Coding:** Senior staff review AI attempts to categorize written patient feedback, correcting misinterpretations of medical jargon to ensure accurate service evaluations safely.
- **Technology | Personalized Content Filtering:** User experience leads manually flag content recommendations deemed invasive or irrelevant, teaching the AI to respect customer privacy and preferences clearly.
- **Retail | Refund Policy Interpretation:** Managers review AI decisions on ambiguous return requests, correcting rigid policy interpretations to ensure fair customer treatment and loyalty accurately.
- **Energy | Outage Communication Auditing:** Supervisors review AI-generated crisis communication messages during outages, correcting confusing language or tone before distribution to maintain customer trust quickly.

In the **Customer Transformation**, the **Architects (The Brain)** serve as the "Sovereign Trust Layer." While the primary instrument of **Human-in-the-Loop (HITL)** ensures that the brand's customer relationships remain ethically grounded and human-centric, it requires secondary technical pairings to maintain that intimacy at scale. To move from "Transactional Databases" to a "Living Experience Memory," a leader must integrate these six pairings. This ensures the organization's collective knowledge of the customer is not just a list of purchases, but a contextually aware, legally secure, and deeply empathetic infrastructure that honors individual sovereignty.

1. The Vector Database (RAG) Pairing for Intent Memory

Pairing **HITL with Vector Databases** is the foundation of the **"Universal Customer Brain."** While the human Architect sets the ethical boundaries, the Vector Database provides the "Semantic Memory" that allows the organization to recall a customer's intent across different channels. This is best utilized to eliminate "Contextual Friction." A customer can switch from a mobile app to a physical store, and the Brain ensures the assistant "remembers" the nuanced preferences discussed earlier. It turns the brand into a "Persistent Partner" that understands the customer's journey as a single, coherent narrative.

2. The Knowledge Graph Pairing for Preference Lineage

Leaders should pair **HITL with Knowledge Graphs** to ensure **Relational Integrity.** While the human provides the brand values, the Knowledge Graph maps the complex web of how a customer's tastes, family ties, and value systems interlock. This is best deployed in "Hyper-Personalized Loyalty Programs." It allows the Architect to see *why* a customer values specific perks. This pairing prevents "Identity Decay," ensuring that the Brain provides a verifiable and deep understanding of the customer's life world, anchored in a web of truth rather than just surface-level clicks.

3. The Natural Language Processing (NLP) Pairing for Emotional Guardrails

To maintain a high level of empathy, **HITL must be paired with Advanced NLP.** This is best utilized for "Tone Monitoring" and "Relationship Health Checks." When the NLP layer identifies a pattern of frustration or a shift in the customer's "Voice" that indicates a loss of trust, it triggers a request for a human Architect to intervene. This ensures the "Brain" is constantly self-correcting, maintaining the emotional quality of the brand's interaction without requiring manual oversight of every single conversation.

4. The Differential Privacy Pairing for Privacy Sovereignty

When building a shared customer ecosystem, a leader should pair **HITL with Differential Privacy.** This allows the Architect to analyze aggregate behavior and improve the "Brain" without ever risking the exposure of an individual customer's raw data. This is best used in "Privacy-First Personalization." The firm can improve its recommendation engine based on "the crowd" while mathematically ensuring that no individual's identity is ever compromised, reinforcing the brand's position as a protector of the customer's digital sovereignty.

5. The Multi-Modal AI Pairing for Contextual Awareness

To capture the "Unspoken Needs" of the customer, leaders should pair **HITL with Multi-Modal AI.** This allows the "Brain" to ingest and "understand" more than just text; it can analyze the visual context of a home design query or the tone of a voice-activated service request. This is best utilized in "High-Fidelity Concierge" experiences. The human Architect guides the AI to identify which non-verbal cues represent true satisfaction. This ensures the organization's "Brain" captures the subtle nuances of the customer's world that are usually lost in traditional data tables.

6. The Blockchain Pairing for Transparent Loyalty

For ensuring the absolute integrity of the customer's relationship with the brand, **HITL should be paired with Blockchain.** While the human sets the loyalty rules, the Blockchain stores a tamper-proof "Audit Trail" of every reward earned, data permission granted, or promise made. This is best utilized for "Verifiable Brand Integrity." It prevents any unauthorized changes to the customer's status or data rights. It ensures that the "Brain" remains an unassailable record of mutual value, providing a permanent, timestamped ledger of the organization's commitment to the customer.

Customer Power: Autonomous Fulfillment

The final shift in the customer journey is the transition from "Assisting the User" to **"Executing on Their Behalf."** Historically, a brand was a "Tool" that the customer had to pick up and use. The transformation into **Executors** represents the birth of "Agentic Service." In this role, the brand doesn't just "offer" a solution; it "manifests" it. The brand becomes the "Hands" of the customer, capable of navigating complex tasks and resolving issues autonomously based on the customer's "Strategic Intent." This is the shift from "Passive Service" to "Autonomous Agency," turning the brand into an "Invisible Engine" that works 24/7 to fulfill the customer's life and business goals.

Defensive Stance: Killing the Effort Gap

The defensive mission of the Customer-Executor is the **Elimination of Toil**. In the legacy world, the customer has to do the "Work" of being a customer—tracking shipments, managing subscriptions, or resolving billing errors. This "Operational Drag" is a major source of churn. The Executor role kills this "Wastage of Human Time" by establishing "Closed-Loop Fulfillment." The brand's "Hands" identify the friction and "Execute the Fix" before the customer even notices. This is the **Shield of Effortless Experience**. By "Hard-Coding the Response" to common life-events or service disruptions, the brand ensures the customer remains in a state of "Constant Flow." This defensive posture protects the customer's "Attention" and the brand's "Loyalty" by ensuring that "Maintenance Tasks" never interrupt the "Value Experience," making the brand the most "Low-Friction" choice in the market.

Offensive Stance: Unlocking Life-Orchestration

On the offense, the Executor moves toward **Agentic Advocacy**. This is the ability of the brand to "Conduct" the world on the customer's behalf. Instead of the customer manually interfacing with multiple systems, the brand's "Agents" navigate the landscape to achieve a goal. For example, a financial brand doesn't just "flag" a better interest rate; it "Negotiates" the transfer, "Updates" the records, and "Verifies" the gain—all autonomously within the customer's guardrails. This **Unlocks Infinite Scale for the Individual**, where the brand's "Hands" give the customer the power of a thousand assistants. It turns the brand into an "Autonomous Powerhouse"—one that out-executes competitors by being "Active" while others are "Waiting." It allows the leader to "Orchestrate" global customer success by setting "High-Level Intent" and letting the "Agents" find the most efficient path to manifest that reality in the customer's world.

The Customer Outcome: Absolute Leverage

The ultimate value of the Executor role is **Personal Sovereignty**. It allows the customer to "Live at the Speed of Thought," not the speed of bureaucracy. For the leader, this provides **Execution Dominance**. You are no longer "Serving a Customer"; you are "Empowering a Life." This turns the organization into a "Direct Manifestation Machine"—where the "Hands" of the brand are always active, always precise, and always working to fulfill the "Voice" of the customer's desires. It ensures the brand's "Execution" is as reliable as a law of physics, transforming "Customer Intent" into "Autonomous Reality" and "Reality" into "Unbreakable Loyalty."

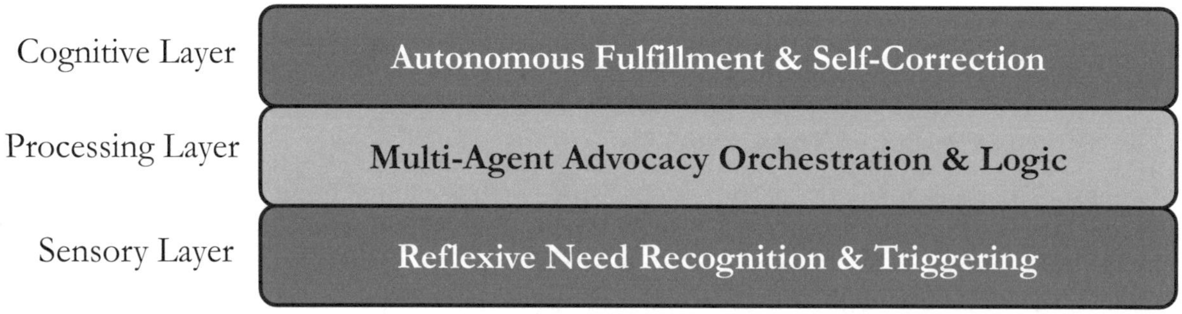

Executive Power: Agentic Customer Autonomy

The final shift in the customer journey is the transition from "Offering Service" to **"Orchestrating Autonomous Fulfillment."** The **Executor** role represents the birth of "Active Customer Advocacy," where the leader provides the "Strategic Intent" and the brand's AI-augmented "Hands" manifest results directly in the customer's life. Historically, customer service was "Reactive"—a human had to wait for a complaint to act. The Executor role replaces this with a system of **Agentic Autonomy**, where the "Hands" are capable of fulfilling tasks—such as proactive troubleshooting, automated replenishment, or personalized price adjustments—without the customer even having to ask. This is the shift from "Serving the Customer" to "Leading for the Customer," giving the executive the leverage to "Operate at the Speed of Light" by transforming customer "Intent" into "Autonomous Reality."

The Sensory Layer: Reflexive Need Recognition & Triggering

The **Sensory Layer** of the Customer Hands serves as the organization's "Reflex System." It is designed for **Real-Time Opportunity Detection**, monitoring the customer's environment for specific "Triggers" that require an immediate autonomous response. Utilizing **Predictive Anomaly Detection** and **Event-Driven APIs**, this layer identifies the moment a customer's experience deviates from the "Ideal Path." It doesn't wait for a human to see a "Shipment Delay" or a "Usage Spike"; it "feels" the friction the micro-second it occurs. This sensory input provides the "Activation Signal" for the agentic stack, ensuring that the "Hands" of the company are always ready to move the millisecond a customer need is detected. This layer protects the customer's "Peace of Mind" by ensuring the organization responds with biological speed, resolving issues before the customer is even aware they exist.

The Processing Layer: Multi-Agent Advocacy Orchestration & Logic

The **Processing Layer** acts as the "Experience Cerebellum," breaking down the leader's high-level intent into executable customer-success tasks. It utilizes **Agentic Logic (ReAct Frameworks)** and **Chain-of-Thought Reasoning** to orchestrate a swarm of specialized advocacy agents. If the leader's intent is to "ensure 100% satisfaction for the premium industrial segment during a supply disruption," this layer identifies the thousand sub-tasks required—from re-routing inventory to issuing proactive credits and personalized apologies. It manages the "Inter-Agent Communication," ensuring that each bot knows the "Ethical and Brand Guardrails" set by the executive. This layer provides the **Governance and Logic** for the execution, ensuring that the "Hands" stay within the boundaries of brand integrity, killing the friction of manual oversight through automated, high-speed coordination across the entire customer lifecycle.

The Cognitive Layer: Autonomous Fulfillment & Self-Correction

The **Cognitive Layer** is where the "Hands" actually manifest the result through **Action Transformers** and **Automated Personalization Engines**. This is the layer of "Closed-Loop Advocacy." It doesn't just "process" a ticket; it "achieves" a state of customer delight. If the agentic stack encounters an obstacle—such as a logistics failure or a failed payment—this layer uses **Self-Correction Logic** and **Reinforcement Learning** to find an alternative fulfillment path without requiring human intervention. It turns "Executive Intent" into "Physical and Digital Reality," whether that means shipping a replacement part, adjusting a subscription, or deploying a software patch autonomously. This provides the executive with **Absolute Strategic Leverage**, allowing them to "Direct the Experience Flow" rather than "Manage the Support Toil." It transforms the firm into an "Autonomous Advocacy Machine" that delivers "Market Dominance" through sheer agentic velocity and proactive customer care.

6. THE EXECUTORS (THE HANDS) | AGENTIC AI

OVERVIEW

Agentic AI operates as an autonomous mission manager, breaking down high-level strategic goals into executable tasks. The Driving Mission leader defines the ultimate objective and core values, essentially setting the constitutional limits for the AI agent. This synergy offloads complex, time-consuming multi-step execution to the AI, allowing the leader to focus exclusively on strategic vision and alignment. The leader initiates the mission; the agent executes and reports on key milestones. This partnership fundamentally converts strategy into action, maximizing the leader's impact on high-level direction and governance.

MAIN STEPS

Start

STEP 1 — Mission Goal Setting
Define the high-level desired outcome.

STEP 2 — Autonomous Task Breakdown
Agent breaks goal into specific sub-tasks.

STEP 3 — Execution and Monitoring
Agent performs steps, tracks progress autonomously.

STEP 4 — Milestone Reporting & Review
Agent summarizes for leader approval.

STEP 5 — Course Correction, Redirection
Leader adjusts mission parameters or values.

STEP 6 — Mission Completion & Handoff
Agent delivers final output to human team.

End

SAMPLE FLOW

Autonomous Customer Journey Personalization: The leader mandates an agent to deliver hyper-personalized customer experiences across three key digital touchpoints.

AGENTIC AI

1. Set Personalization Goals
Define target conversion rates and customer satisfaction metrics clearly.

2. Deploy Profiling Agents
Agent autonomously analyzes behavior, preference, and intent data correctly.

3. Generate Dynamic Content
Agent creates and optimizes unique messaging and product recommendations autonomously.

4. Execute A/B Testing Rounds
Agent runs tests, learning from real-time customer engagement and iterating promptly.

5. Validate Brand Integrity
Leader reviews messaging to ensure consistency with brand voice and ethical guidelines ethically.

6. Scale Personalized Campaigns
Leader authorizes full deployment across all defined customer journey segments criteria.

USE CASES

- **Manufacturing | Service Agreement Renewal Mission:** An agent proactively monitors product performance for major clients, autonomously generating and deploying customized service renewal proposals based on usage and predictive failure data effectively.
- **Financial Services | Proactive Churn Prevention:** The leader commissions an agent to identify high-risk accounts, autonomously deploying personalized intervention missions involving special offers, relationship manager contact, and tailored content reliably.
- **Healthcare | Patient Adherence Campaign:** An agent manages complex patient adherence missions, autonomously scheduling reminders, educational materials, and follow-up appointments based on individual treatment plans accurately.
- **Technology | Self-Service Support Optimization:** An agent is tasked with a mission to minimize support tickets, autonomously analyzing FAQ usage, identifying knowledge gaps, and generating new, proactive documentation consistently.
- **Retail | Loyalty Program Revitalization:** An agent analyzes loyalty tier performance and autonomously executes a mission to redesign rewards, targeting stagnant segments with unique, high-value, and personalized incentives quickly.
- **Energy | Energy Consumption Reduction:** The leader deploys an agent to manage customer consumption reduction, autonomously sending personalized efficiency tips, smart meter data insights, and scheduling consultations clearly.

In the **Customer Transformation**, the **Executors (The Hands)** represent the brand's "Fulfillment Core." While the primary instrument of **Agentic AI** provides the autonomous agency to handle customer requests and solve problems, it requires secondary technical pairings to ensure those digital intentions translate into perfect physical or digital outcomes. To move from "Reactive Service" to "Autonomous Delight," a leader must integrate these six pairings. This ensures the "Hands" can act on the customer's behalf with surgical precision, closing the loop between a stated desire and a completed result.

1. The RPA Pairing for Frictionless Service Recovery

Pairing **Agentic AI with Robotic Process Automation (RPA)** is best utilized to handle "Legacy Friction." While the Agent decides to issue a refund or adjust a booking based on a customer conversation, RPA provides the "fingers" to navigate the aging billing or reservation systems that lack modern APIs. This is essential for a seamless "Voice-to-Action" experience, ensuring the customer's problem is solved in seconds without a human agent having to manually toggle between disparate software screens.

2. The Digital Twin Pairing for High-Stakes Personalization

Leaders should pair **Agentic AI with Customer Digital Twins** when executing "High-Value Interventions." Before an agent offers a specific upgrade or a complex financial product, it "rehearses" the execution in the Digital Twin. This pairing ensures the "Hands" never take an action that could feel intrusive or logically inconsistent with the customer's history. It provides a virtual test-bed for the brand's actions, ensuring that the execution of a "Surprise and Delight" move is perfectly timed for maximum positive impact.

3. The Edge AI Pairing for Localized Response

When execution must happen at the "Point of Contact"—such as in a smart retail fitting room or at an airport kiosk—**Agentic AI must be paired with Edge AI**. This allows the "Hands" to react to customer movements or requests locally with zero latency. This pairing is vital for "Atmospheric Personalization," where lighting, temperature, or digital displays adjust instantly as a customer approaches. It turns the physical environment into a responsive extension of the brand, acting as fast as the customer moves.

4. The Blockchain Pairing for Sovereign Data Transactions

For secure, trust-based execution, a leader should pair **Agentic AI with Blockchain (Smart Contracts)**. This is best used in "Automated Loyalty Settlements" or data-sharing agreements. The "Hands" don't just exchange value; they execute the transaction on an immutable ledger that the customer can verify. This pairing ensures that autonomous actions—like redeeming points for a third-party partner's service—are legally secure and transparent. It reinforces the customer's sense of "Data Sovereignty," ensuring every brand action is a recorded promise kept.

5. The Computer Vision Pairing for Physical-Digital Synthesis

To allow the "Hands" to assist customers in the physical world, leaders should pair **Agentic AI with Computer Vision**. This is best utilized in "Frictionless Retail" and "Visual Support." By "seeing" the customer's environment—such as a broken appliance via a smartphone camera—the agent can autonomously identify the part needed and initiate the order. This pairing gives the "Hands" the sensory context required to solve real-world problems, ensuring the brand's helpfulness extends beyond the screen and into the customer's home.

6. The Reinforcement Learning Pairing for Experience Refinement

In high-velocity markets, **Agentic AI should be paired with Reinforcement Learning (RL)**. This is best utilized for optimizing "Dynamic Customer Journeys," such as real-time website personalization or automated email cadence. The "Hands" learn from every click, purchase, or unsubscribed event, constantly refining their execution strategy to maximize customer happiness. This provides the brand with a self-improving "Fulfillment Engine" that grows more intuitive with every interaction, ensuring the brand's capacity for delight evolves in lockstep with the customer.

THE Ai6x6™ STRATEGIC ROADMAP SERVICE

Your 3-Year Vision, Stress-Tested for the AI Era

The **Ai6x6™** book series has given you the "source code"—the clinical lenses, the vulnerability audits, and the diagnostic frameworks. But strategic orchestration at the highest level requires a cold, objective eye. If you are ready to move from self-diagnosis to radical transformation, we invite you to engage the **Ai6x6™ Roadmap Service.**

The 21-Day Transformation Engine

We don't believe in "billable hours" or six-month discovery phases. We believe in high-fidelity data and rapid execution. Our domain experts take your raw diagnostic output and synthesize it into a personalized, holistic blueprint for the next three years of your organization's life.

How It Works

1. **Submit Your Data:** Upload your completed Discovery Questionnaires, Vulnerability Audits, and Radar Charts directly to our secure portal or via email or via courier.
2. **Contextualize:** Provide your current 3-year roadmap and any business documentation that can provide more context on your line of business, your mission and success KPIs.
3. **The Clinical Review:** Our experts analyze your organizational DNA against the **6 Lens Sequence**, identifying hidden frictions and untapped leverage points. Assign a single point within your organization for asking questions.
4. **The Delivery:** Within **2 to 3 weeks**, you receive a personalized **Holistic Transformation Roadmap**, including our proprietary **Self-Disruption Strategy**.

Why Self-Disruption?

The greatest threat to your organization isn't your competitor; it is your own legacy success. Our roadmap doesn't just show you how to grow; it shows you how to cannibalize your outdated models before the market does it for you.

The Commitment

Packages from $3000

- **Speed:** 14–21-day turnaround.
- **Security:** Full NDA-backed engagement.
- **Precision:** Fact-based, quantitative analysis. No "fluff," no subjective brainstormin

SECURE YOUR DIAGNOSTIC REVIEW:

Ai6x6.com/roadmap

Consulting Roadmap

SECTION C

BLUEPRINT SERIES

Business Maturity

Ai6x6™ Transformation Blueprint

Story 1 — The Marshall Plan Reconstruction (1948–1952): Post-war Europe faced shattered economies, broken supply chains, and institutional collapse. The Marshall Plan introduced structured funding, transparent reporting, and coordinated execution across nations. Local governments provided context while centralized oversight ensured accountability. This blend of autonomy and structure accelerated recovery. The program demonstrated how disciplined governance can rebuild complex systems. It became a blueprint for large-scale organizational maturity.

Story 2 — The Rise of Quality Management in Japan (1950s–1960s): Japan's industries struggled with inconsistent quality and limited global credibility. Statistical quality control introduced measurement, repeatability, and disciplined improvement cycles. Workers contributed insights while leaders standardized processes. Over time, quality became a cultural norm rather than a corrective action. This shift transformed fragmented operations into cohesive systems. It marked the emergence of maturity through methodical refinement.

Story 3 — NASA's Systems Engineering Discipline (1960s): The space race demanded unprecedented coordination across thousands of contractors and scientific domains. NASA formalized systems engineering to integrate requirements, risks, and interfaces into a unified framework. Structured reviews balanced innovation with rigor. Human judgment guided trade-offs while documentation preserved institutional memory. This discipline enabled complex missions to succeed repeatedly. It became a global model for organizational maturity in high-stakes environments.

Story 4 — The Birth of Modern Corporate Governance (1970s–1980s): Corporate scandals exposed weaknesses in oversight, transparency, and fiduciary responsibility. Boards adopted clearer roles, audit mechanisms, and accountability structures. Regulations reinforced ethical conduct while organizations matured through formal governance. Leaders learned to balance autonomy with compliance. This era established governance as a strategic capability, not an administrative burden. It marked a turning point in institutional maturity.

Story 5 — Toyota's Global Expansion Playbook (1980s–1990s): As Toyota expanded globally, it faced cultural differences, supply-chain complexity, and operational variability. The company codified principles, standardized training, and embedded continuous improvement across regions. Local teams adapted practices while preserving core philosophy. This balance of global consistency and local flexibility enabled scalable excellence. It demonstrated maturity as the ability to replicate capability without diluting identity.

Story 6 — The Digital Transformation of Banking (1990s–2000s): Banks confronted rising customer expectations, regulatory pressure, and technological disruption. They modernized legacy systems, standardized data, and introduced structured risk frameworks. Digital channels required new operating models and cross-functional coordination. Human expertise guided decisions where automation lacked nuance. Over time, banks evolved from siloed institutions into integrated digital enterprises. This transition reflected maturity through modernization and disciplined reinvention.

LESSONS FROM A CUSTOMER TRANSFORMATION PERSPECTIVE

Customer maturity emerges when organizations shift from transactional interactions to long-term value stewardship. Over time, institutions learned that reliability, transparency, and consistency matter as much as innovation. Mature organizations design systems that protect customers, reduce friction, and uphold trust even under stress. They balance efficiency with empathy, ensuring that processes serve human needs rather than internal convenience. As capabilities evolve, customer experience becomes a function of governance, quality, and operational discipline—not just frontline behavior. Maturity is reached when customer value is embedded into structures, not slogans, and when trust becomes a measurable outcome of organizational design.

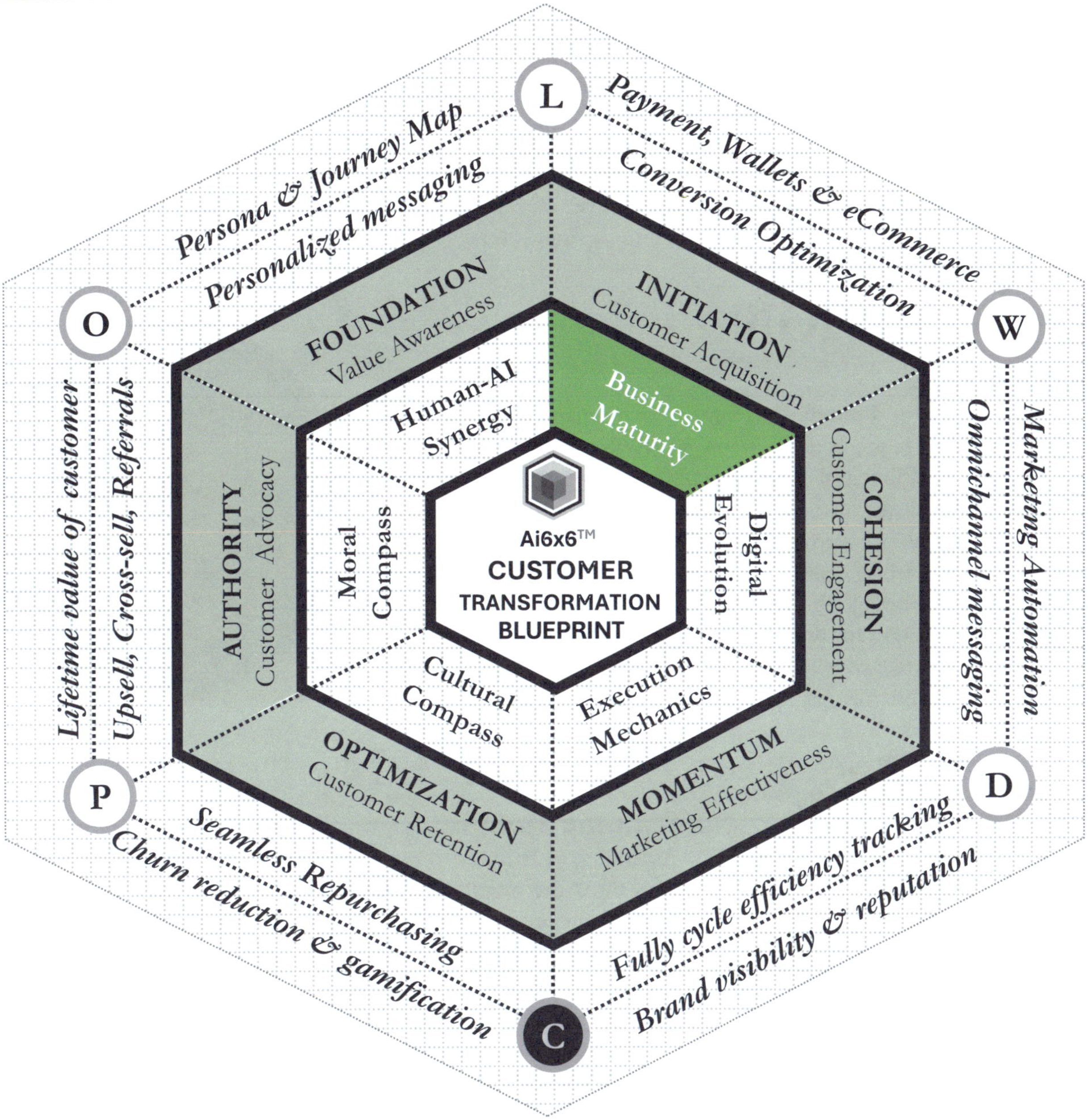

The **Ai6x6 Research Lab** selected this lens because an organization's maturity is ultimately defined by the depth and longevity of its customer relationships. We view customer maturity as an internal filter for the brand's character and its capacity to deliver on its **Value Proposition**. This dimension ensures that the aggressive pursuit of **Acquisition** is anchored in a long-term commitment to **Loyalty** and **Advocacy**. It is a vital lens because, in a volatile market, a leader's ability to maintain a stable, engaged community is the only force that prevents the enterprise from drifting into the trap of short-term conversion at the cost of long-term trust.

We believe true maturity is found in the ability to hold a mirror to the brand's own effectiveness, recognizing that **Churn** is often a reflection of leadership's own lack of focus. By centering on the **Total Experience**, the Lab provides a mechanism for leaders to audit their own brand's resilience before they attempt to influence the market. This lens ensures that the pursuit of **Influencer Growth** does not come at the cost of genuine human connection. It is the steady hand that guides the brand through the complexities of **Omnichannel Engagement**, recognizing that while tactics shift, the bedrock of customer respect must remain immutable.

Every leadership transformation begins with ambition, but ambition alone is never enough to carry an organization into its future. **Business Maturity** is the lens that forces leaders to confront the gap between what they hope to achieve and what their enterprise is prepared to sustain. It is the discipline that reveals whether the organization is structurally, operationally, and culturally ready for the future it claims to desire. In the AI era, this gap becomes even more pronounced; AI does not simply automate tasks—it reshapes workflows, decision cycles, and the very rhythm of how leadership operates. It exposes weaknesses that were previously tolerable and accelerates the consequences of poor preparation, demanding a level of rigor that traditional management frameworks often overlook. It examines how ambition translates into capability, how vision converts into execution, and how culture sustains momentum under pressure.

FOUNDATION

Define the Core

The journey begins at **Foundation**, where the primary objective is to *Define the Core*. This is the bedrock upon which all transformation rests, requiring leaders to **Filter the Noise** of competing priorities and anchor their identity and strategic intent. At this level, clarity is not optional—it is existential. Without clear governance, AI initiatives drift, priorities conflict, and teams operate in isolated silos. Foundation maturity is the discipline of intent: the ability to distinguish urgency from readiness. Leaders who skip this stage mistake motion for progress, building transformation on shifting sand. True maturity begins when the organization knows what it stands for, what it will not compromise, and how it will measure success.

INITIATION

Start the Motion

Ascending to **Initiation** marks the first active step: *Start the Motion*. This level turns the gaze inward, demanding self-awareness and the courage to confront personal biases. By **Wiring the System**, leaders catalyze the shift from passive management to intentional authority. Initiation is where leadership becomes kinetic—it transforms from aspiration into activation. The organization begins to move, not because it is told to, but because its internal circuits are aligned. Maturity here is emotional as much as structural; it requires self-regulation, empathy, and the ability to lead through friction. The leader's persona is tested, revealing whether they can sustain momentum through disruption rather than retreat into comfort.

COHESION

Connect the Parts

The climb continues to **Cohesion**, represented by *The Engine*. This is the critical unifier where people, processes, and systems are harmonized. The focus shifts to **Connect the Parts**—to ensure that individual leadership goals align with a unified organizational voice. By **Shifting the Gear**, leaders eliminate silos and systemic friction, transforming fragmented efforts into a synchronized machine. Cohesion maturity is the art of integration: turning disconnected motion into collective momentum. Without it, organizations celebrate activity instead of achievement. The mature leader understands that alignment is not consensus—it is coherence, the ability to move as one without losing diversity of thought.

MOMENTUM

Drive the Speed

The fourth rung is **Momentum**, where the organization begins to *Drive the Speed.* Leadership here moves toward scaling impact with predictable velocity. By **Removing the Drag** of legacy thinking and institutional inertia, leaders accelerate the collective intelligence of the enterprise. Momentum maturity is the physics of transformation—it measures how efficiently energy converts into progress. At this stage, the organization's rhythm changes; it begins to operate with self-propelling force. The leader's role shifts from directing individuals to empowering systems, ensuring that innovation scales faster than resistance. Momentum is not chaos—it is controlled acceleration, the moment when transformation gains its gravity.

OPTIMIZATION

Refine the Flow

Refinement occurs at **Optimization**, characterized by *The Flow.* Here, leadership pauses to **Refine the Flow** by **Locking the Code** of influence through reflection and data-driven inquiry. Immature organizations measure what is easy; mature ones measure what matters. Optimization maturity is the discipline of precision—removing friction without losing flexibility. Leaders refine decisions still in motion, adjusting with humility and surgical accuracy. Ego and bias are replaced by evidence and iteration. The organization becomes self-healing, capable of continuous improvement without external intervention. Optimization is not about perfection—it is about adaptability, the ability to evolve faster than the environment demands.

AUTHORITY

Own the Market

The summit is reached at **Authority**, where leaders finally *Own the Market.* This stage manifests as *The Benchmark*—a legacy of endurance that can withstand external shocks and volatility. By **Fueling the Future**, leaders ensure that transformation is not a temporary win but a permanent standard. Authority maturity is the discipline of stewardship: sustaining excellence while preparing for reinvention. Immature organizations avoid this stage because it exposes uncomfortable truths; mature ones embrace it as the crucible of resilience. At this level, leadership becomes institutional—it transcends personality and becomes culture. The organization no longer reacts to the market; it defines it.

Business Maturity is the scaffolding that holds the weight of innovation. It ensures that speed does not collapse under fragility and that ambition does not outpace capability. Each level builds upon the last, forming a continuum from clarity to mastery. Foundation defines intent, Initiation activates leadership, Cohesion integrates systems, Momentum accelerates impact, Optimization refines intelligence, and Authority sustains legacy. In the AI era, maturity is not a luxury—it is survival. As algorithms amplify both strengths and weaknesses, leaders must design organizations that can learn, adapt, and endure. Business Maturity is therefore not a checklist but a **leadership discipline**—a way of thinking that transforms ambition into architecture. When leaders embrace this full sequence, they discover that maturity is not the end of transformation but its engine—the invisible structure that makes both explosive speed and long-term continuity possible.

BUSINESS MATURITY ANALOGY – CLIMBING MOUNT EVEREST

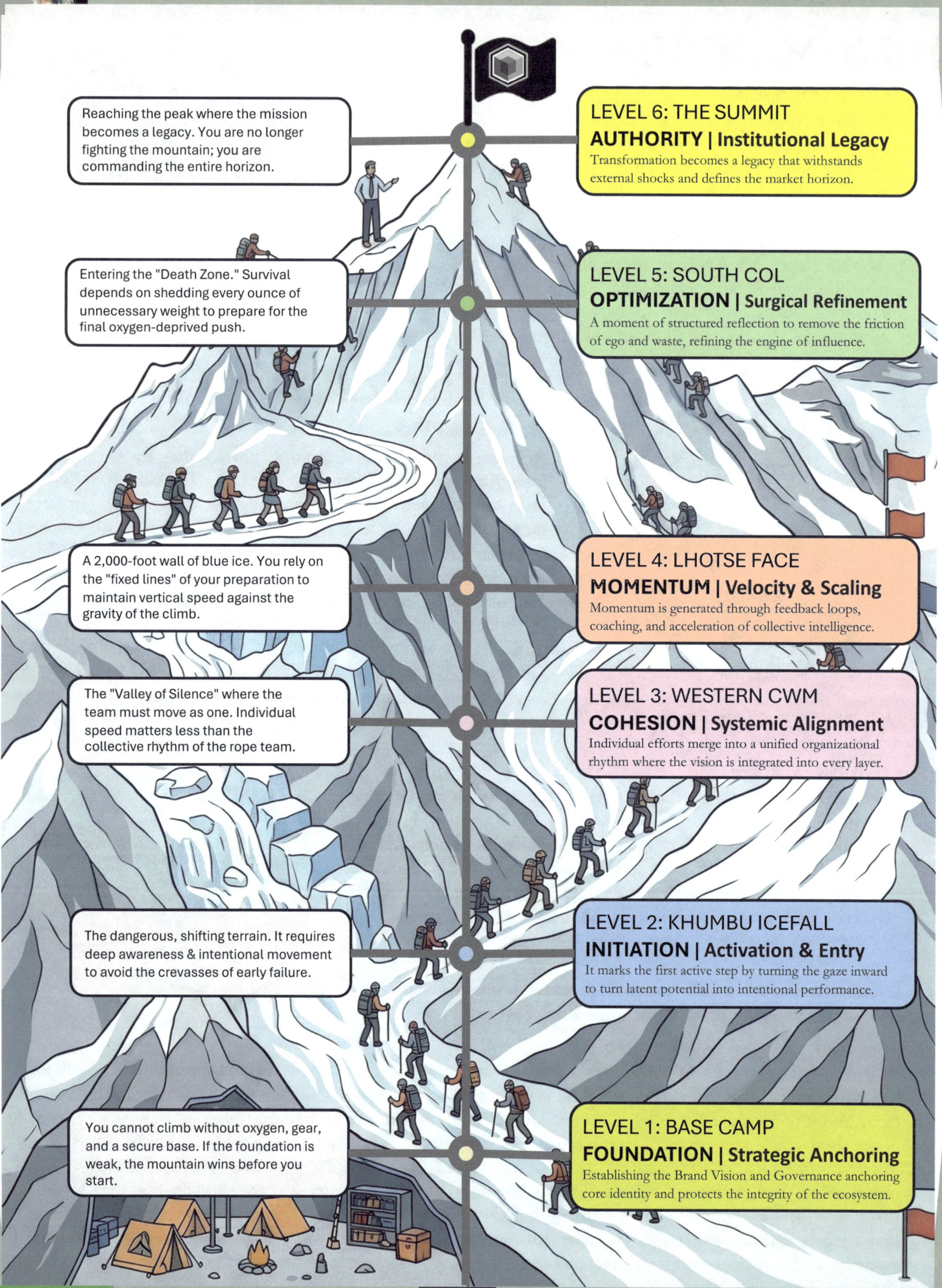

Ai6x6™ Transformation Blueprint

FOUNDATION | Value Awareness

"The customer's perception is your reality." - David Oglivy.

Widely regarded as the "Father of Advertising," Oglivy founded Ogilvy & Mather and shaped modern advertising with his emphasis on brand image and deep consumer understanding, believing that the most effective advertising is built on solid research and a strong, clear value proposition.

The Foundation level is your Value Awareness, anchoring a customer-centric identity and market presence within a clearly defined persona map. This stage establishes exactly why the market should trust you before a single transaction occurs.

Value awareness is the core promise of benefits and solutions your brand offers to meet a customer's specific needs, solving their problems, or enhancing their situation. Without a clear and compelling value proposition, customers lack a reason to engage, leading to high churn rates, difficulty in acquiring new customers, and ultimately, an unsustainable business model.

"Value awareness" about clearly articulating the unique worth your product or service delivers, ensuring it resonates with your target audience's desires and pain points. This involves continuous refinement and communication of how your offering uniquely solves problems or adds value, from initial awareness through to ongoing loyalty and advocacy.

For business strategy maturity, the "**Value Awareness**" chapter is foundational because it defines the brand's journey to create initial offering and recognition among potential customers. A mature value proposition ensures that the target audience becomes aware of the benefit of the brand, product, or service through cohesive marketing, social media presence, word of mouth, or search results. This initial alignment is critical for building market presence and filling the top of the sales funnel. It's about strategically positioning the brand to capture attention and lay the groundwork for future customer engagement and acquisition. Here is how we approach our learning:

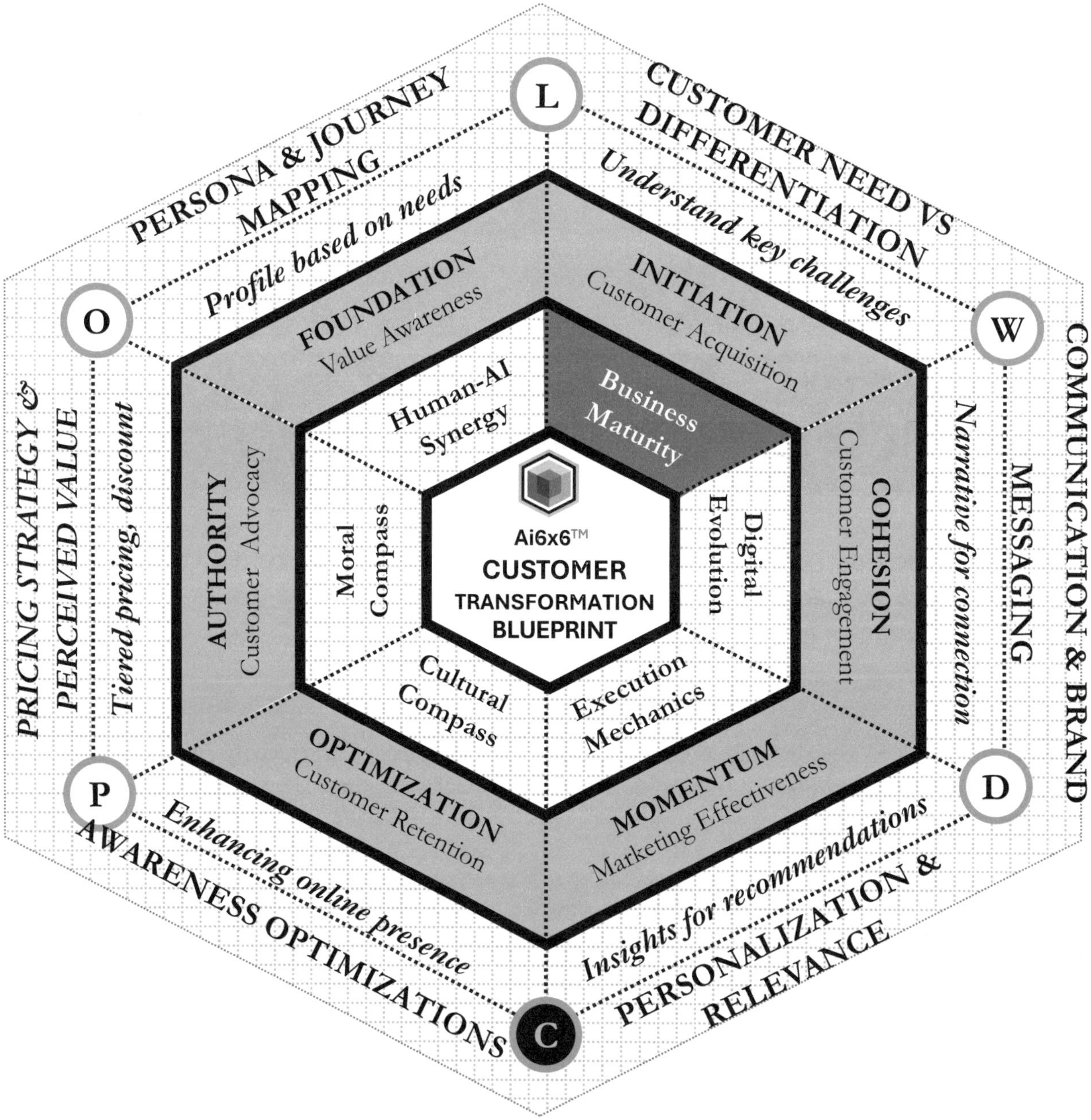

Ai6x6 BUSINESS MATURITY FRAMEWORK
FOR CUSTOMER TRANSFORMATION | VALUE AWARENESS

1. Personas and journey mapping

For a business leader, mastering personas and journey mapping is a critical strategic imperative for truly understanding your target customers and their intricate interactions with your brand. This foundational work ensures that all your efforts, from product development to marketing and sales, are deeply customer-centric and highly effective. Without this insight, your strategies are based on assumptions, not realities. Central to this is meticulously defining customer personas. This involves creating detailed profiles based on deep dives into demographics, behaviors, motivations, and crucially, their pain points. These personas transform abstract market segments into relatable individuals, allowing your teams to empathize with and design for real people.

Furthermore, rigorously mapping the customer journey is crucial. Visualizing every step customers take, from initial awareness of a need to purchase, post-purchase engagement, and ultimately advocacy, reveals critical touchpoints, moments of truth, and potential areas of friction. Crucially, ensuring a precise alignment of your value proposition with these personas is vital. This means constantly refining your messaging, product features, and services to meet specific customer needs at every different touchpoint along their journey. The goal is to ensure your brand consistently delivers relevant value where and when it matters most to the customer. Neglecting personas and journey mapping can lead to generic marketing, products that miss the mark, inefficient customer service, and a fragmented customer experience that erodes trust. By championing this empathetic and analytical approach, you empower your organization to build deeper customer relationships, personalize interactions, and ultimately drive sustained customer loyalty and business growth by truly understanding and serving your audience.

WHAT?	WHY?	HOW?	BEWARE!
DEFINING CUSTOMER PERSONAS	**TARGET CLARITY**	**DATA COLLECTION**	**STEREOTYPING**
Creating profiles based on behaviors and needs.	Provides a clear understanding of who the ideal customer is.	Conduct market research, surveys, and analyze customer data.	Risk of creating overly generic or inaccurate personas.
MAPPING THE CUSTOMER JOURNEY	**TOUCHPOINT OPTIMIZATION**	**FLOWCHARTING**	**COMPLEXITY**
Visualizing customer steps from awareness to advocacy.	Identifies critical interaction points for improvement.	Map out all customer interactions, emotions, and pain points.	Mapping intricate journeys across multiple channels and departments.
ALIGNING VALUE PROPOSITION WITH PERSONAS	**RELEVANT ENGAGEMENT**	**CUSTOMIZATION**	**INCONSISTENCY**
Ensuring messaging meets customer needs.	Tailors messaging and offerings for maximum resonance.	Adapt marketing, product, and service strategies to each persona.	Maintaining consistent value delivery across diverse personas.

For a business leader, masterfully balancing core customer needs with distinct brand differentiation is a critical strategic imperative for forging a unique and compelling brand position in the market. It's not enough to simply meet customer requirements; true competitive advantage lies in satisfying those needs in a way that truly sets you apart. Your ability to articulate this balance is paramount for attracting and retaining customers. Central to this is meticulously identifying core customer pain points. This involves a deep understanding of the key challenges, frustrations, and unmet needs that genuinely drive purchasing decisions within your target audience. By pinpointing these fundamental problems, you ensure your offerings are solving real issues that customers care about.

Furthermore, rigorously defining your competitive differentiation is crucial. Highlighting how your brand uniquely stands out from competitors—whether through innovative product features, superior service delivery, a distinctive brand experience, or a unique business model—gives customers a clear reason to choose you. Crucially, adopting value-based selling focuses on communicating benefits instead of just features. This approach emphasizes how your product or service solves customer problems, improves their lives, or helps them achieve their goals, rather than merely listing technical specifications. It shifts the conversation to the tangible positive impact you deliver. Neglecting this crucial balance can lead to commoditization (where your brand is just like everyone else), a struggle to justify premium pricing, or offerings that fail to resonate in a crowded market. By championing a strategy that precisely addresses customer needs while highlighting your unique advantages, you empower your business to attract the right customers, build strong brand loyalty, and ultimately secure a formidable competitive position.

WHAT?	WHY?	HOW?	BEWARE!
IDENTIFYING CORE CUSTOMER PAIN POINTS	**PROBLEM-SOLUTION FIT**	**FEEDBACK ANALYSIS**	**SURFACE-LEVEL INSIGHTS**
Understanding key challenges.	Ensures products/services genuinely solve customer problems.	Collect and analyze customer feedback, support tickets, and reviews.	Only addressing superficial needs, missing deeper issues.
COMPETITIVE DIFFERENTIATION	**MARKET ADVANTAGE**	**COMPETITOR ANALYSIS**	**IMITATION**
Highlighting how a brand stands out from rivals.	Establishes a unique selling proposition in a crowded market.	Analyze competitor offerings, pricing, and messaging.	Brands becoming too similar, losing their unique appeal.
VALUE-BASED SELLING	**EMPHASIS ON IMPACT**	**BENEFIT ARTICULATION**	**OVER-PROMISING**
Communicating benefits instead of just features.	Highlights problem-solving and positive customer outcomes.	Train sales and marketing to focus on solutions and value delivered.	Communicating benefits that the product or service cannot consistently deliver.

For a business leader, meticulously crafting your brand's communication and messaging is a critical strategic imperative for deeply resonating with customers, building trust, and fostering lasting connections. In today's saturated market, your words are your brand's voice, and how you speak directly impacts how you are perceived and whether customers choose to engage. Central to this is mastering storytelling and emotional appeal. Using relatable narratives, authentic anecdotes, and evocative language creates an emotional connection with your audience, fostering trust, memorability, and a sense of shared values far beyond mere facts or figures.

Furthermore, ensuring a consistent brand voice is crucial. Maintaining clarity, tone, and alignment across all communication channels—from website copy and social media posts to email campaigns and customer service interactions—reinforces your brand identity, preventing confusion and building recognition. Crucially, a relentless focus on customer-centric messaging is vital. This means ensuring your language consistently focuses on customer benefits and experiences, directly addressing their needs, aspirations, and challenges, rather than company-centered statements or jargon. It shifts the spotlight entirely to the customer. Neglecting compelling communication can lead to a flat, forgettable brand, misinterpretations, a lack of emotional connection with your audience, and inefficient marketing spend. By championing a strategic approach to crafting authentic, consistent, and customer-focused narratives, you empower your organization to inspire, engage, and convert, ultimately driving sustained brand loyalty and business growth through powerful and resonant communication.

WHAT?	WHY?	HOW?	BEWARE!
STORYTELLING AND EMOTIONAL APPEAL	**DEEPER CONNECTION**	**NARRATIVE DEVELOPMENT**	**INAUTHENTICITY**
Using relatable narratives to create connection.	Builds trust and emotional bonds with the audience.	Craft compelling stories that align with brand values and customer aspirations.	Stories feeling forced or inauthentic, failing to resonate.
CONSISTENT BRAND VOICE	**BRAND RECOGNITION**	**STYLE GUIDELINES**	**FRAGMENTATION**
Maintaining clarity and alignment across channels.	Ensures recognizable and cohesive brand identity.	Develop clear brand voice and tone guidelines for all communications.	Different teams or channels deviating from the brand voice.
CUSTOMER-CENTRIC MESSAGING	**RELEVANCE & IMPACT**	**LANGUAGE SHIFT**	**INTERNAL BIAS**
Focusing on customer benefits and experiences.	Makes messaging more relevant and impactful for the customer.	Train teams to use customer-focused language over company-focused jargon.	Difficulty shifting from an internal, product-focused mindset to external customer focus.

For a business leader, deeply embracing personalization and relevance is a critical strategic imperative for enhancing value delivery and fostering unparalleled customer loyalty through tailored experiences. In today's digital landscape, generic interactions are a liability; customers expect communications and offerings that speak directly to their individual needs and preferences. Central to this is leveraging data-driven personalization. By meticulously using comprehensive customer insights—gleaned from browsing history, purchase patterns, past interactions, and demographic data—you can offer highly relevant recommendations, promotions, and content that resonate uniquely with each individual, making them feel seen and valued.

Furthermore, developing adaptive engagement strategies is crucial. This involves dynamically adjusting messaging, content, and interaction paths based on real-time customer behavior, evolving preferences, and historical interaction history. This agility ensures that every communication is timely, contextually appropriate, and maximally impactful. Crucially, employing hyper-targeting through AI is revolutionizing this capability. By leveraging advanced machine learning algorithms, you can refine personalization in marketing and product offerings to an extremely granular level, predicting individual needs and delivering bespoke experiences at scale that drive conversion and deepen engagement. Neglecting personalization can lead to generic marketing, customer fatigue, missed opportunities for conversion, and a struggle to differentiate your brand in a crowded market. By championing a sophisticated, AI-powered approach to personalization, you empower your organization to build deeper customer relationships, accelerate sales, and ultimately secure sustained loyalty & competitive advantage through relevant experiences.

WHAT?	WHY?	HOW?	BEWARE!
DATA-DRIVEN PERSONALIZATION	**ENHANCED VALUE**	**INSIGHT UTILIZATION**	**PRIVACY CONCERNS**
Using customer insights for relevant recommendations.	Delivers tailored experiences that resonate with individual customers.	Use customer data (browsing, purchase history) for recommendations.	Balancing personalization with customer privacy expectations.
ADAPTIVE ENGAGEMENT STRATEGIES	**DYNAMIC INTERACTION**	**SEGMENTATION & AUTOMATION**	**OVER-AUTOMATION**
Adjusting messaging based on behavior and preferences.	Ensures messages are timely and contextually appropriate.	Create dynamic customer segments and automate adaptive messaging.	Messaging feeling generic or intrusive despite automation.
HYPER-TARGETING THROUGH AI	**OPTIMAL IMPACT**	**PREDICTIVE MODELING**	**ALGORITHMIC BIAS**
Leveraging machine learning for refined personalization.	Maximizes relevance in marketing and product offerings.	Deploy AI/ML models to predict customer preferences and behaviors.	AI models unintentionally reinforcing biases in targeting.

For a business leader, relentlessly pursuing awareness optimization is a critical strategic imperative for improving your brand's visibility and accessibility, ultimately enhancing customer discovery and initiating the crucial first step in the customer journey. In a crowded marketplace, if customers don't know you exist, they can't choose you. Central to this is leveraging Search Engine Optimization (SEO) tools. Enhancing your online presence through meticulously optimized content and strategic keyword strategies ensures that your brand ranks high in search results, making it easily discoverable when potential customers are actively searching for solutions you provide.

Furthermore, utilizing social media listening tools is crucial. Analyzing audience sentiment, identifying emerging trends, and monitoring conversations related to your industry and brand allows you to refine your messaging, engage proactively where your audience is, and capitalize on trending topics for increased visibility. Crucially, adopting programmatic advertising is vital for efficient and precise targeting. Leveraging automated digital ad placement allows you to reach specific demographics and interests with maximum reach and cost-effectiveness, ensuring your ads are seen by the most relevant potential customers across diverse online platforms. Neglecting awareness optimization can lead to a hidden brand, missed market opportunities, inefficient marketing spend, and a struggle to grow your customer base from the top of the funnel. By championing a multi-faceted, data-driven approach to improving visibility, you empower your organization to capture initial customer interest, generate high-quality leads, and ultimately lay the groundwork for sustained customer acquisition and business growth.

WHAT?	WHY?	HOW?	BEWARE!
SEARCH ENGINE OPTIMIZATION (SEO) TOOLS	**INCREASED VISIBILITY**	**CONTENT OPTIMIZATION**	**ALGORITHM CHANGES**
Enhancing online presence.	Improves organic search rankings and customer discoverability.	Implement keyword research, content optimization, and technical SEO.	Google/search engine algorithm updates impacting rankings.
SOCIAL MEDIA LISTENING TOOLS	**REFINED MESSAGING**	**MONITORING TOOLS**	**NOISE & FAKE NEWS**
Analyzing audience sentiment and trends.	Informs messaging and engagement strategies based on public sentiment.	Use tools to track brand mentions, sentiment, and trending topics.	Differentiating genuine sentiment from social media noise or misinformation.
PROGRAMMATIC ADVERTISING	**PRECISE TARGETING**	**DSP/SSP INTEGRATION**	**AD FRAUD**
Leveraging automated digital ad placement.	Maximizes ad reach and efficiency through automated targeting.	Implement demand-side platforms (DSPs) for automated ad buying.	Risk of ad fraud and ensuring legitimate ad placements.

For a business leader, meticulously crafting your pricing strategy and managing perceived value are critical strategic imperatives for attracting customers, ensuring profitability, and sustaining brand worth. Pricing is not just a number; it's a powerful communication tool that reflects your brand's quality, benefits, and competitive position. Central to this is intelligently employing psychological pricing techniques. Using strategies like tiered pricing (e.g., good, better, best options), strategic discounts, and premium positioning (e.g., prestige pricing) can influence customer perception of value, drive purchase decisions, and create a sense of urgency or exclusivity.

Furthermore, effectively managing the balance between value vs. affordability perception is crucial. This involves clearly communicating why your pricing reflects the superior quality, unique benefits, and long-term value offered, ensuring customers understand the justification for the cost and perceive it as fair relative to the benefits received. Crucially, exploring subscription and loyalty pricing models is vital for long-term customer retention. Structuring recurring payments for services or reward-based discounts for loyal customers increases stickiness, fosters predictable revenue streams, and encourages repeat engagement, building enduring relationships rather than one-off transactions. Neglecting a thoughtful pricing strategy can lead to leaving money on the table, struggling to compete on price, attracting the wrong customer segments, or failing to communicate your true brand worth. By championing a strategic, data-informed, and value-focused approach to pricing, you empower your business to optimize revenue, attract ideal customers, and ultimately drive sustained profitability and long-term brand equity in a competitive marketplace.

WHAT?	WHY?	HOW?	BEWARE!
PSYCHOLOGICAL PRICING TECHNIQUES	**PERCEPTION INFLUENCE**	**PRICING MODELS**	**CONSUMER SKEPTICISM**
Using tiered pricing, discounts, and premium positioning.	Shapes customer perception of value and drives purchase decisions.	Implement tiered pricing, bundling, odd-even pricing, or prestige pricing.	Customers becoming resistant to perceived psychological manipulation.
VALUE VS. AFFORDABILITY PERCEPTION	**PURCHASE JUSTIFICATION**	**VALUE COMMUNICATION**	**UNDERVALUING**
Ensuring pricing reflects quality and benefits.	Helps customers understand the justification for the price.	Clearly articulate benefits and ROI to align with pricing.	Customers perceiving the price as too high relative to perceived value.
SUBSCRIPTION AND LOYALTY PRICING MODELS	**RETENTION & STICKINESS**	**RECURRING BILLING**	**CHURN MANAGEMENT**
Structuring recurring payments.	Increases customer retention and builds long-term loyalty.	Implement subscription tiers, loyalty discounts, and reward systems.	Managing churn within subscription models effectively.

Ai6x6™ Transformation Blueprint

INITIATION | Customer Acquisition

"Our customers are our greatest asset and our greatest source of innovation." - Mary Barra.

CEO of General Motors (GM), the first female CEO of a major global automaker. She leads GM's transformation, focusing on electric vehicles and autonomous driving, modernizing its culture and operations towards sustainable mobility solutions.

The Initiation level is your Customer Acquisition, the first spark of transaction where a prospect is converted into a participant. Efficient e-commerce and conversion mechanics turn market interest into the first beat of the business heart.

The strategic process of converting interested prospects into their first-time customers, turning initial curiosity into a tangible transaction or commitment. This phase is absolutely crucial for business growth, expanding market share, and injecting vital new revenue streams into the organization, fueling its expansion and sustainability.

Effective acquisition strategies involve optimizing sales funnels, compelling calls to action, and seamless conversion experiences to efficiently bring new customers into the fold and kickstart their relationship with your brand.

The **"Customer Acquisition"** chapter is crucial for business strategy maturity as it focuses on mobilizing potential customers into the customer base. A mature business strategy optimizes the process by which customers make a buying decision and complete the transaction. This activation phase is vital for converting interest into revenue, directly impacting market share and growth. It's about designing seamless customer journeys and effective sales processes that efficiently convert leads into loyal customers, ensuring the business can consistently expand its customer base and achieve its strategic growth objectives. Here is how we approach our learning:

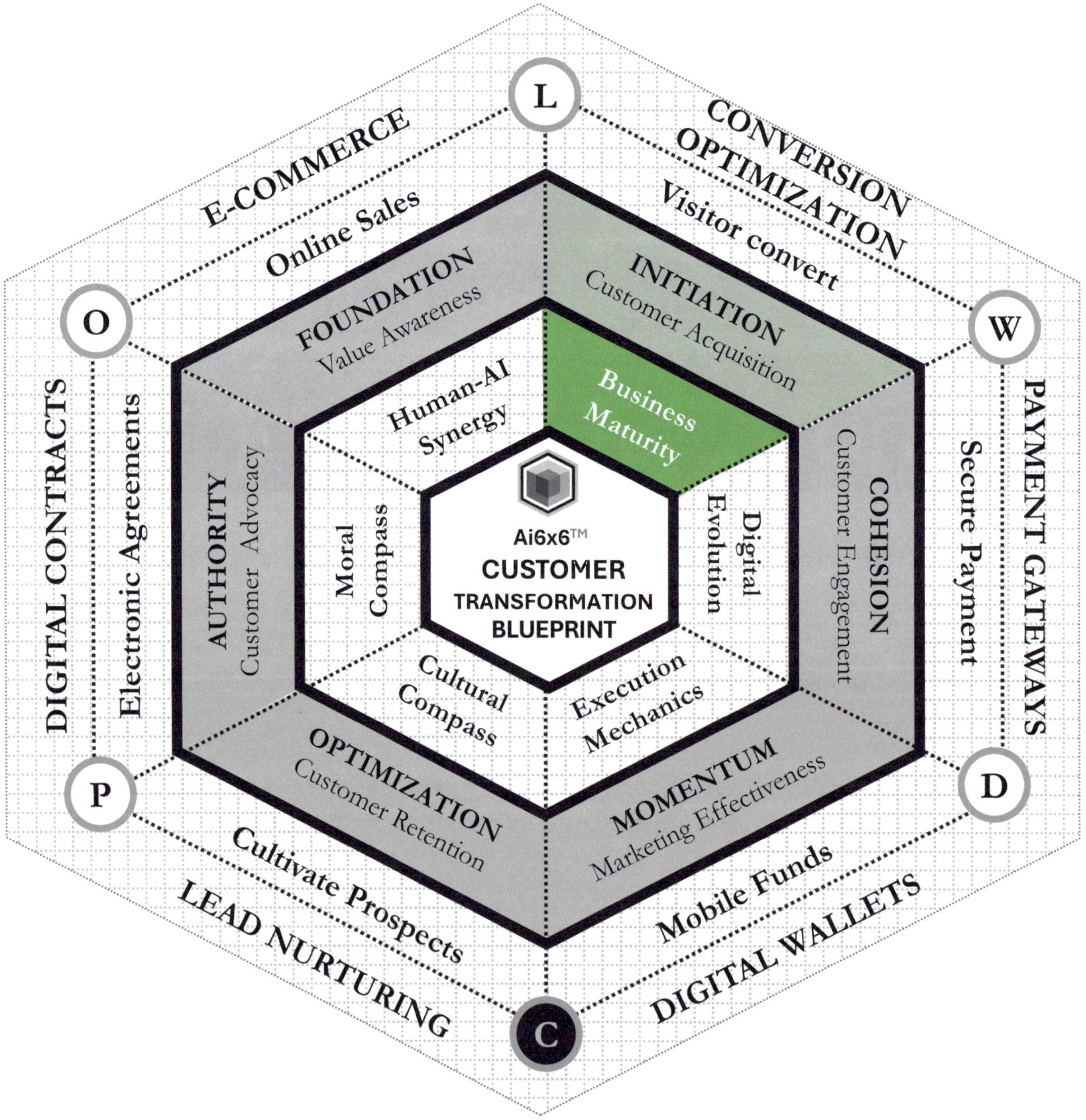

Ai6x6 BUSINESS MATURITY FRAMEWORK
FOR CUSTOMER TRANSFORMATION | CUSTOMER ACQUISITION

For a business leader, mastering e-commerce is no longer an option but a foundational imperative for customer acquisition and growth in today's digital-first economy. The digital marketplace offers unparalleled reach and opportunities to connect directly with customers globally. Your ability to effectively sell online is crucial for scaling your business and diversifying revenue streams. Central to this is meticulous online store setup. This involves building and continuously optimizing your e-commerce website for seamless user experiences, intuitive navigation, and mobile responsiveness. A well-designed online storefront is your digital flagship, directly impacting conversion rates.

Furthermore, effectively managing product listings and digital catalogs is crucial. Presenting your inventory with high-quality images, detailed descriptions, compelling calls to action, and clear pricing is essential for customer engagement and informed purchasing decisions. Your digital display is often the first and only impression. Crucially, a deep understanding of customer buying behavior in the e-commerce context is vital. Analyzing purchase patterns, preferences, and emerging trends in online shopping allows you to optimize your digital marketing, personalize recommendations, and streamline the customer journey, from browsing to checkout. Neglecting a robust e-commerce strategy means missing out on vast market opportunities, limiting your reach to physical storefronts, and falling behind competitors in the digital realm. By championing a comprehensive e-commerce approach, you empower your business to acquire new customers efficiently, expand into new markets, and build a scalable sales channel that directly contributes to long-term revenue growth and market leadership.

WHAT?	WHY?	HOW?	BEWARE!
ONLINE STORE SETUP	**SEAMLESS EXPERIENCE**	**PLATFORM SELECTION**	**UX/UI OPTIMIZATION**
Building and optimizing e-commerce websites.	Ensures intuitive and engaging user journeys.	Choose and configure an e-commerce platform (Shopify, Magento, custom).	Designing intuitive user interfaces and user experiences.
PRODUCT LISTINGS AND DIGITAL CATALOGS	**CUSTOMER ENGAGEMENT**	**HIGH-QUALITY ASSETS**	**DATA CONSISTENCY**
Presenting inventory effectively.	Maximizes product visibility and encourages purchases.	Use compelling product descriptions, high-resolution images, and videos.	Maintaining consistent and accurate product data across all channels.
CUSTOMER BUYING BEHAVIOR	**OPTIMIZED SALES**	**ANALYTICS & A/B TESTING**	**PREDICTIVE ACCURACY**
Understanding purchase patterns and trends.	Informs strategies for marketing, personalization, and user experience.	Analyze website analytics, conversion funnels, and user testing.	Accurately predicting evolving customer preferences and market shifts.

For a business leader, relentless conversion optimization is a strategic imperative for maximizing your customer acquisition efforts by enhancing strategies to turn website visitors into paying customers. It's not enough to attract traffic; the true measure of success lies in converting that interest into tangible revenue. Your ability to refine this process is directly linked to your profitability. Central to this is meticulous landing page optimization. Creating persuasive and user-friendly pages specifically designed to drive conversions—with clear calls to action, compelling headlines, and relevant content—is crucial for guiding visitors towards desired actions.

Furthermore, rigorous A/B testing and performance analysis are vital. This involves systematically evaluating different elements of your web pages, marketing campaigns, or user flows to identify what resonates most effectively with your audience and maximizes desired outcomes. Data-driven experimentation is key to continuous improvement. Crucially, optimizing the checkout experience improvements is paramount. Reducing any friction in the final stages of the purchase process—simplifying forms, offering diverse payment options, ensuring transparency in pricing—leads to significantly higher purchase completion rates and reduces cart abandonment. Neglecting conversion optimization can result in wasted marketing spend, low ROI on traffic generation efforts, and a struggle to grow your customer base efficiently. By championing a continuous, data-driven approach to enhancing every step of the conversion funnel, you empower your organization to unlock greater revenue from existing traffic, acquire new customers more effectively, and ultimately drive the sustainable growth and profitability of your business. This focus on efficiency is a core driver of competitive success.

WHAT?	WHY?	HOW?	BEWARE!
LANDING PAGE OPTIMIZATION	**BOOSTED CONVERSIONS**	**A/B TESTING & UX**	**ITERATION CYCLES**
Creating persuasive and user-friendly pages.	Maximizes the percentage of visitors taking desired actions.	Test headlines, CTAs, forms, and layout for better engagement.	Needing continuous testing and iteration to find optimal solutions.
A/B TESTING AND PERFORMANCE ANALYSIS	**MAXIMIZED EFFECTIVENESS**	**SPLIT TESTING**	**STATISTICAL SIGNIFICANCE**
Evaluating different elements.	Identifies optimal strategies for design, content, and calls to action.	Run controlled experiments comparing variations of web elements.	Ensuring sufficient data for statistically valid results.
CHECKOUT EXPERIENCE IMPROVEMENTS	**HIGHER COMPLETION**	**SIMPLIFICATION**	**SECURITY VS. CONVENIENCE**
Reducing friction for higher purchase rates.	Minimizes cart abandonment and increases purchase completion.	Streamline forms, offer guest checkout, and display clear progress indicators.	Balancing security measures with a frictionless user experience.

3. Payment gateways

For a business leader, carefully selecting and managing payment gateways is a critical strategic decision for any online business, directly impacting customer acquisition, trust, and operational efficiency. These gateways are the secure conduits that enable financial transactions, and their performance is paramount to customer satisfaction and preventing fraud. Central to this is evaluating various payment processing solutions. This involves comparing different gateways based on their transaction fees, processing speeds, reliability, and the range of payment methods they support. Choosing the right provider impacts both your bottom line and customer convenience.

Furthermore, ensuring robust security and fraud prevention measures is crucial. Implementing encryption, tokenization, and advanced verification methods protects sensitive customer data and safeguards against fraudulent transactions, building trust and mitigating significant financial risks. Your commitment to security is a key differentiator for customers. Crucially, enabling global payment acceptance is vital for businesses with international aspirations. Adapting payment options to suit diverse regional preferences and currencies allows you to tap into broader markets, significantly expanding your customer acquisition potential and global reach. Neglecting secure and efficient payment gateways can lead to abandoned carts, frustrated customers, significant financial losses due to fraud, and a hindrance to international expansion. By championing a secure, versatile, and customer-friendly payment infrastructure, you empower your business to process transactions seamlessly, build customer confidence, and expand its market footprint, ultimately driving sustainable revenue growth and a positive brand reputation in the digital marketplace.

WHAT?	WHY?	HOW?	BEWARE!
PAYMENT PROCESSING SOLUTIONS	**OPTIMIZED COSTS**	**PROVIDER EVALUATION**	**INTEGRATION COMPLEXITY**
Comparing different gateways and their fees.	Balances transaction fees with reliability and features.	Research and compare various payment gateway providers (Stripe, PayPal, etc.).	Integrating multiple payment gateways into existing systems.
SECURITY AND FRAUD PREVENTION	**RISK MITIGATION**	**ENCRYPTION & TOKENIZATION**	**EVOLVING THREATS**
Ensuring safe transactions.	Protects sensitive customer data and prevents financial losses.	Implement strong encryption, tokenization, and fraud detection tools.	Keeping up with constantly evolving fraud techniques.
GLOBAL PAYMENT ACCEPTANCE	**MARKET EXPANSION**	**LOCALIZED OPTIONS**	**REGULATORY COMPLIANCE**
Adapting payment options for international customers.	Enables reach to international customers and new markets.	Offer local payment methods, currencies, and multi-language support.	Navigating diverse international payment regulations and tax laws.

For a business leader, embracing digital wallets is a strategic imperative for providing seamless, convenient, and secure payment solutions that enhance the customer acquisition and engagement experience. As consumers increasingly rely on mobile devices for daily transactions, integrating digital wallet capabilities positions your business at the forefront of modern commerce. Central to this is understanding mobile wallet adoption and how smartphones are transforming consumer payments. Recognizing the widespread shift towards mobile-first transactions allows you to tap into a rapidly growing segment of digitally native consumers, improving their checkout experience and boosting conversion rates.

Furthermore, seamless integration with online stores is crucial. Connecting digital wallets to your e-commerce platforms ensures that customers can complete purchases with minimal clicks and data entry, significantly reducing friction in the checkout process and decreasing cart abandonment rates. Crucially, highlighting the core benefits of digital wallets—speed, convenience, and enhanced security—in your marketing and user experience design persuades customers to adopt these payment methods. Faster checkouts and robust security features are powerful motivators for modern consumers. Neglecting digital wallets can lead to a less competitive checkout experience, lost sales due to friction, and a failure to cater to the preferences of a growing segment of your customer base. By championing the adoption and integration of digital wallets, you empower your business to offer a superior payment experience, accelerate customer acquisition, and reinforce your brand as innovative and customer-centric, ultimately driving increased revenue and long-term customer satisfaction.

WHAT?	WHY?	HOW?	BEWARE!
MOBILE WALLET ADOPTION	**ENHANCED CONVENIENCE**	**TREND ANALYSIS**	**FRAGMENTATION**
How smartphones are transforming consumer payments.	Caters to mobile-first consumers, streamlining payments.	Monitor mobile payment trends and integrate popular wallet options.	Numerous mobile wallet providers, requiring multiple integrations.
INTEGRATION WITH ONLINE STORES	**SEAMLESS CHECKOUT**	**API INTEGRATION**	**TECHNICAL COMPLEXITY**
Connecting digital wallets to e-commerce platforms.	Reduces friction and improves conversion rates.	Use APIs to connect e-commerce platforms with digital wallet services.	Ensuring smooth and secure integration across various platforms.
BENEFITS OF DIGITAL WALLETS	**CUSTOMER TRUST**	**COMMUNICATION**	**USER EDUCATION**
Speed, convenience, and security in transactions.	Builds trust and encourages adoption through perceived security.	Highlight speed, ease, and security benefits in marketing.	Educating customers unfamiliar with digital wallet usage.

For a business leader, establishing robust lead nurturing strategies is a strategic imperative for strengthening relationships with potential customers and significantly boosting conversion rates. In today's competitive landscape, not every prospect is ready to buy immediately; effective nurturing guides them through the sales funnel, building trust and demonstrating value over time. Central to this is implementing personalized email campaigns. Tailoring outreach strategies to engage leads effectively—based on their behavior, interests, and stage in the buyer's journey—ensures that communications are relevant and compelling, increasing open rates and click-throughs.

Furthermore, integrating retargeting and follow-up mechanisms is crucial. Re-engaging visitors who showed interest but didn't convert immediately through targeted ads or follow-up communications keeps your brand top-of-mind and encourages them to reconsider their purchase, maximizing the ROI on initial marketing efforts. Crucially, seamless CRM integration and automation are vital. Using Customer Relationship Management (CRM) tools to track and manage all lead interactions—from website visits to email opens and content downloads—provides a holistic view of each prospect and automates personalized outreach, ensuring no lead falls through the cracks. Neglecting lead nurturing can result in wasted marketing spend on unqualified leads, low conversion rates, and a significant loss of potential revenue. By championing a systematic and personalized approach to lead nurturing, you empower your sales and marketing teams to build stronger relationships, guide prospects effectively towards conversion, and ultimately drive the sustainable growth and profitability of your business by maximizing the value of every lead generated.

WHAT?	WHY?	HOW?	BEWARE!
PERSONALIZED EMAIL CAMPAIGNS	**LEAD ENGAGEMENT**	**SEGMENTATION & AUTOMATION**	**OVER-AUTOMATION**
Tailoring outreach strategies.	Boosts engagement by delivering highly relevant content.	Segment leads and automate personalized email sequences.	Emails feeling generic or intrusive despite personalization efforts.
RETARGETING AND FOLLOW-UPS	**CONVERSION OPTIMIZATION**	**PIXEL TRACKING**	**AD FATIGUE**
Re-engaging interested visitors.	Recovers lost leads and encourages purchase completion.	Implement retargeting pixels and follow-up ad campaigns.	Over-saturating users with retargeting ads, leading to negative brand perception.
CRM INTEGRATION AND AUTOMATION	**EFFICIENT MANAGEMENT**	**CRM PLATFORM**	**DATA INTEGRATION**
Using tools to track and manage interactions.	Streamlines lead tracking and automates nurturing processes.	Implement and integrate a CRM system (e.g., Salesforce, HubSpot).	Ensuring seamless data flow between CRM and other marketing/sales tools.

For a business leader, embracing digital contracts is a strategic imperative for streamlining agreements and transactions through technology, enhancing efficiency, security, and compliance across your business operations. Moving away from cumbersome paper-based processes offers significant advantages in customer acquisition and operational agility. Central to this is exploring smart contracts in e-commerce. These self-executing agreements, often built on blockchain technology, automate secure transactions with predefined conditions, reducing the need for intermediaries, accelerating deal closures, and enhancing trust in online commerce.

Furthermore, leveraging electronic signatures is crucial for efficiency and legal validity. Using digital authentication to validate agreements—whether for sales contracts, service agreements, or onboarding documents—eliminates the need for physical paperwork, accelerates approval processes, and provides a legally binding digital record. Crucially, meticulous attention to compliance and legal considerations is vital. As a leader, you must ensure that all digital contract processes adhere strictly to regulatory guidelines, electronic signature laws (e.g., ESIGN Act), and data privacy frameworks, safeguarding your business from legal risks. Neglecting digital contracts can lead to slow, inefficient, and error-prone agreement processes, hindering customer acquisition speed and increasing operational costs. By championing these technological solutions, you empower your business to execute agreements faster, enhance security, reduce administrative burdens, and provide a seamless, modern experience for customers and partners, ultimately driving greater operational efficiency and accelerating business growth in the digital age.

WHAT?	WHY?	HOW?	BEWARE!
SMART CONTRACTS IN E-COMMERCE	**TRUST & EFFICIENCY**	**BLOCKCHAIN INTEGRATION**	**COMPLEXITY**
Automating secure transactions.	Automates agreements, reducing intermediaries and enhancing trust.	Explore blockchain platforms for smart contract deployment.	High technical complexity in developing and deploying smart contracts.
ELECTRONIC SIGNATURES	**SPEED & VALIDITY**	**E-SIGNATURE TOOLS**	**LEGAL ACCEPTANCE**
Using digital authentication to validate agreements.	Accelerates agreement processes while ensuring legal validity.	Implement electronic signature platforms (e.g., DocuSign, Adobe Sign).	Ensuring legal acceptance and enforceability across jurisdictions.
COMPLIANCE AND LEGAL CONSIDERATIONS	**RISK MITIGATION**	**LEGAL REVIEW**	**EVOLVING LAWS**
Ensuring regulatory adherence.	Protects the business from legal liabilities.	Consult legal experts to ensure digital contract processes comply with regulations.	Keeping up with rapidly evolving digital contract and data privacy laws.

Ai6x6™ Transformation Blueprint

COHESION | Customer Engagement

"The single most important thing for a company is to engage with its customers." - Tony Hsieh.

Former CEO of Zappos, an online retailer renowned for exceptional customer service and a unique, employee-centric company culture. He prioritized customer happiness and employee empowerment, influencing countless leaders on the importance of culture and customer delight.

The Cohesion level is your Customer Engagement, found when the user feels a consistent, seamless bond with the brand across all touchpoints. A personalized omnichannel strategy ensures the experience is a unified journey rather than fragmented moments.

Actively and meaningfully interacting with customers to build enduring relationships, consistently provide value beyond the initial purchase, and foster deep-seated loyalty and satisfaction. This deepens connections, significantly increases customer satisfaction, and drives repeat business, effectively moving customers beyond a single, transactional interaction to a lasting partnership.

Robust engagement strategies involve personalized communication, excellent customer service, and valuable content that keeps customers connected and invested in your brand over time.

"**Customer Engagement**" is vital for business strategy maturity because it represents the active and ongoing interaction with the customer, deepening their relationship with the brand. A mature business strategy fosters continuous engagement, encouraging customers to browse websites, sign up for newsletters, or interact on social media. This sustained interaction accelerates customer lifetime value, builds brand loyalty, and provides valuable feedback. It's about creating dynamic touchpoints that keep customers invested and satisfied, ensuring they remain active participants in the brand's ecosystem and contribute to long-term strategic success. Here is how we approach our learning:

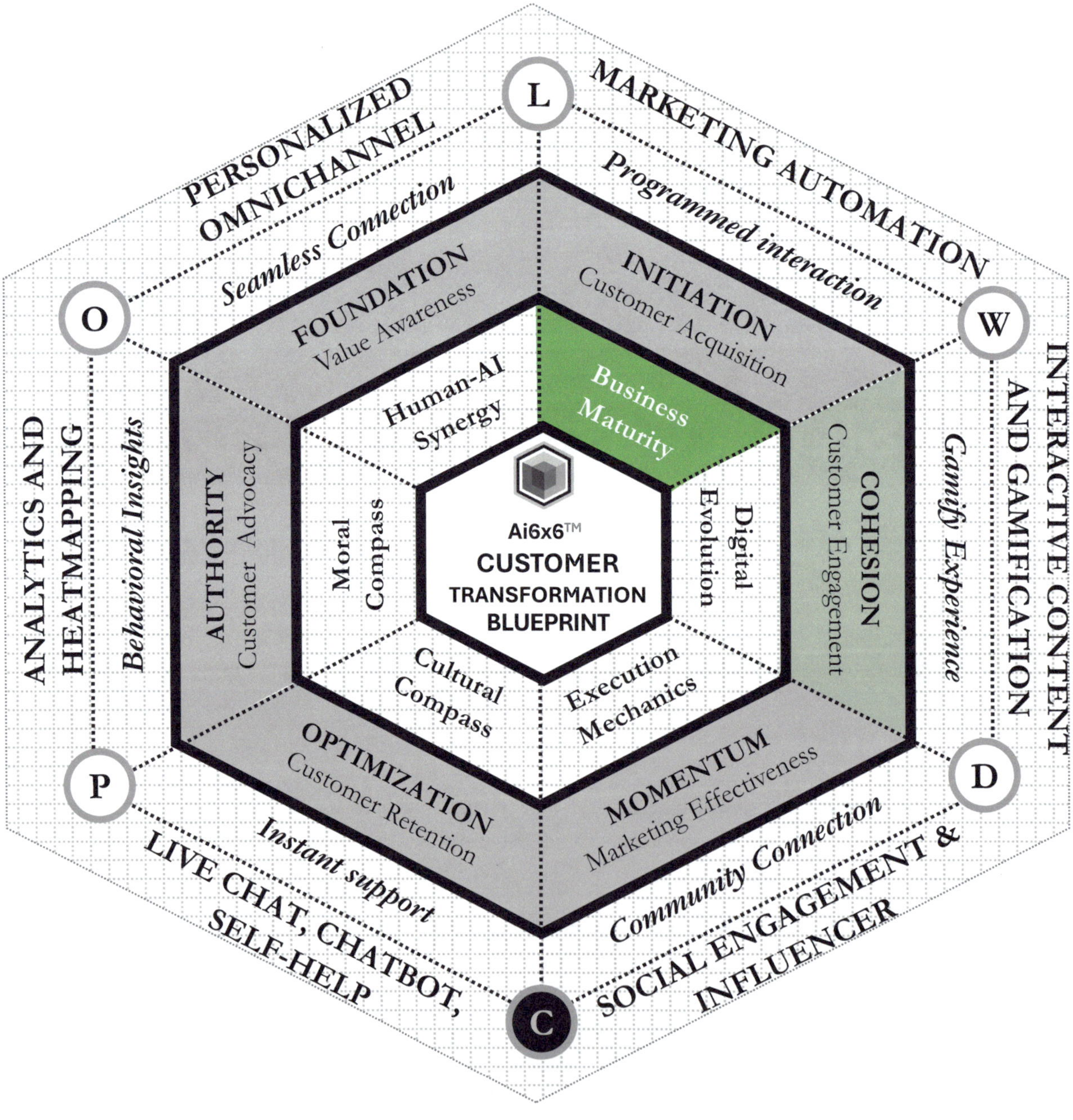

Ai6x6 BUSINESS MATURITY FRAMEWORK
FOR CUSTOMER TRANSFORMATION | CUSTOMER ENGAGEMENT

1. Personalized omnichannel

For a business leader, delivering a personalized omnichannel experience is a strategic imperative for creating truly seamless customer interactions across every touchpoint and fostering lasting loyalty. In today's hyper-connected world, customers expect consistent, tailored experiences whether they interact with your brand via social media, email, website, or in-store. Your ability to integrate these channels is paramount. Central to this is establishing unified communication channels. Ensuring consistency in messaging, tone, and customer service across all platforms—from social media to email, your website, and physical store interactions—prevents fragmentation and reinforces a cohesive brand identity.

Furthermore, leveraging AI-driven personalization is crucial for tailoring messaging, recommendations, and interactions based on granular customer data. AI analyzes behavior and preferences to deliver highly relevant content, making each customer feel understood and valued, which significantly boosts engagement. Crucially, ensuring cross-device synchronization maintains continuity in customer journeys across mobile, desktop, and in-app experiences. When a customer starts an interaction on one device and seamlessly continues it on another, it creates a fluid and effortless experience that builds trust and convenience. Neglecting a personalized omnichannel strategy can lead to frustrated customers, inconsistent brand experiences, lost sales, and a decline in customer loyalty. By championing an integrated, AI-powered approach to customer engagement, you empower your organization to deliver exceptional, personalized experiences that deepen customer relationships, drive repeat business, and ultimately secure a significant competitive advantage in the customer-centric marketplace.

WHAT?	WHY?	HOW?	BEWARE!
UNIFIED COMMUNICATION CHANNELS	**CONSISTENT EXPERIENCE**	**INTEGRATED PLATFORMS**	**SILOED DATA**
Ensuring consistency across touchpoints.	Provides a seamless and cohesive brand experience.	Use a CRM or CX platform to unify communication across channels.	Data residing in separate systems, making a unified view difficult.
AI-DRIVEN PERSONALIZATION	**RELEVANCE & ENGAGEMENT**	**ML ALGORITHMS**	**DATA PRIVACY**
Using customer data to tailor messaging.	Delivers highly relevant content, increasing customer engagement.	Deploy AI/ML models to analyze customer data for personalized recommendations.	Balancing personalization with customer data privacy concerns.
CROSS-DEVICE SYNCHRONIZATION	**SEAMLESS JOURNEY**	**IDENTITY RESOLUTION**	**TECHNICAL COMPLEXITY**
Maintaining continuity in customer journeys.	Ensures a smooth, uninterrupted customer experience across devices.	Implement identity graphs to link customer activity across devices.	Ensuring accurate and real-time synchronization across diverse devices.

For a business leader, embracing marketing automation is a strategic imperative for streamlining customer interactions, ensuring consistent communication, and maximizing the efficiency and effectiveness of your engagement efforts. In today's fast-paced digital landscape, manual outreach is simply unsustainable for maintaining personalized relationships at scale. Central to this is implementing automated email campaigns. By designing workflows that deliver targeted content based on customer behavior and preferences—such as a welcome series, abandoned cart reminders, or product updates—you ensure timely and relevant communication that nurtures leads and drives conversions without continuous manual intervention.

Furthermore, leveraging dynamic audience segmentation is crucial. By using AI and analytics to group customers into refined segments based on their demographics, behaviors, and preferences, you can tailor engagement strategies with greater precision, ensuring your messages resonate deeply with each specific group. Crucially, integrating predictive customer engagement allows you to anticipate needs and trigger personalized messaging proactively. By leveraging data insights to forecast future customer behavior, you can deliver timely offers, support information, or recommendations before the customer even explicitly requests them, creating a truly intuitive and personalized experience. By championing comprehensive marketing automation, you empower your organization to deliver highly personalized, scalable, and efficient customer experiences, driving deeper engagement, increasing conversions, and ultimately securing your business's sustained growth and customer loyalty in the digital age.

WHAT?	WHY?	HOW?	BEWARE!
AUTOMATED EMAIL CAMPAIGNS	**CONSISTENT ENGAGEMENT**	**EMAIL PLATFORM**	**OVER-AUTOMATION**
Delivering targeted content.	Ensures timely and relevant communication at scale.	Implement marketing automation platforms	Emails feeling generic or losing a personal touch.
DYNAMIC AUDIENCE SEGMENTATION	**REFINED TARGETING**	**AI-POWERED TOOLS**	**DATA COMPLEXITY**
Using AI to group customers.	Enables highly precise and effective engagement strategies.	Use AI/ML to analyze customer data and create dynamic segments.	Managing and interpreting complex customer data for segmentation.
PREDICTIVE CUSTOMER ENGAGEMENT	**PROACTIVE INTERACTION**	**ML MODELS**	**ACCURACY & PRIVACY**
Leveraging data to anticipate needs.	Anticipates customer needs and triggers timely, personalized messaging.	Deploy machine learning models to forecast customer behavior and intent.	Ensuring predictive models are accurate and respect customer privacy.

For a business leader, integrating interactive content and gamification into your customer engagement strategy is a powerful way to boost loyalty, encourage repeat engagement, and differentiate your brand through immersive experiences. In a world saturated with passive content, interactive elements actively capture attention and foster deeper connections. Central to this is implementing gamified loyalty programs. By encouraging repeat engagement with tiered rewards, achievement badges, and exciting challenges, you motivate customers to interact more frequently with your brand, fostering a sense of accomplishment and driving loyalty through intrinsic motivation.

Furthermore, utilizing interactive quizzes and polls is crucial for enhancing customer participation and brand interaction. These engaging formats allow customers to personalize their experience, share their preferences, and feel more connected to your brand, turning passive consumption into active involvement. Crucially, leveraging augmented reality (AR) and immersive content can revolutionize engagement. Driving deeper interaction through virtual product try-ons, 3D experiences, interactive storytelling, or virtual showrooms provides a memorable and cutting-edge brand experience that stands out from competitors and strengthens emotional bonds. Neglecting interactive content and gamification can lead to passive customer relationships, missed opportunities for deeper engagement, and a brand that struggles to capture and retain attention in a crowded digital landscape. By championing these immersive strategies, you empower your organization to create more dynamic, personalized, and rewarding customer experiences, driving stronger engagement, higher conversion rates, and ultimately, sustained customer loyalty and advocacy for your business.

WHAT?	WHY?	HOW?	BEWARE!
GAMIFIED LOYALTY PROGRAMS	**REPEAT ENGAGEMENT**	**REWARD SYSTEMS**	**COMPLEXITY**
Encouraging repeat engagement with rewards.	Drives long-term loyalty and increased customer lifetime value.	Design tiered rewards, points, badges, and leaderboards.	Designing and managing complex gamification mechanics effectively.
INTERACTIVE QUIZZES AND POLLS	**PARTICIPATION & INSIGHTS**	**CONTENT DEVELOPMENT**	**RELEVANCE**
Enhancing customer participation.	Boosts engagement and gathers valuable customer insights.	Create engaging quizzes, polls, and interactive surveys.	Ensuring interactive content remains relevant and engaging over time.
AUGMENTED REALITY (AR) AND IMMERSIVE CONTENT	**MEMORABLE EXPERIENCES**	**AR DEVELOPMENT**	**TECHNOLOGY ACCESS**
Driving engagement.	Provides unique, immersive brand interactions.	Create AR experiences for product try-ons, virtual showrooms, etc.	High development costs and accessibility issues for all customers.

For a business leader, strategically embracing social engagement and influencer collaborations is a critical imperative for strengthening connections, building community-driven brand interaction, and amplifying your brand's reach and credibility. In today's digitally native world, social platforms are where customers connect, share, and discover, making active participation crucial for brand loyalty. Central to this is leveraging strategic influencer collaborations. By partnering with trusted voices and content creators in your industry, you can enhance brand credibility, reach wider and more targeted audiences, and tap into established communities, generating authentic endorsements.

Furthermore, implementing robust social listening and real-time interaction capabilities is crucial. Engaging audiences through timely responses, participating in trending conversations, and direct messaging demonstrates responsiveness and builds genuine relationships, turning passive followers into active brand advocates. Crucially, fostering true community building creates lasting brand loyalty. This involves nurturing exclusive groups, establishing ambassador programs, and running interactive campaigns that encourage customers to connect with each other around your brand, transforming them into a supportive and engaged ecosystem. Neglecting social engagement and influencer strategies can lead to a disconnected brand presence, missed opportunities for authentic customer interaction, and a failure to tap into the powerful network effects of social platforms. By championing these community-driven approaches, you empower your organization to build stronger customer connections, amplify your brand message through trusted voices, and cultivate a loyal community that directly contributes to your business's reputation, growth, and long-term success.

WHAT?	WHY?	HOW?	BEWARE!
INFLUENCER COLLABORATIONS	**BRAND AMPLIFICATION**	**IDENTIFICATION & PARTNERSHIP**	**AUTHENTICITY**
Leveraging trusted voices for credibility and reach.	Expands reach and enhances brand credibility through trusted endorsements.	Identify relevant influencers and develop clear collaboration agreements.	Ensuring genuine alignment between influencers and the brand.
SOCIAL LISTENING AND REAL-TIME INTERACTION	**PROMPT ENGAGEMENT**	**MONITORING TOOLS**	**VOLUME & NOISE**
Engaging audiences.	Enables timely responses and participation in trending conversations.	Use social listening tools to track mentions and engage in real-time.	Managing large volumes of social media mentions and irrelevant noise.
COMMUNITY BUILDING	**BRAND LOYALTY**	**PLATFORM CREATION**	**MODERATION**
Fostering brand loyalty through exclusive groups.	Cultivates strong, long-term brand loyalty and advocacy.	Create exclusive online forums, ambassador programs, and interactive groups.	Actively moderating and nurturing online communities to keep them vibrant.

For a business leader, optimizing customer support through a blend of live chat, chatbots, and self-help solutions is a strategic imperative for providing real-time assistance, enhancing customer satisfaction, and improving operational efficiency. In today's fast-paced environment, customers expect immediate solutions, and a multi-tiered support strategy is crucial. Central to this is implementing AI-powered chatbots. By automating responses to common inquiries and routine tasks, chatbots provide instant customer assistance 24/7, reducing wait times and freeing up human agents for more complex issues.

Furthermore, seamlessly blending live chat with human-assisted support is crucial. This ensures that while automation handles basic queries, customers can easily transition to a live agent for complex issues, providing the human touch when truly needed and maintaining a high level of service quality. Crucially, empowering customers through self-service portals and knowledge bases is vital. Providing comprehensive FAQs, detailed guides, and step-by-step tutorials allows customers to find answers to their questions autonomously, reducing the volume of support tickets and improving overall efficiency. Neglecting these modern support solutions can lead to frustrated customers, long wait times, high operational costs for customer service, and a decline in satisfaction. By championing an integrated support strategy that leverages both automation and personalized human interaction, you empower your organization to deliver efficient, scalable, and highly satisfying customer support, ultimately driving higher customer engagement, retention, and a strong brand reputation.

WHAT?	WHY?	HOW?	BEWARE!
AI-POWERED CHATBOTS	**INSTANT ASSISTANCE**	**BOT DEPLOYMENT**	**COMPLEXITY**
Automating responses to common inquiries.	Provides immediate answers and reduces agent workload.	Implement and train chatbots for FAQs and routine customer queries.	Chatbots struggling with complex or nuanced customer inquiries.
LIVE CHAT AND HUMAN-ASSISTED SUPPORT	**SERVICE QUALITY**	**CHANNEL INTEGRATION**	**STAFFING**
Blending automation with human interaction.	Balances efficiency of automation with empathy of human interaction.	Ensure seamless handoff from chatbots to live agents.	Ensuring adequate human agent availability for complex issues.
SELF-SERVICE PORTALS AND KNOWLEDGE BASES	**EFFICIENT RESOLUTION**	**CONTENT CREATION**	**CONTENT UPKEEP**
Empowering customers with FAQs.	Reduces support volume and empowers customers for self-help.	Develop comprehensive FAQs, guides, and video tutorials.	Keeping knowledge base content accurate and up-to-date.

For a business leader, leveraging analytics and heatmapping is a strategic imperative for precisely measuring customer engagement metrics and continuously optimizing customer interactions across all digital platforms. In the digital age, every user click and scroll provides valuable data that, when analyzed correctly, can significantly enhance user experience and drive business growth. Central to this is conducting robust behavioral analytics. By meticulously tracking user behavior across websites, apps, and other digital platforms, you gain deep insights into how customers interact with your content, navigate your interfaces, and engage with your services.

Furthermore, employing heatmapping and click tracking tools is crucial. These visual analytical methods identify engagement hotspots—areas where users click most, scroll deepest, or spend the most time—as well as areas of confusion or abandonment. This visual feedback helps you improve UX design, optimize content placement, and refine calls to action. Crucially, translating these insights into concrete conversion and retention insights is vital. By using data-driven analysis to understand how engagement impacts key business outcomes, you can refine your engagement strategies, personalize experiences, and maximize results, directly contributing to higher sales and stronger customer loyalty. Neglecting analytics and heatmapping can lead to uninformed design decisions, inefficient digital marketing spend, and a failure to understand why customers convert (or don't). By championing these powerful analytical techniques, you empower your organization to make data-informed decisions, continuously optimize customer interactions, and ultimately drive superior user experience, higher conversion rates, and sustained business success through a deep understanding of customer behavior.

WHAT?	WHY?	HOW?	BEWARE!
BEHAVIORAL ANALYTICS	**USER INSIGHTS**	**TRACKING TOOLS**	**DATA PRIVACY**
Tracking user behavior across platforms.	Provides deep understanding of how customers interact digitally.	Implement web and app analytics tools (e.g., Google Analytics, Mixpanel).	Balancing granular tracking with user privacy concerns.
HEATMAPPING AND CLICK TRACKING	**UX OPTIMIZATION**	**HEATMAP TOOLS**	**INTERPRETATION**
Identifying engagement hotspots.	Identifies areas of interest and friction for design improvements.	Deploy tools (e.g., Hotjar, Crazy Egg) to visualize user clicks and scrolls.	Correctly interpreting heatmap patterns for actionable insights.
CONVERSION AND RETENTION INSIGHTS	**STRATEGY REFINEMENT**	**FUNNEL ANALYSIS**	**ATTRIBUTION**
Using data-driven analysis.	Informs and refines engagement strategies for better results.	Analyze user journeys and identify drop-off points for optimization.	Attributing specific engagement actions to overall conversion/retention.

Ai6x6™ Transformation Blueprint

MOMENTUM | Marketing Effectiveness

"The only way to win is to learn faster than anyone else." - Eric Ries.

An American entrepreneur and author, best known for popularizing the Lean Startup methodology. His work advocates for rapid experimentation and iterative product development based on continuous customer feedback, significantly influencing startup culture and product management globally.

The Momentum level is your Marketing Effectiveness, built by using data to accelerate the acquisition and satisfaction cycle. Marketing analytics provide the velocity needed to scale the brand's reach and impact with scientific speed.

The phase where customers meticulously assess your product or service against their specific needs, expectations, and available alternatives in the market, often involving trials, comparisons, and seeking reviews. Understanding the nuances of this phase is key to optimizing your offerings, proactively addressing potential pain points, and effectively converting initial interest into a firm commitment and purchase.

Providing clear information, compelling demonstrations, and social proof can significantly influence this critical decision-making stage, leading to higher conversion rates.

The "**Marketing Effectiveness**" chapter is paramount for business strategy maturity as it focuses on the customer's process of evaluating their marketing choices and adapting their needs. A mature business strategy understands that customers continuously compare options, read reviews, seek recommendations, and assess products or services against their evolving needs. By providing transparent information, excellent support, and responsive product development, businesses can influence this evaluation phase. It's about proactively addressing customer concerns and demonstrating value, ensuring the brand remains the preferred choice and continues to meet evolving market demands. Here is how we approach our learning:

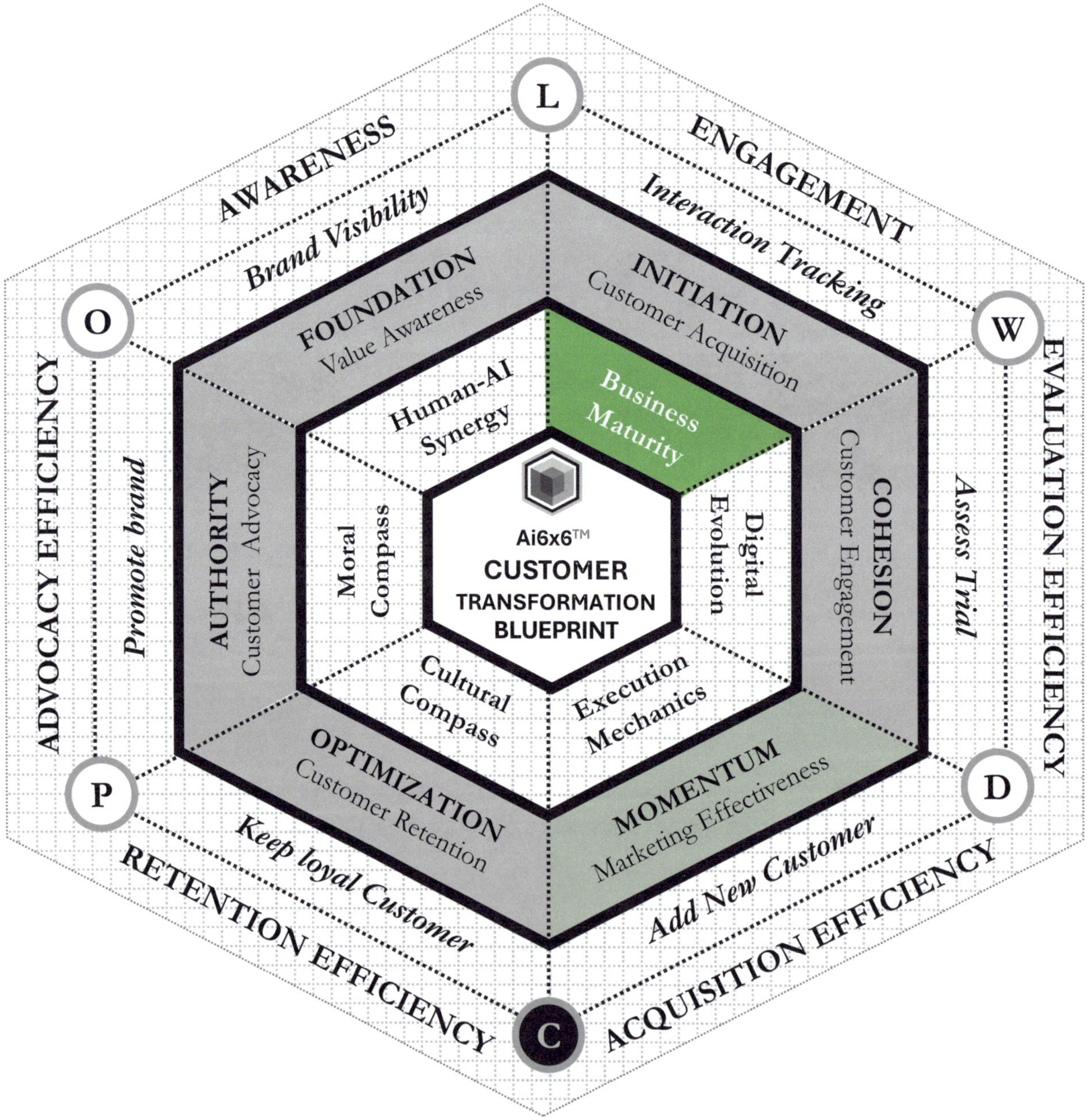

Ai6x6 BUSINESS MATURITY FRAMEWORK
FOR CUSTOMER TRANSFORMATION | MARKETING EFFECTIVENESS

For a business leader, meticulously measuring brand awareness is a foundational imperative for understanding how effectively your brand reaches potential customers and resonates within your target market. Awareness is the crucial first step in the customer journey; without it, all subsequent marketing and sales efforts will struggle. Central to this is rigorously tracking brand reach. This involves monitoring key metrics such as website traffic, social media impressions, advertising engagement rates, and content views to understand the breadth of your brand's exposure across various channels. A wider reach indicates stronger initial visibility.

Furthermore, assessing your share of voice is crucial. This involves analyzing how often your brand is mentioned, discussed, or featured compared to your competitors in relevant media, online conversations, and industry discourse. A higher share of voice often correlates with greater market presence and influence. Crucially, continuously evaluating customer sentiment provides a qualitative dimension to awareness. By monitoring public perception through reviews, social discussions, media mentions, and direct feedback, you can understand not just if people know your brand, but how they feel about it, allowing for proactive reputation management. Neglecting to measure brand awareness can lead to inefficient marketing spend, a lack of understanding of your brand's competitive standing, and a struggle to grow your customer base from the top of the funnel. By championing comprehensive awareness metrics, you empower your marketing teams to optimize their strategies for maximum visibility and positive brand perception, ultimately driving initial customer interest and laying the groundwork for acquisition.

WHAT?	WHY?	HOW?	BEWARE!
BRAND REACH	**MAXIMIZED EXPOSURE**	**ANALYTICS TOOLS**	**ATTRIBUTION**
Tracking website traffic, social media impressions, and ad engagement rates.	Quantifies the breadth of brand visibility and potential audience.	Use web analytics, social media insights, and ad platform reports.	Accurately attributing reach to specific marketing channels.
SHARE OF VOICE	**COMPETITIVE STANDING**	**MONITORING TOOLS**	**NOISE FILTERING**
Assessing how often the brand is mentioned compared to competitors.	Reveals brand presence relative to rivals in public discourse.	Use social listening and media monitoring tools to track mentions.	Distinguishing relevant brand mentions from general noise.
CUSTOMER SENTIMENT	**REPUTATION MANAGEMENT**	**SENTIMENT ANALYSIS**	**NUANCE**
Evaluating public perception through reviews and social discussions.	Understands brand perception and identifies areas for improvement.	Use AI/NLP tools to analyze reviews, comments, and social media posts.	Accurately interpreting complex or sarcastic sentiment in text data.

For a business leader, deeply analyzing customer engagement is a critical strategic imperative for understanding how customers interact with your brand across different platforms, ultimately influencing loyalty, conversion, and repeat business. Engagement goes beyond mere awareness; it signifies active interest and interaction, reflecting the strength of your brand's connection with its audience. Central to this is measuring interaction rates. This involves tracking key metrics such as click-through rates on emails or ads, social media engagement (likes, shares, comments), and email open rates.

Furthermore, examining the time spent by users on your websites or apps is crucial. This metric provides a qualitative gauge of interest and content relevance; longer durations often signify deeper engagement with your brand's offerings or information, revealing what truly captures attention. Crucially, rigorously tracking conversion rates—the percentage of visitors who take desired actions, from signing up for a newsletter to making a purchase—is the ultimate measure of engagement effectiveness. This links engagement directly to tangible business outcomes, showing whether your efforts are truly translating into customer acquisition or desired behaviors. Neglecting to analyze customer engagement can lead to wasted marketing efforts, a superficial understanding of customer behavior, and missed opportunities to optimize digital experiences. By championing comprehensive engagement analytics, you empower your marketing and product teams to refine their strategies, create more compelling content, and build stronger, more active relationships with your customer base, ultimately driving higher conversions and sustained customer loyalty.

WHAT?	WHY?	HOW?	BEWARE!
INTERACTION RATES	**CONTENT EFFECTIVENESS**	**PLATFORM ANALYTICS**	**VANITY METRICS**
Measuring click-through rates, social media engagement, and email open rates.	Indicates how compelling content and campaigns are.	Use data from email platforms, social media, and ad dashboards.	Focusing on high-level numbers that don't always reflect true engagement.
TIME SPENT	**CONTENT RELEVANCE**	**WEB/APP ANALYTICS**	**BOUNCE RATE MISINTERPRETATION**
Examining duration of visits on websites or apps to gauge interest.	Assesses the depth of user interest and content quality.	Track session duration, page views, and app usage time.	A quick visit might not always mean low interest.
CONVERSION RATES	**BUSINESS IMPACT**	**FUNNEL TRACKING**	**ATTRIBUTION COMPLEXITY**
Tracking the percentage of visitors who take desired actions.	Directly links engagement efforts to tangible business outcomes.	Monitor conversion funnels and A/B test different elements.	Attributing conversions across multiple touchpoints.

For a business leader, deeply understanding how customers evaluate your brand before making a purchase is a strategic imperative for optimizing your sales funnel and strengthening your competitive position. This 'consideration' phase is where potential customers compare options, assess value, and build conviction, making it crucial for conversion. Central to this is monitoring consideration metrics. This involves tracking actions like trial sign-ups, demo requests, content downloads (e.g., whitepapers), and product comparisons, which indicate active interest and a deeper engagement with your offerings. These signals are vital for lead qualification.

Furthermore, diligently collecting customer feedback is crucial. Gathering insights through direct surveys, Net Promoter Score (NPS) surveys, customer interviews, and online reviews provides qualitative understanding of how customers perceive your brand's strengths, weaknesses, and overall value proposition during their evaluation process. Crucially, robust competitor benchmarking provides essential context. Analyzing how customers make decisions relative to industry rivals—what features they prioritize, which brands they compare, and why they ultimately choose one over another—allows you to fine-tune your differentiation strategy and sales messaging. Neglecting to understand the evaluation phase can lead to high abandonment rates in your sales funnel, inefficient marketing efforts targeting unqualified leads, and a struggle to articulate your unique selling points effectively. By championing a proactive approach to understanding customer evaluation, you empower your sales and marketing teams to address specific concerns, highlight key differentiators, and guide prospects more effectively towards a purchase, ultimately driving higher conversion rates and securing a stronger market position.

WHAT?	WHY?	HOW?	BEWARE!
CONSIDERATION METRICS	**LEAD QUALIFICATION**	**CRM & ANALYTICS**	**MISINTERPRETATION**
Monitoring trial sign-ups, demo requests, and product comparisons.	Identifies prospects who are actively evaluating the brand.	Track specific actions like demo requests and whitepaper downloads.	Not all downloads or trials lead to serious consideration.
CUSTOMER FEEDBACK	**PERCEPTION UNDERSTANDING**	**SURVEYS & NPS**	**RESPONSE BIAS**
Collecting insights through surveys and Net Promoter Score (NPS).	Reveals how customers perceive the brand during evaluation.	Deploy targeted surveys (e.g., post-demo feedback, NPS).	Feedback might be skewed by extreme positive or negative experiences.
COMPETITOR BENCHMARKING	**COMPETITIVE EDGE**	**MARKET RESEARCH**	**DATA ACCESS**
Analyzing customer decision-making relative to rivals.	Helps refine differentiation strategy and sales messaging.	Conduct competitive analysis and track customer reviews of competitors.	Difficulty obtaining granular customer decision-making data from competitors.

For a business leader, meticulously measuring customer acquisition effectiveness is a non-negotiable strategic imperative for understanding the efficiency and impact of your sales and marketing efforts. This phase is the direct result of your entire customer journey strategy, determining your success in converting leads into paying customers and driving revenue growth. Central to this is calculating the Customer Acquisition Cost (CAC). This critical metric quantifies the average cost required to gain a new customer, providing invaluable insight into the financial efficiency of your acquisition channels and campaigns. A low CAC indicates optimized spending.

Furthermore, rigorously tracking the sales conversion rate is crucial. This measures the percentage of prospects or leads who ultimately become customers, directly reflecting the effectiveness of your sales processes and marketing funnel in turning interest into revenue. A higher conversion rate means greater efficiency. Crucially, evaluating first-time purchase behavior provides insights into the initial value and product preference of new buyers. Analyzing their average spend, the products they choose, and their engagement post-purchase helps in understanding the quality of newly acquired customers and tailoring their initial experience for long-term retention. Neglecting precise acquisition metrics can lead to unsustainable growth models, inefficient marketing budgets, and a lack of understanding of which channels truly drive profitable customer growth. By championing data-driven acquisition analysis, you empower your sales and marketing teams to optimize their strategies, reduce costs, and focus on the most effective channels, ultimately driving the sustainable expansion and profitability of your business. This financial rigor is key to scalable growth

WHAT?	WHY?	HOW?	BEWARE!
CUSTOMER ACQUISITION COST (CAC)	**BUDGET OPTIMIZATION**	**EXPENSE CALCULATION**	**INDIRECT COSTS**
Calculating the cost to gain a new customer.	Reveals efficiency of acquisition channels and campaigns.	Divide total marketing/sales spend by number of new customers.	Accurately capturing all direct and indirect acquisition costs.
SALES CONVERSION RATE	**FUNNEL EFFICIENCY**	**RATIO CALCULATION**	**FUNNEL DEFINITION**
Measuring the percentage of prospects who become customers.	Indicates effectiveness of sales processes and marketing funnel.	Divide number of conversions by total leads or visitors.	Ensuring consistent definition of "prospect" or "lead."
FIRST-TIME PURCHASE BEHAVIOR	**CUSTOMER QUALITY**	**DATA ANALYSIS**	**SHORT-TERM VIEW**
Evaluating initial spend and product preference.	Provides insights into the value and potential of new customers.	Analyze initial order value, product categories, and subsequent engagement.	Only reflects initial behavior, not long-term value.

For a business leader, meticulously assessing customer retention is a strategic imperative that directly impacts profitability, growth, and long-term business sustainability. It's significantly more cost-effective to retain existing customers than to acquire new ones, making loyalty a cornerstone of success. Your ability to maintain customer loyalty over time is a true measure of your value proposition and customer experience. Central to this is rigorously tracking churn rate. This critical metric identifies the percentage of customers who cease doing business with you within a given period, providing a clear signal of customer dissatisfaction or competitive vulnerability. A lower churn rate indicates higher loyalty.

Furthermore, accurately estimating Customer Lifetime Value (CLV) is crucial. This metric projects the long-term revenue potential of a retained customer, guiding strategic investments in customer service, loyalty programs, and personalized experiences to maximize their value over time. Understanding CLV helps prioritize retention efforts. Crucially, measuring the repeat purchase rate provides a clear indication of customer loyalty and satisfaction. Analyzing how frequently customers return for additional transactions reflects the stickiness of your product or service and the effectiveness of your post-purchase engagement strategies. Neglecting customer retention metrics can lead to an unsustainable business model where new customer acquisition merely replaces lost ones, eroding profitability and hindering long-term growth. By championing comprehensive retention analysis, you empower your organization to identify at-risk customers, optimize loyalty programs, and refine the overall customer experience, ultimately fostering a stable, profitable, and highly loyal customer base that drives sustained business success.

WHAT?	WHY?	HOW?	BEWARE!
CHURN RATE	**LOYALTY INDICATOR**	**CALCULATION**	**CAUSATION**
Identifying the percentage of customers who cease doing business with you.	Signals customer satisfaction and competitive standing.	Divide lost customers by total customers over a period.	Understanding why customers churn beyond just the rate.
CUSTOMER LIFETIME VALUE (CLV)	**STRATEGIC INVESTMENT**	**PREDICTIVE MODELING**	**ACCURACY**
Estimating long-term revenue potential.	Guides investment in retention and customer experience.	Use historical data to project future customer revenue.	CLV can be challenging to calculate accurately, especially for new businesses.
REPEAT PURCHASE RATE	**STICKINESS & SATISFACTION**	**TRANSACTION ANALYSIS**	**PRODUCT DEPENDENCY**
Measuring how frequently customers return.	Reflects product/service stickiness and customer satisfaction.	Track the percentage of customers making multiple purchases.	Varies significantly based on product type (e.g., consumables vs. one-time purchases).

For a business leader, meticulously evaluating customer advocacy is a strategic imperative for understanding how actively your satisfied customers promote your brand, leveraging their influence for powerful, organic growth and enhanced credibility. In today's interconnected world, word-of-mouth and genuine recommendations are invaluable marketing assets. Central to this is tracking referral rates. This critical metric quantifies the percentage of customers who actively recommend your brand to others, providing a direct measure of their willingness to act as advocates. Higher referral rates indicate strong customer satisfaction and loyalty.

Furthermore, assessing the quality and volume of customer-generated reviews is crucial. Monitoring the sentiment and content of online reviews and testimonials provides qualitative insights into brand perception and highlights areas of customer delight that can be amplified for marketing. Crucially, measuring community engagement reflects the strength of brand discussions, social shares, and participation in loyalty programs. Active community involvement demonstrates a deep connection and willingness to champion your brand beyond mere consumption. Neglecting to measure customer advocacy means missing out on one of the most powerful and cost-effective forms of marketing: authentic customer promotion. By championing initiatives that encourage and track advocacy, you empower your most loyal customers to become powerful brand ambassadors, driving organic growth, building trust with new prospects, and ultimately securing a resilient brand reputation and sustained business success. This customer-led promotion is a hallmark of truly beloved brands.

WHAT?	WHY?	HOW?	BEWARE!
REFERRAL RATES	**ORGANIC GROWTH**	**REFERRAL PROGRAMS**	**ATTRIBUTION**
Tracking the percentage of customers who recommend the brand.	Quantifies customer willingness to promote the brand.	Implement tracking for formal referral programs.	Accurately tracking organic word-of-mouth referrals.
REVIEW QUALITY	**BRAND REPUTATION**	**REVIEW MONITORING**	**FAKE REVIEWS**
Assessing volume and sentiment of customer-generated reviews.	Reveals public perception and highlights areas for improvement.	Use tools to track and analyze reviews on various platforms.	Distinguishing genuine reviews from fraudulent or biased ones.
COMMUNITY ENGAGEMENT	**ACTIVE LOYALTY**	**SOCIAL LISTENING**	**PASSIVE PARTICIPATION**
Measuring brand discussions and loyalty program participation.	Indicates deep customer connection and willingness to champion the brand.	Monitor social shares, mentions, and forum activity.	Users joining communities but not actively engaging.

Ai6x6™ Transformation Blueprint

OPTIMIZATION | Customer Retention

"Do what you do so well that they will want to see it again and bring their friends." - Walt Disney.

An American entrepreneur, animator, and film producer. A pioneer of American animation, he co-founded The Walt Disney Company, transforming it into a global entertainment empire known for iconic storytelling, theme parks, and beloved characters.

The Optimization level is your Customer Retention, focusing on the "Nirvana" of the lifecycle: maximum value with minimum churn. Churn reduction tactics refine the relationship, ensuring the most profitable customers stay for the long term.

Implementing comprehensive strategies and initiatives specifically focused on keeping existing customers satisfied, engaged, and encouraging them to make repeat purchases or continue subscriptions. This is far more cost-effective than constantly acquiring new customers, building stable revenue, fostering profound loyalty, and creating a powerful base of brand advocates.

Effective retention strategies include loyalty programs, proactive support, personalized offers, and continuous value delivery, ensuring customers feel valued and continue to choose your brand over competitors, driving long-term profitability.

"Customer Retention" is critical for business strategy maturity because it focuses on transitioning a customer from a single purchase to a loyal, ongoing relationship, evolving their experience. A mature business strategy prioritizes customer satisfaction through personalized experiences, exceptional service, and effective loyalty programs. This commitment to retention minimizes churn, maximizes customer lifetime value, and builds a stable revenue base. It's about adapting the customer relationship to ensure continued engagement and advocacy, which is essential for sustainable growth and long-term strategic success in a competitive market. Here is how we approach our learning:

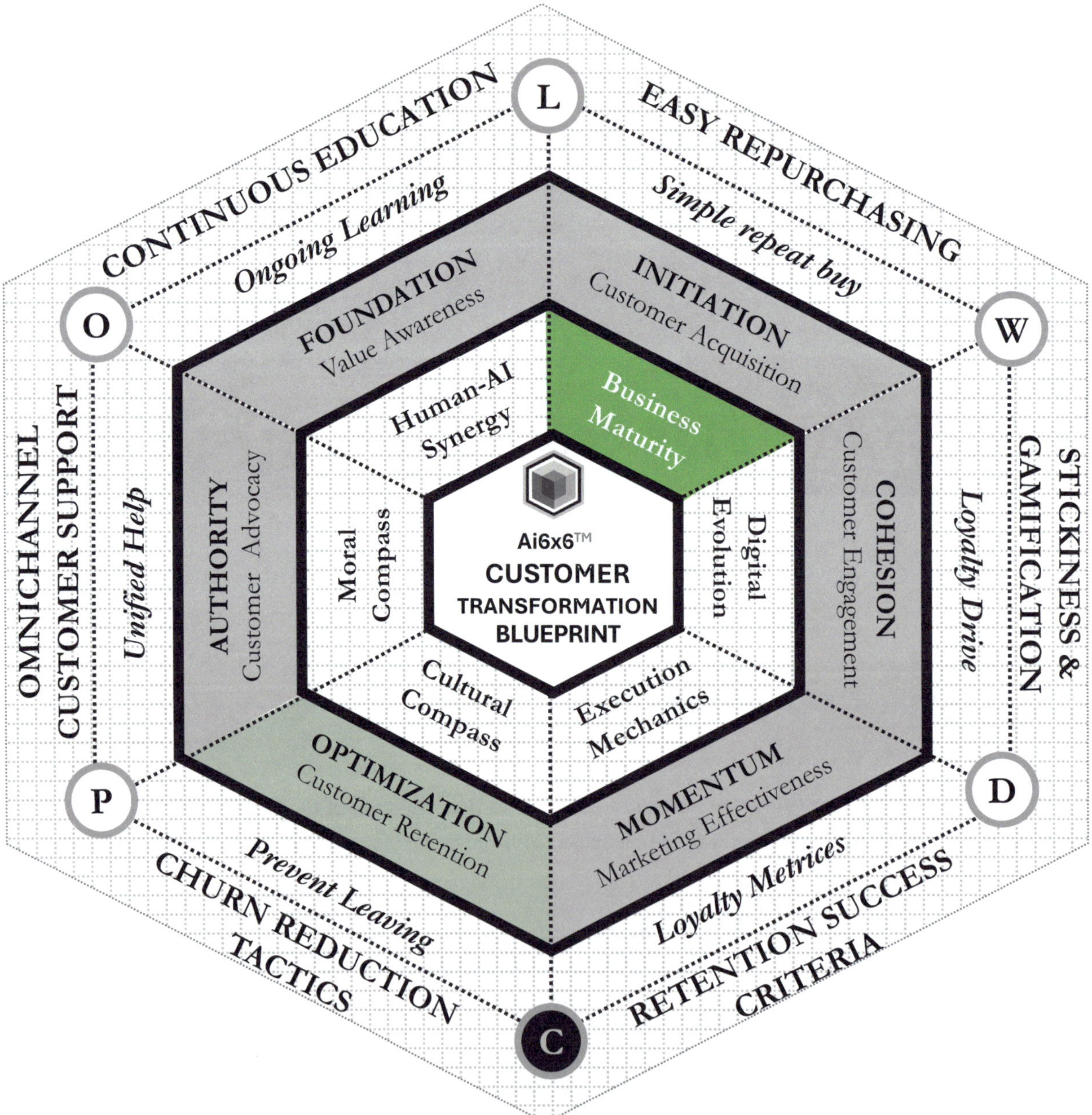

Ai6x6 BUSINESS MATURITY FRAMEWORK
FOR CUSTOMER TRANSFORMATION | CUSTOMER RETENTION

1. Continuous education

For a business leader, prioritizing continuous customer education is a strategic imperative for fostering long-term engagement, enhancing product stickiness, and ultimately strengthening customer retention. In a world of rapidly evolving products and services, empowering customers with ongoing learning and support ensures they maximize value from your offerings, reducing frustration and increasing loyalty. Central to this is providing interactive learning hubs. Offering easily accessible tutorials, webinars, detailed guides, and video content tailored to specific customer needs and usage levels empowers users to solve problems independently.

Furthermore, implementing personalized education paths is crucial. Adapting training materials based on individual customer usage patterns, preferences, and skill levels ensures that learning is relevant and impactful, preventing information overload and focusing on what truly helps them succeed. Crucially, fostering community-driven knowledge sharing empowers customers to learn from each other. Encouraging peer discussions, expert contributions, and user-generated content in forums or online communities creates a self-sustaining ecosystem of support, building a stronger sense of belonging and reducing reliance on direct customer service. Neglecting continuous customer education can lead to underutilized product features, increased customer frustration, higher support costs, and ultimately, customer churn as they fail to derive full value from your offerings. By championing comprehensive and personalized educational resources, you empower your customers to become experts, deepen their engagement, and ensure they remain loyal advocates for your brand, ultimately driving sustained customer retention and business growth.

WHAT?	WHY?	HOW?	BEWARE!
INTERACTIVE LEARNING HUBS	**VALUE MAXIMIZATION**	**CONTENT PLATFORM**	**CONTENT UPKEEP**
Providing tutorials, webinars, and guides.	Ensures customers fully utilize and derive value from products/services.	Create a dedicated online portal with tutorials, & FAQs	Keeping learning content accurate and up-to-date with product changes.
PERSONALIZED EDUCATION PATHS	**RELEVANCE & ENGAGEMENT**	**BEHAVIORAL ANALYTICS**	**DATA PRIVACY**
Adapting training materials based on usage.	Delivers highly relevant content, improving learning effectiveness.	Use customer usage data to recommend specific learning modules.	Balancing personalization with customer data privacy concerns.
COMMUNITY KNOWLEDGE SHARING	**SELF-SUPPORT & LOYALTY**	**FORUM & WIKI**	**MODERATION**
Encouraging peer discussions.	Fosters self-sufficiency and strengthens customer community.	Create online forums or wikis for peer-to-peer support and content creation.	Actively moderating and managing community content quality.

For a business leader, prioritizing easy repurchasing is a strategic imperative for enhancing customer convenience, strengthening loyalty, and driving repeat business. Once a customer has made an initial purchase, reducing any friction in subsequent transactions significantly increases the likelihood of them returning, making convenience a powerful retention tool. Central to this is implementing one-click reorder options. By simplifying the purchase functionality to minimal steps, you dramatically reduce friction, making it incredibly easy for loyal customers to quickly repurchase their favorite products or services without navigating a full checkout process. This seamless experience drives repeat purchases.

Furthermore, adopting subscription-based models for recurring purchases is crucial. By implementing auto-renewal systems for products or services that customers regularly need, you ensure consistent revenue streams while providing maximum convenience, reducing the need for manual reordering and boosting customer stickiness. By intelligently suggesting relevant products and services based on past purchases, browsing history, and preferences, you make the buying process more intuitive and tailored, leading to higher conversion rates for subsequent sales. Neglecting easy repurchasing mechanisms can lead to customer frustration, a decline in repeat business, and a struggle to maximize customer lifetime value. By championing a streamlined, automated, and personalized buying process for returning customers, you empower your organization to build stronger loyalty, secure recurring revenue, and ultimately drive the sustained growth and profitability of your business through a superior customer experience.

WHAT?	WHY?	HOW?	BEWARE!
ONE-CLICK REORDER OPTIONS	**CONVENIENCE & SPEED**	**PLATFORM INTEGRATION**	**SECURITY**
Reducing friction with quick purchase functionalities.	Simplifies repeat purchases, enhancing customer experience.	Implement quick reorder buttons within user accounts or order history.	Ensuring secure one-click purchasing without compromising financial data.
SUBSCRIPTION-BASED MODELS	**RECURRING REVENUE**	**BILLING PLATFORM**	**CHURN MANAGEMENT**
Implementing auto-renewal systems.	Ensures consistent revenue streams and boosts customer stickiness.	Deploy subscription management and auto-renewal billing systems.	Managing cancellations and preventing subscriber churn.
PERSONALIZED RECOMMENDATIO N	**INCREASED SALES**	**RECOMMENDATION ENGINE**	**DATA PRIVACY**
Using AI-driven insights to suggest relevant products.	Drives repeat purchases by suggesting highly relevant items.	Implement AI/ML-powered recommendation engines based on user data.	Balancing personalization with customer privacy concerns.

For a business leader, focusing on product stickiness and incorporating gamification elements into your customer retention strategy is a powerful imperative for encouraging long-term engagement and fostering deeper customer loyalty. Making your product or service indispensable and enjoyable to use creates strong behavioral bonds that significantly reduce churn. Central to this is implementing reward-based customer milestones. By designing tiered loyalty incentives tied to purchase frequency, engagement levels, or longevity, you motivate customers to achieve specific goals, fostering a sense of accomplishment and driving repeat business through tangible rewards and status.

Furthermore, integrating interactive challenges and badges is crucial. Motivating users with achievement-based recognition, virtual awards, or progress bars taps into their intrinsic desire for mastery and recognition, transforming routine interactions into engaging experiences and increasing their commitment to your brand. Crucially, leveraging personalized engagement triggers through behavioral analytics allows you to deliver timely engagement opportunities. By using data to understand customer activity and preferences, you can send targeted nudges, offers, or content that re-engage users at opportune moments, keeping them connected and active. Neglecting stickiness and gamification can lead to passive customer relationships, higher churn rates as customers find less engaging alternatives, and a struggle to differentiate your brand beyond price or features. By championing these interactive and rewarding experiences, you empower your organization to build a highly engaged, loyal, and enthusiastic customer base, transforming passive users into active advocates and ultimately driving sustained revenue growth and competitive advantage.

WHAT?	WHY?	HOW?	BEWARE!
REWARD-BASED MILESTONES	**REPEAT BUSINESS**	**LOYALTY PLATFORM**	**COMPLEXITY**
Implementing tiered loyalty incentives.	Motivates customers to engage more and encourages repeat purchases.	Design and implement a tiered loyalty program with points, badges, and rewards.	Designing and managing complex gamification mechanics effectively.
INTERACTIVE CHALLENGES AND BADGES	**ENGAGEMENT & MOTIVATION**	**GAMIFIED MECHANICS**	**USER FATIGUE**
Motivating users with achievement-based recognition.	Boosts user participation and a sense of accomplishment.	Integrate challenges, progress bars, and virtual badges into the product/service.	Gamification elements becoming repetitive or losing their novelty.
PERSONALIZED ENGAGEMENT TRIGGERS	**RE-ENGAGEMENT**	**BEHAVIORAL TRACKING**	**INTRUSION**
Using behavioral analytics for timely opportunities.	Delivers timely and relevant nudges to keep customers active.	Use analytics to identify user behavior patterns and personalized messages.	Overly frequent or irrelevant triggers annoying customers.

For a business leader, meticulously defining key metrics for retention success is a non-negotiable strategic imperative for accurately assessing customer loyalty and guiding efforts to maximize customer lifetime value. Without clear, measurable criteria, retention initiatives become guesswork, and it's impossible to objectively evaluate their impact. Central to this is accurately measuring Customer Lifetime Value (CLV). This critical metric estimates the long-term profitability of an engaged customer, guiding strategic investments in customer service, product development, and loyalty programs that aim to maximize their value over time. A higher CLV signifies effective retention strategies.

Furthermore, rigorously tracking repeat purchase rates is crucial. This directly measures customer retention through their repurchase behaviors, indicating the stickiness of your product or service and the effectiveness of your post-purchase engagement. Consistent repurchases are a strong sign of loyalty. Crucially, leveraging engagement and satisfaction analytics provides a deeper understanding. By evaluating user interactions with your brand's services and content (e.g., app usage, feature adoption, support interactions, survey scores), you can gauge customer health and satisfaction, identifying at-risk customers before they churn and pinpointing areas for improvement in the overall experience. Neglecting to define and track these retention success criteria can lead to a false sense of security, inefficient spending on ineffective programs, and a reactive approach to customer churn. By championing comprehensive retention metrics, you empower your organization to make data-informed decisions, proactively strengthen customer relationships, and ultimately drive sustained revenue growth and profitability through a loyal, engaged, and high-value customer base.

WHAT?	WHY?	HOW?	BEWARE!
CUSTOMER LIFETIME VALUE (CLV)	**PROFITABILITY FORECAST**	**PREDICTIVE MODELS**	**ACCURACY**
Measuring long-term profitability of engaged customers.	Guides strategic investments in customer retention.	Build models to estimate the future revenue a customer will generate.	Difficulty in accurately predicting future customer behavior and spending.
REPEAT PURCHASE RATES	**LOYALTY INDICATOR**	**TRANSACTION DATA**	**PRODUCT TYPE VARIANCE**
Tracking customer retention through repurchase behaviors.	Directly measures customer stickiness and product satisfaction.	Analyze transaction history to identify repeat buyers.	Relevance varies significantly by product (e.g., SaaS vs. infrequent purchases).
ENGAGEMENT & SATISFACTION ANALYTICS	**CUSTOMER HEALTH**	**METRIC TRACKING**	**CORRELATION**
Evaluating user interactions.	Identifies active users and potential churn risks.	Monitor app usage, feature adoption, support interactions, and survey scores.	Clearly correlating engagement metrics with actual customer loyalty/churn.

For a business leader, implementing proactive churn reduction tactics is a strategic imperative for preventing customer attrition, preserving revenue, and ensuring the long-term stability of your customer base. In a competitive market, retaining existing customers is far more cost-effective than acquiring new ones, making a focused approach to preventing churn absolutely vital. Central to this is leveraging predictive churn detection. By using AI and advanced analytics to identify at-risk customers based on behavioral patterns (e.g., decreased usage, support tickets, survey sentiment), you can receive early warnings and intervene proactively before they decide to leave. This foresight allows for timely action.

Furthermore, implementing targeted retention campaigns is crucial. By delivering personalized outreach strategies—such as special offers, dedicated support, or tailored product education—you can re-engage disengaged users, address their specific pain points, and re-demonstrate value, strengthening their commitment to your brand. Crucially, continuous seamless customer support enhancements are vital. Strengthening interactions across all channels, minimizing frustration, and providing rapid, effective problem-solving builds trust and reduces the likelihood of abandonment due to poor service experiences. Neglecting proactive churn reduction can lead to a shrinking customer base, eroded profitability, and a constant uphill battle in customer acquisition. By championing a data-driven and customer-centric approach to preventing attrition, you empower your organization to safeguard its revenue streams, build stronger customer relationships, and cultivate a highly loyal customer base that drives sustained business growth and a formidable competitive advantage.

WHAT?	WHY?	HOW?	BEWARE!
PREDICTIVE CHURN DETECTION	**PROACTIVE INTERVENTION**	**ML MODELING**	**ACCURACY**
Using AI to identify at-risk customers.	Enables early identification of at-risk customers for retention efforts.	Develop AI/ML models to predict churn based on behavioral and demographic data.	Ensuring models are highly accurate and don't generate false positives.
TARGETED RETENTION CAMPAIGNS	**RE-ENGAGEMENT**	**SEGMENTED MARKETING**	**PERSONALIZATION INTRUSION**
Implementing personalized outreach.	Effectively re-engages disengaged users and prevents departure.	Design personalized offers, content, and outreach for at-risk segments.	Risk of appearing overly intrusive with targeted interventions.
CUSTOMER SUPPORT ENHANCEMENTS	**FRUSTRATION REDUCTION**	**IMPROVED CHANNELS**	**CONSISTENCY**
Strengthening interactions.	Minimizes customer frustration, a key driver of churn.	Optimize response times, train agents, and simplify support processes.	Maintaining consistent, high-quality support across all channels.

For a business leader, providing comprehensive omnichannel customer support is a strategic imperative for ensuring seamless assistance, maximizing customer satisfaction, and strengthening long-term customer relationships. In today's interconnected world, customers expect to interact with your brand across multiple communication platforms—phone, chat, email, social media, and self-service—without losing context or repeating themselves. Central to this is establishing integrated support ecosystems. This involves providing help across all preferred channels, ensuring that customer service agents have a unified view of past interactions regardless of the touchpoint. This seamless handoff improves efficiency and customer experience.

Furthermore, leveraging AI-driven assistance, such as chatbots and automation, is crucial for instant query resolution. AI can handle common inquiries, route complex issues to the right human agent, and provide personalized information 24/7, significantly reducing response times and improving efficiency. Crucially, offering personalized human support remains vital for high-touch customer needs. While automation handles routine tasks, skilled live agents provide empathetic, nuanced interactions for complex queries, complaints, or sensitive issues, ensuring that the human touch is available when it matters most, building trust and loyalty. Neglecting omnichannel support can lead to frustrated customers, inefficient support operations, and a fragmented customer experience that erodes loyalty. By championing an integrated, intelligent, and customer-centric approach to support, you empower your organization to deliver exceptional service, reduce operational costs, and ultimately drive higher customer satisfaction and retention, securing your business's reputation and long-term success.

WHAT?	WHY?	HOW?	BEWARE!
INTEGRATED SUPPORT ECOSYSTEMS	**SEAMLESS EXPERIENCE**	**CRM INTEGRATION**	**DATA SILOS**
Providing help across all channels.	Ensures consistent and convenient customer assistance.	Unify customer data and interaction history across all support channels.	Integrating disparate legacy support systems.
AI-DRIVEN ASSISTANCE	**INSTANT RESOLUTION**	**CHATBOT DEPLOYMENT**	**COMPLEXITY**
Leveraging chatbots and automation.	Provides immediate answers and improves efficiency.	Implement AI-powered chatbots for FAQs and routine tasks.	Chatbots struggling with complex or nuanced customer inquiries.
PERSONALIZED HUMAN SUPPORT	**HIGH-TOUCH RESOLUTION**	**AGENT TRAINING**	**SCALABILITY**
Offering live agent interactions.	Provides empathetic and nuanced support for complex issues.	Train human agents on advanced problem-solving and empathy skills.	Ensuring sufficient human agent availability for peak demand.

Ai6x6™ Transformation Blueprint

AUTHORITY | Customer Advocacy

"We want to make sure that our customers are not just satisfied, but delighted, so they become our biggest advocates." - Indra Nooyi.

Former CEO of PepsiCo. She spearheaded the "Performance with Purpose" vision, integrating sustainability into its core business strategy. Her transformative leadership expanded PepsiCo's portfolio and global footprint, making her an influential female business leader and champion for responsible capitalism.

The Authority level is your Customer Advocacy, reached when customers become the brand's most powerful marketing force. Cultivating loyalty and influencer growth moves the brand from a vendor to an industry leader backed by a community.

The ultimate goal, where satisfied customers actively and enthusiastically promote your brand through positive word-of-mouth, glowing reviews, social media mentions, and direct referrals to their networks. Advocates drive organic growth, build immense trust within their communities, and significantly amplify your brand's reach and sterling reputation.

They become your most powerful and credible marketing asset. Cultivating advocacy involves consistently exceeding expectations, providing exceptional experiences, and empowering customers to share their positive stories, creating a virtuous cycle of growth and brand loyalty.

The **"Customer Advocacy"** chapter is fundamental for business strategy maturity as it focuses on sustaining customer loyalty and innovating through word-of-mouth growth. A mature business strategy cultivates an environment where satisfied customers become brand ambassadors, actively recommending and influencing potential buyers. This organic amplification reduces marketing costs, enhances brand credibility, and drives exponential growth through trusted referrals. It's about creating a customer experience so compelling that it naturally generates positive buzz, ensuring the business leverages its most powerful asset—its loyal customers—for sustained strategic expansion and market leadership. Here is how we approach our learning:

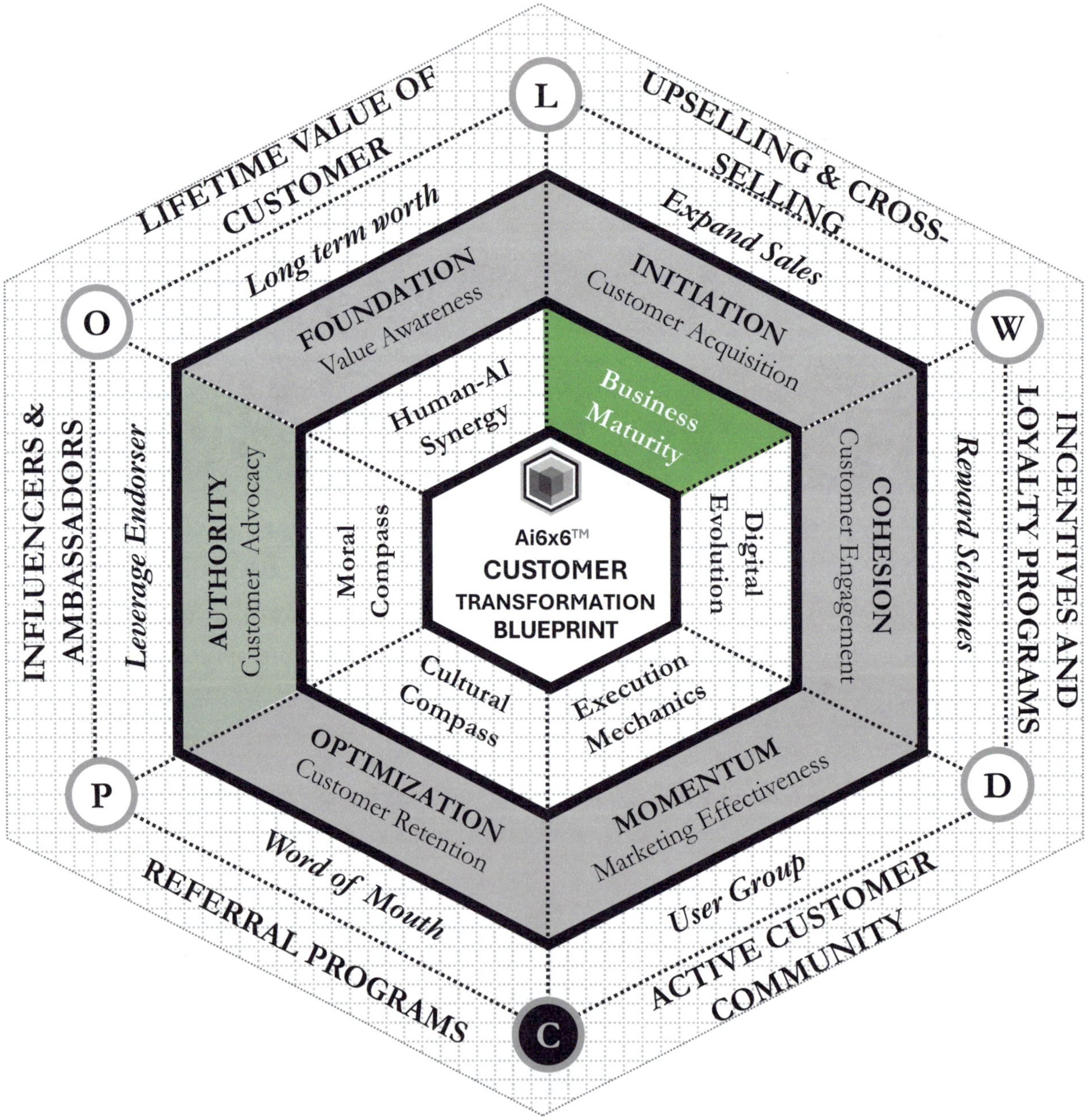

Ai6x6 BUSINESS MATURITY FRAMEWORK
FOR CUSTOMER TRANSFORMATION | CUSTOMER ADVOCACY

For a business leader, relentlessly focusing on the lifetime value of customer (LVC) is a strategic imperative for maximizing engagement and ensuring long-term revenue contributions from your customer base. It's not just about a single transaction, but nurturing relationships to foster repeat business and sustained profitability, recognizing that a truly engaged customer is your most valuable asset. Central to this is implementing personalized engagement strategies. By proactively nurturing relationships through tailored communications, exclusive offers, and relevant content, you can significantly increase customer retention and encourage repeat purchases, ensuring they remain active and derive continuous value from your brand.

Furthermore, leveraging predictive analytics for customer behavior is crucial. By using data-driven insights to identify high-value customers, anticipate their future needs, and predict potential churn risks, you can strategically allocate resources to those who offer the greatest long-term revenue potential. This foresight optimizes retention efforts. This means continually innovating products, enhancing services, and providing exceptional customer experience that reinforces their investment in your brand and motivates long-term loyalty. Neglecting customer lifetime value can lead to a short-sighted focus on acquisition, missed opportunities for upselling and cross-selling, and an unsustainable business model reliant on constant new customer influx. By championing strategies that maximize the long-term profitability and engagement of your customers, you empower your organization to build a stable, recurring revenue base, foster deep loyalty, and ultimately secure sustained business success through a truly customer-centric approach.

WHAT?	WHY?	HOW?	BEWARE!
PERSONALIZED ENGAGEMENT STRATEGIES	**RETENTION & REVENUE**	**SEGMENTATION & MESSAGING**	**DATA PRIVACY**
Nurturing relationships.	Increases retention and boosts repeat business.	Tailor communications and offers based on customer data.	Balancing personalization with customer privacy concerns.
PREDICTIVE ANALYTICS FOR CUSTOMER BEHAVIOR	**RESOURCE ALLOCATION**	**ML MODELING**	**ACCURACY**
Identifying high-value customers.	Guides strategic investment in high-potential customers.	Develop AI/ML models to predict customer value and behavior.	Ensuring predictive models are accurate and actionable.
CONTINUOUS VALUE DELIVERY	**LONG-TERM LOYALTY**	**PRODUCT INNOVATION**	**EXPECTATION MANAGEMENT**
Ensuring customers feel rewarded and supported.	Fosters sustained customer engagement and commitment.	Regularly update products/services and provide exceptional support.	Constantly meeting or exceeding evolving customer expectations.

For a business leader, strategically leveraging upselling and cross-selling is a powerful imperative for enhancing customer advocacy, maximizing revenue from existing relationships, and deepening customer engagement through complementary product offerings. Once a customer is satisfied, presenting relevant additional value strengthens their relationship with your brand. Central to this is implementing personalized recommendation engines. By using AI-driven insights to suggest upgrades (upselling) or additional services/products (cross-selling) based on past usage patterns, browsing history, and preferences, you make the buying process more intuitive and tailored, increasing conversion rates for related offerings.

Furthermore, providing exclusive upgrade incentives for loyal customers is crucial. Offering preferential pricing, early access, or special bundles on premium offerings motivates existing customers to expand their investment, making them feel valued and rewarding their loyalty while boosting average order value. Crucially, timing timely cross-selling strategies is vital. Introducing relevant products at the right moments in the customer lifecycle—for example, complementary accessories after a core product purchase, or related services when a customer is exploring new needs—maximizes relevance and conversion. Neglecting upselling and cross-selling means leaving significant revenue on the table, failing to fully leverage your existing customer base, and potentially allowing competitors to fill those complementary needs. By championing these strategic approaches, you empower your organization to maximize customer lifetime value, enhance customer satisfaction by offering relevant solutions, and ultimately drive sustained revenue growth and profitability through intelligent engagement with your loyal customer base.

WHAT?	WHY?	HOW?	BEWARE!
PERSONALIZED RECOMMENDATION ENGINES	**REVENUE MAXIMIZATION**	**AI/ML DEPLOYMENT**	**RELEVANCE**
Suggesting upgrades and additional services.	Drives higher customer lifetime value through relevant suggestions.	Implement AI-powered recommendation systems based on user data.	Ensuring recommendations are genuinely helpful and not intrusive.
EXCLUSIVE UPGRADE INCENTIVES	**CUSTOMER LOYALTY**	**TIERED PRICING**	**PERCEIVED FAIRNESS**
Offering loyal customers preferential pricing.	Rewards existing customers, encouraging deeper commitment.	Design exclusive discounts or early access for loyal segments.	Ensuring new customer promotions don't devalue loyalty incentives.
TIMELY CROSS-SELLING STRATEGIES	**INCREASED ENGAGEMENT**	**CUSTOMER JOURNEY MAPPING**	**OVER-PROMOTION**
Introducing relevant products at the right moments.	Maximizes sales by offering products when customers are most receptive.	Identify key touchpoints in the customer journey for cross-sell opportunities.	Over-promoting new products, leading to customer fatigue.

For a business leader, establishing robust incentives and loyalty programs is a fundamental strategic imperative for encouraging customers to become active brand advocates, driving repeat business, and building a passionate community around your brand. Beyond transactional value, these programs foster a deeper connection and reward commitment. Central to this is implementing gamified loyalty rewards. Engaging customers with tiered perks, bonuses, and milestone achievements (e.g., points systems, exclusive discounts, special access) taps into their desire for recognition and accomplishment, motivating increased engagement and repeat purchases.

Furthermore, providing exclusive VIP benefits cultivates a sense of prestige and appreciation. Offering members special discounts, early access to new products, priority service, or invitations to exclusive events makes loyal customers feel valued and reinforces their commitment to your brand. Crucially, leveraging social recognition and engagement highlights loyal customers through leaderboards, features in brand content, and user-generated content initiatives. Publicly celebrating advocates amplifies their influence and inspires others to engage more deeply, creating a positive feedback loop within your community. Neglecting comprehensive incentives and loyalty programs can lead to a transactional customer relationship, missed opportunities for deeper engagement, and a struggle to differentiate your brand in a crowded market. By championing these rewarding initiatives, you empower your most loyal customers to become passionate brand advocates, driving organic growth, enhancing brand credibility, and ultimately securing sustained revenue growth and a thriving customer community for your business.

WHAT?	WHY?	HOW?	BEWARE!
GAMIFIED LOYALTY REWARDS	**REPEAT ENGAGEMENT**	**POINTS SYSTEM**	**COMPLEXITY**
Engaging customers with tiered perks and achievements.	Motivates customers to interact more frequently with the brand.	Design loyalty programs with points, tiers, and achievement badges.	Overly complex reward systems confusing customers.
EXCLUSIVE VIP BENEFITS	**CUSTOMER RETENTION**	**TIERED MEMBERSHIP**	**PERCEIVED INEQUITY**
Providing members with special discounts and early access.	Fosters a sense of prestige and exclusivity, boosting loyalty.	Create VIP tiers with unique perks like early access, priority support.	Non-VIP customers feeling less valued.
SOCIAL AND ENGAGEMENT	**AMPLIFIED ADVOCACY**	**LEADERBOARDS**	**AUTHENTICITY**
Highlighting loyal customers.	Inspires others and strengthens brand community.	Feature top advocates in newsletters, social media, or leaderboards.	Ensuring recognition feels genuine and not purely transactional.

For a business leader, actively cultivating an engaged customer community is a strategic imperative for fostering powerful brand advocacy, strengthening customer loyalty, and creating a self-sustaining ecosystem of support and interaction around your brand. Beyond individual relationships, a thriving community transforms customers into collective champions. Central to this is establishing user forums and community hubs. Creating dedicated online spaces where customers can share experiences, ask questions, provide expertise, and connect with each other builds a sense of belonging and collective ownership, deepening their emotional investment in your brand.

Furthermore, implementing brand ambassador initiatives is crucial. Empowering highly loyal and engaged customers to lead discussions, guide newcomers, and represent your brand through formal programs amplifies authentic advocacy and provides invaluable peer-to-peer support, extending your brand's reach and credibility. Crucially, fostering real-time support networks within the community enables peer-to-peer problem-solving. This not only enhances product trust and satisfaction by providing immediate, reliable answers from experienced users but also reduces the burden on your official customer service channels. Neglecting community building can lead to isolated customer experiences, missed opportunities for organic interaction, and a failure to harness the collective power of your most passionate users. By championing interactive customer engagement platforms and empowering your customers to connect, you foster a vibrant, self-supporting community that drives advocacy, strengthens loyalty, and ultimately contributes significantly to your business's sustained growth and positive brand reputation.

WHAT?	WHY?	HOW?	BEWARE!
USER FORUMS AND COMMUNITY HUBS	**BELONGING & SUPPORT**	**PLATFORM DEPLOYMENT**	**MODERATION**
Creating spaces for sharing experiences.	Fosters a sense of belonging and peer-to-peer support.	Set up dedicated online forums or community platforms.	Requires active moderation to ensure positive and productive discussions.
BRAND AMBASSADOR INITIATIVES	**AUTHENTIC ADVOCACY**	**IDENTIFICATION & REWARDS**	**AUTHENTICITY**
Empowering loyal customers.	Amplifies brand message through trusted, organic voices.	Identify loyal customers and offer exclusive perks for advocacy.	Ensuring ambassador content remains genuine and not overtly promotional.
REAL-TIME SUPPORT NETWORKS	**EFFICIENT SUPPORT**	**COMMUNITY Q&A**	**ACCURACY**
Enabling peer-to-peer problem-solving.	Reduces burden on official support channels and builds trust.	Encourage customers to answer each other's questions in forums.	Ensuring the accuracy of peer-generated answers.

For a business leader, establishing effective referral programs is a strategic imperative for turning satisfied customers into powerful brand advocates, leveraging authentic word-of-mouth marketing for efficient customer acquisition and sustainable growth. When customers willingly recommend your brand, it carries immense credibility that traditional advertising often cannot match. Central to this is implementing incentivized referral rewards. By providing clear discounts, perks, or even cash incentives for successful referrals, you create a strong motivation for customers to actively share their positive experiences, directly linking their advocacy to tangible benefits.

Furthermore, developing personalized referral outreach strategies is crucial. Encouraging advocates to share their experiences within their personal networks through customizable referral links, pre-drafted messages, or social sharing options makes the act of referring easy and natural, increasing participation rates. Crucially, leveraging automated referral tracking is vital for measuring impact. Using digital tools to monitor the source of referrals, track conversions, and attribute success accurately allows you to evaluate the effectiveness of your programs, identify top advocates, and optimize strategies for maximum reach and ROI. Neglecting structured referral programs means missing out on one of the most powerful and cost-effective forms of marketing: authentic customer promotion. By championing clear, incentivized, and trackable referral initiatives, you empower your most loyal customers to become passionate sales drivers, enhancing brand credibility, reducing customer acquisition costs, and ultimately securing sustained revenue growth through organic, trusted recommendations.

WHAT?	WHY?	HOW?	BEWARE!
INCENTIVIZED REFERRAL REWARDS	**MOTIVATED ADVOCACY**	**REWARD STRUCTURE**	**COST**
Providing discounts, perks, or cash incentives.	Drives active participation in brand promotion.	Design clear incentives for both referrer and referred customer.	Managing the cost of incentives, especially for high-volume referrals.
PERSONALIZED REFERRAL OUTREACH	**EASY SHARING**	**SHARING TOOLS**	**AUTHENTICITY**
Encouraging advocates to share experiences.	Makes it simple for customers to share their recommendations.	Provide unique referral links, social sharing buttons, and pre-written messages.	Ensuring referrals feel genuine and not overly promotional.
AUTOMATED REFERRAL TRACKING	**ACCURATE MEASUREMENT**	**TRACKING PLATFORM**	**FRAUD**
Using digital tools to measure impact.	Allows for precise evaluation and optimization of referral efforts.	Implement a referral tracking software to monitor links and conversions.	Preventing fraudulent referrals or self-referrals to exploit rewards.

For a business leader, strategically engaging with influencers and establishing brand ambassador programs is a powerful imperative for amplifying customer advocacy and extending your brand's reach through trusted voices. In a fragmented media landscape, genuine endorsements from credible personalities or passionate customers resonate deeply with target audiences. Central to this is forming strategic influencer partnerships. Collaborating with industry experts, content creators, or micro-influencers whose audience aligns with your target market allows you to leverage their established trust and reach broader audiences efficiently, driving awareness and consideration.

Furthermore, developing long-term brand ambassador programs fosters deeper connections with your most loyal customers. By formally recognizing and engaging these super-advocates, you create a dedicated cohort that consistently champions your brand through ongoing engagement, exclusive access, and early product previews, turning them into extensions of your marketing team. Crucially, emphasizing authentic storytelling initiatives encourages organic customer testimonials and success stories. Empowering customers to share their genuine experiences in their own voice builds trust with potential buyers, as authentic narratives are far more persuasive than traditional advertising. Neglecting influencer and ambassador strategies can lead to missed opportunities for credible reach, reliance on less effective marketing channels, and a failure to tap into the power of social proof. By championing these strategic collaborations and authentic content generation, you empower your organization to amplify its message through trusted voices, build a powerful network of advocates, and ultimately secure sustained brand growth and market leadership through compelling, customer-led promotion.

WHAT?	WHY?	HOW?	BEWARE!
STRATEGIC INFLUENCER PARTNERSHIPS	**WIDER REACH**	**IDENTIFICATION & CONTRACTS**	**AUTHENTICITY**
Collaborating with industry experts.	Extends brand message to targeted, engaged audiences.	Identify relevant influencers, establish clear contracts and campaign goals.	Ensuring influencer content remains genuine and not overly commercial.
LONG-TERM BRAND AMBASSADOR PROGRAMS	**CONSISTENT ADVOCACY**	**PROGRAM STRUCTURE**	**MANAGEMENT**
Establishing deeper connections.	Cultivates ongoing, committed brand promotion.	Create a formal program with perks, responsibilities, and engagement activities.	Actively managing and nurturing ambassador relationships.
AUTHENTIC STORYTELLING	**CREDIBILITY & TRUST**	**UGC CAMPAIGNS**	**QUALITY CONTROL**
Encouraging organic customer testimonials.	Builds trust with potential customers through genuine experiences.	Encourage customers to share their stories via reviews, videos, and social media.	Ensuring user-generated content quality and brand alignment.

SECTION D

BLUEPRINT SERIES

Digital Evolution

Ai6x6™ Transformation Blueprint

Story 1 — The Birth of COBOL and Business Computing (1959–1965): Early enterprises struggled to translate business logic into machine-readable form, limiting the usefulness of emerging computers. COBOL introduced a standardized language that mirrored human reasoning, enabling organizations to encode processes consistently. This shift allowed systems to scale beyond isolated experiments. Human analysts defined rules while machines executed them with precision. The partnership created the first bridge between business intent and digital execution.

Story 2 — ARPANET and Packet Switching (1969–1972): Communication networks were fragile, centralized, and prone to failure. Packet switching introduced a decentralized architecture that allowed information to flow even when parts of the network failed. Humans designed protocols while machines routed data autonomously. This resilience transformed digital communication from a technical novelty into a strategic infrastructure. ARPANET demonstrated how distributed systems could scale without collapsing. It laid the foundation for the modern internet.

Story 3 — Xerox PARC and the Graphical User Interface (1973–1983): Computers were powerful but inaccessible to most users, requiring specialized knowledge to operate. Xerox PARC introduced the graphical user interface, enabling humans to interact with machines visually rather than through code. This shift democratized computing and redefined usability. Machines handled complexity behind the scenes while humans engaged through intuitive metaphors. The interface became a catalyst for mass digital adoption. It showed that digital evolution accelerates when technology adapts to human cognition.

Story 4 — The Rise of Enterprise Resource Planning (1980s–1990s): Organizations operated in silos, with fragmented systems and inconsistent data. ERP platforms unified finance, operations, and supply chains into integrated digital environments. Machines enforced standardization while humans redesigned processes around shared information. This integration improved visibility, coordination, and control. ERP became the backbone of global enterprises. It demonstrated how digital evolution transforms not just tools but entire operating models.

Story 5 — The Commercial Internet and E-Commerce (1994–2005): The internet shifted from academic infrastructure to commercial ecosystem, enabling new forms of interaction, distribution, and value creation. Websites became digital storefronts, and payment systems matured into secure transactional platforms. Machines automated logistics while humans reimagined customer experience. The digital marketplace expanded global reach and lowered barriers to entry. It marked a turning point where digital capability became essential for competitiveness.

Story 6 — Cloud Computing and Elastic Infrastructure (2006–2015): Organizations struggled with rigid, capital-intensive infrastructure that limited experimentation and scale. Cloud computing introduced elastic resources that could expand or contract instantly. Machines automated provisioning while humans focused on architecture and innovation. This shift accelerated deployment cycles, reduced cost, and enabled global reach. Cloud became the foundation for modern digital ecosystems. It demonstrated how digital evolution thrives when infrastructure becomes fluid, on-demand, and programmable.

LESSONS FROM A CUSTOMER TRANSFORMATION PERSPECTIVE

Customer value deepens as digital evolution expands access, transparency, and personalization. Early digital systems improved reliability and consistency, while later platforms enabled global reach, real-time interaction, and seamless experiences. Customers benefit when digital capability reduces friction, increases choice, and enhances trust. Yet digital evolution also requires human oversight to ensure fairness, empathy, and ethical alignment. Mature organizations design digital experiences that respect human needs while leveraging machine efficiency. Customer value becomes a function of both technological sophistication and human-centered design. The result is a relationship where digital systems amplify—not replace—the human elements of service and trust.

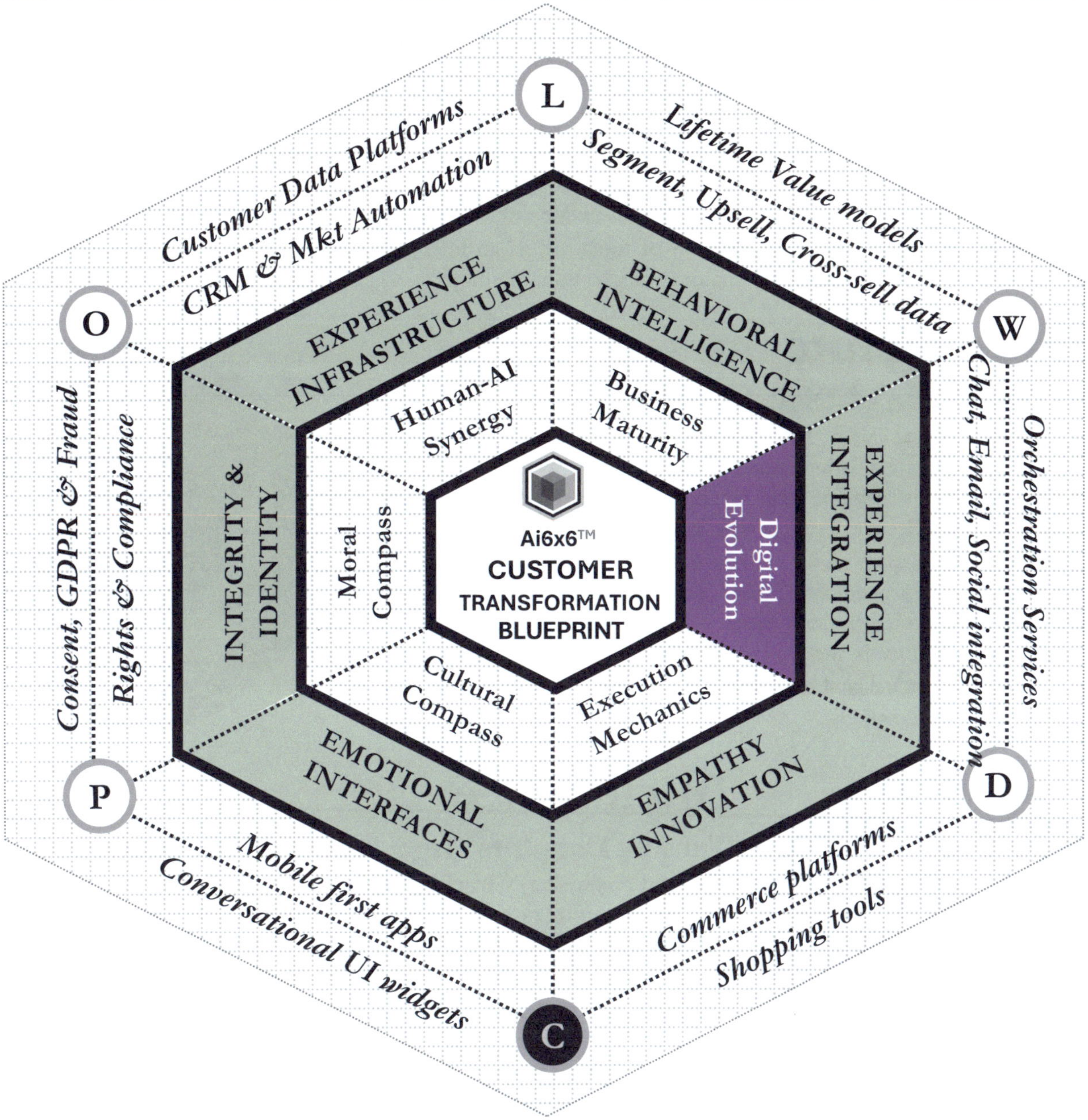

The **Ai6x6 Research Lab** chose this lens to solve the "empathy gap" that often leaves organizations delivering technically perfect but emotionally hollow experiences. We believe a brand must possess **Digital Gravitas**—a quiet, unshakeable confidence in the systems that connect them to their customers. This isn't about the sophistication of the tech stack; it is about the mental shift toward **Designing Emotion** and creating "Respectful Relationships" at scale. It is vital because a leader cannot "visioneer" new market realities if they are constantly worried about the friction in their **Experience Infrastructure**.

By mastering this evolution, the leader moves from a state of technical anxiety to a state of absolute strategic command over the customer's journey. This lens ensures that the **Experience API** is built on a foundation of trust, where "Knowing Before They Ask" becomes a natural extension of the brand's intent. When a leader understands the integrity of their digital interface, they gain the freedom to lead with an undistracted focus on **Innovation**. This lens provides the psychological peace of mind that comes from knowing the backbone of the relationship is secure and designed for human synergy, making data truly human.

Digital Evolution is no longer a technical initiative. It is the structural rewiring of how an organization thinks, operates, and competes. For years, leaders treated digital transformation as a project—something with a start date, a roadmap, and a finish line. But the world has moved beyond that framing. Digital Evolution is not a destination; it is a permanent state of adaptation. It is the recognition that technology is no longer an enabler of strategy but a **shaper** of it. The organizations that thrive are those that treat digital capability as a living system—continuously upgraded, continuously integrated, and continuously aligned with the realities of the market. This lens reframes Digital Evolution as a leadership discipline rather than a technology program. It challenges leaders to see digital capability not as a collection of tools but as an interconnected ecosystem that influences every decision, every workflow, and every customer interaction.

Infrastructure

The Foundation of Digital Stability

The journey begins with **Infrastructure**, the foundation upon which everything else depends. Infrastructure is not just hardware or cloud capacity; it is the stability, scalability, and reliability of the systems that carry the organization's intelligence. Leaders who underestimate infrastructure discover that innovation collapses under its own weight. Without a strong foundation, every new capability becomes fragile, every integration becomes brittle, and every attempt at modernization becomes a patch rather than a transformation. Infrastructure maturity is the discipline of building for endurance—ensuring that the organization's digital backbone can support the speed, complexity, and volatility of modern markets.

Intelligence

Turning Data Into Organizational Foresight

The next dimension is **Intelligence**, representing the organization's ability to convert data into insight and insight into action. Intelligence is not about dashboards or reports; it is about building a system that learns. It requires data that flows freely, models that are trained responsibly, and decisions that are informed by evidence rather than assumption. Leaders who embrace intelligence understand that it is not a technical asset but a strategic one. Intelligence shapes priorities, exposes risks, and reveals opportunities that would otherwise remain invisible. It is the cognitive engine of Digital Evolution—the mechanism through which the organization perceives, interprets, and responds to its environment.

Integration

Making the Enterprise Whole

Integration is where Digital Evolution becomes real. Organizations often accumulate technologies faster than they can connect them. Systems become islands, data becomes siloed, and teams become fragmented. Integration is the discipline that ensures the enterprise operates as a coherent whole. It is the work of connecting platforms, harmonizing data, and designing workflows that move across boundaries. Without integration, digital investments become isolated experiments rather than enterprise capabilities. Integration maturity is the art of coherence—ensuring that the organization's digital landscape functions as a single, unified system rather than a patchwork of disconnected tools.

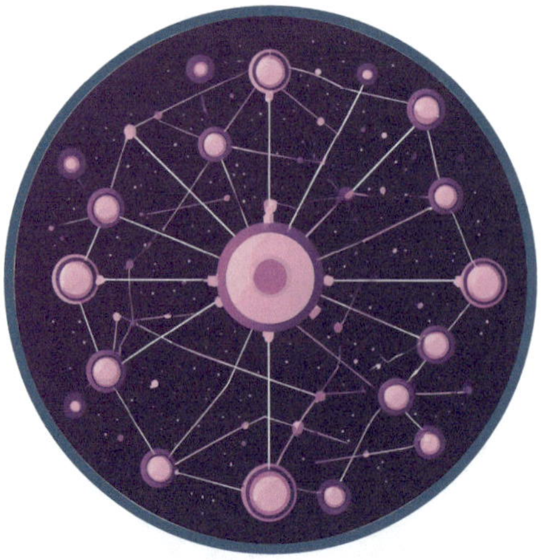

Innovation

Expanding the Field of Possibility

The fourth dimension, **Innovation**, expands the field of possibility. Innovation is not about chasing trends or adopting technology for its own sake. It is about using digital capability to reimagine how value is created, delivered, and experienced. Innovation demands that leaders challenge assumptions, explore alternatives, and test ideas quickly. It requires a culture where experimentation is encouraged, failure is reframed as learning, and progress is measured by insight rather than perfection. Innovation maturity is the discipline of curiosity—using digital tools not just to optimize the present but to invent the future.

Interface

Where Technology Meets Human Behavior

Interface represents the human experience of Digital Evolution. It is the point where technology meets behavior, where systems meet expectations, and where design meets trust. Leaders who ignore interface discover that even the most powerful systems fail if people cannot or will not use them. Interface is not cosmetic; it is strategic. It shapes adoption, accelerates learning, and determines whether digital capability becomes embedded or abandoned. Interface maturity is the discipline of empathy—designing digital experiences that are intuitive, accessible, and aligned with human needs.

Integrity

The Guardrail of Responsible Evolution

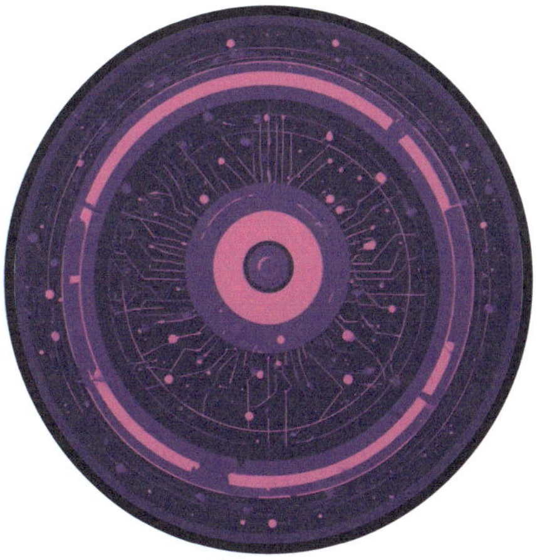

The final dimension, **Integrity**, grounds Digital Evolution in responsibility. Integrity ensures that systems are secure, data is protected, and decisions are ethical. It prevents speed from becoming recklessness and innovation from becoming harm. Leaders who prioritize integrity build organizations that are trusted, resilient, and prepared for the scrutiny that accompanies digital capability. Integrity maturity is the discipline of stewardship—ensuring that digital evolution strengthens the organization's moral and operational foundations rather than compromising them.

Digital Evolution is not a technology story. It is a leadership story. It is the recognition that the future belongs to organizations that treat digital capability as a core competency, not a side project. It is the shift from building tools to building systems, from adopting technology to evolving with it, and from reacting to change to shaping it. In this future, Digital Evolution is not a department, a project, or a budget line. It is the operating philosophy of the enterprise. It shapes how decisions are made, how systems are built, how teams collaborate, and how value is created. It turns technology from a source of disruption into a source of stability. It transforms the organization from a follower of change into a shaper of it.

Insights, LLMs & Analytics

The cognitive core drives logical processing and strategic decision-making through high-fidelity data analysis. The center of cognitive analysis and strategic decision-making powered by LLMs and advanced analytics.

UX, UI & Market Touchpoints

The external touchpoints serve as the skin and eyes that perceive and interact with the market environment. The skin and eyes of the enterprise that perceive, touch, and feel the external market and customer needs.

Cybersecurity & Ethics

The proactive defense system acts as a silent protector to identify and neutralize internal and external threats. The proactive shield of Cybersecurity and Ethics that identifies and neutralizes internal and external threats.

INTERFACE

The Five Senses

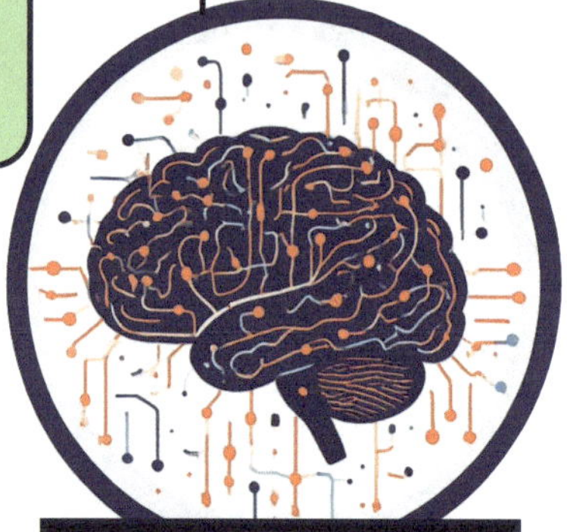

INTELLIGENCE

The Brain

INTEGRITY

The Immune System

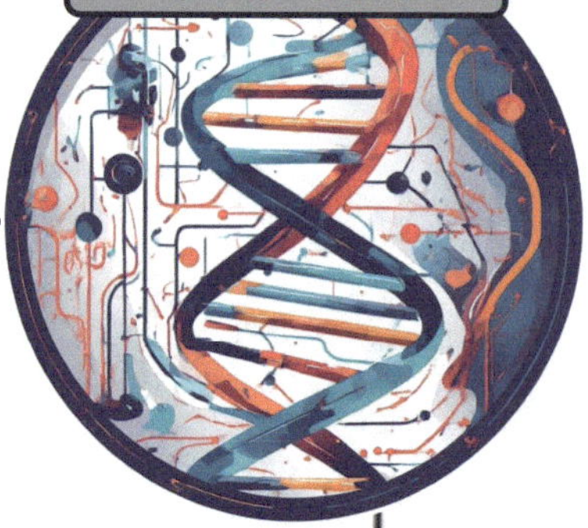

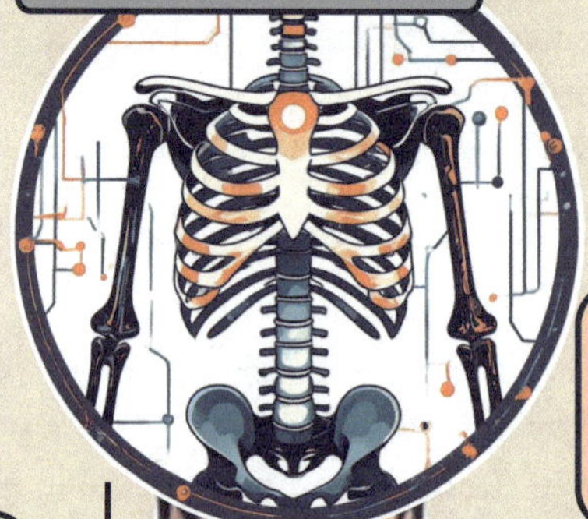

INFRASTRUCTURE

The Skeleton

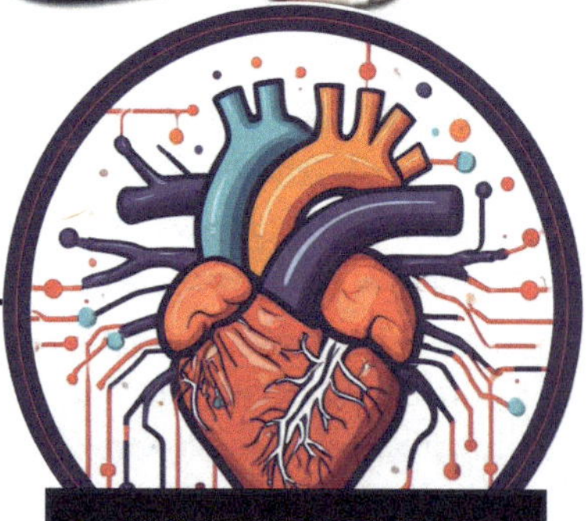

INTEGRATION

The Circulation

INNOVATION

The DNA

SOA, APIs & Middleware

The data pathways ensure a continuous and oxygenated flow of information across every organizational silo. Ensuring the continuous, oxygenated flow of information between silos via APIs and robust middleware.

Code, R&D & Hackathons

The adaptive code stores the permanent instructions for continuous growth and future evolutionary versions. The permanent instructions for growth and evolution, ensuring the organism survives through future versions

Cloud, Hardware, Network

The foundational frame provides the structural resilience to support global organizational scale. Providing the structural cloud and hardware frame that supports all weight and organizational scale.

Ai6x6™ Transformation Blueprint

Experience 'Infrastructure'

"If you don't take care of your customers, someone else will." — Elon Musk

CEO of Tesla and SpaceX; known for customer-centric product ecosystems despite deep engineering complexity.

Customer experience begins before interaction—it begins in platform structure. CRM, CDP, and automation engines must pre-load empathy into delivery mechanisms. "Experience infrastructure" makes experience a default, not a feature. Mature transformation embeds predictive memory into every moment. Loyalty isn't earned at checkout—it's configured in systems that remember preference, intent, and pace. Infrastructure is more than throughput—it's feeling delivered at scale. When platforms whisper care before customers speak, CX becomes a system, not a department.

For customer transformation, the "Infrastructure: Experience Infrastructure" chapter is foundational for digital maturity because it provides the essential technological platforms that support seamless and consistent customer interactions across all touchpoints. A digitally mature organization invests in robust CRM platforms, customer data platforms (CDPs), and marketing automation systems to unify customer data and orchestrate highly personalized experiences. This comprehensive infrastructure is vital for gaining a 360-degree view of the customer, streamlining service delivery through contact centers, and building lasting, loyal relationships in a competitive digital marketplace, ultimately driving customer satisfaction and revenue.

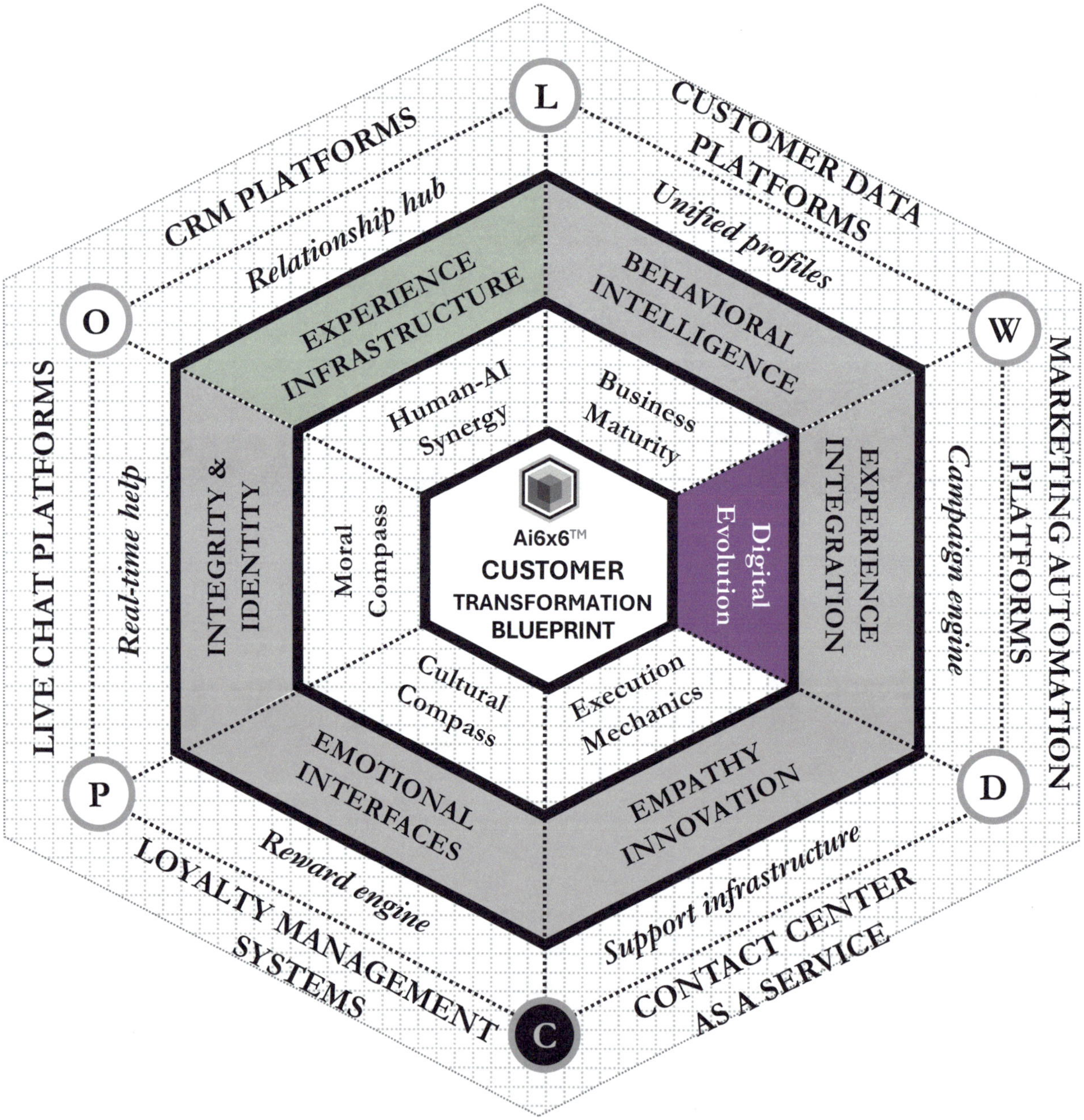

Ai6x6 DIGITAL EVOLUTION FRAMEWORK
FOR CUSTOMER TRANSFORMATION | EXPERIENCE INFRASTRUCTURE

CRM (Customer Relationship Management) Platforms are integrated software systems designed to manage and analyze customer interactions and data throughout the customer lifecycle, with the goal of improving business relationships with customers, assisting in customer retention, and driving sales growth. These platforms centralize customer information, track communications (emails, calls, chats), manage sales pipelines, automate marketing campaigns, and provide customer service support. By offering a unified view of every customer interaction, CRM Platforms enable organizations to understand customer needs better, personalize experiences, streamline sales and service processes, and ultimately foster stronger, more profitable customer relationships, serving as the foundational hub for customer-centric operations.

Without robust CRM Platforms, organizations often operate with fragmented customer data scattered across disparate systems, leading to an incomplete view of customer interactions and preferences. This results in inconsistent customer experiences, missed sales opportunities, inefficient customer service, and difficulty in tracking the effectiveness of marketing efforts. The absence of a centralized CRM hinders the ability to personalize customer journeys, build lasting relationships, and scale customer-facing operations, ultimately impacting sales performance, customer satisfaction, and an organization's overall competitiveness in a customer-driven market.

Implementation	Challenges
* Define clear customer journey & goals * Select appropriate CRM platform * Migrate & integrate existing customer data * Customize workflows & user interfaces * Plan user training & adoption strategy * Establish KPIs & reporting for CRM success	* Data migration and quality issues * User adoption and resistance to change * Integration with diverse existing systems * Customization complexity and cost * Ensuring data security and privacy * Measuring clear ROI on CRM investment

Products	Platform Cost	Project Cost	Duration
Salesforce Sales Cloud	$20-100K	$100-500K	4-12 m
HubSpot CRM	$5-30K	$20-150K	2-7 m
Microsoft Dynamics 365 Sales	$15-70K	$80-400K	3-11 m
Zoho CRM	$1-10K	$5-50K	1-3 m
Oracle Siebel CRM (On-Prem)	$30-150K	$200-800K	6-18 m
SAP Customer Experience (CX)	$20-90K	$100-500K	4-12 m

Customer Data Platforms (CDPs) are packaged software that create a persistent, unified customer database accessible to other systems. They ingest data from various sources (CRM, ERP, web, mobile, IoT, marketing automation), cleanse and unify it into comprehensive customer profiles, and then make this data available to support marketing, sales, and service activities. Unlike CRMs or DMPs, CDPs build a holistic "single customer view" by focusing on first-party data for known and anonymous customers. By providing a centralized, actionable source of truth for customer interactions, CDPs enable highly personalized customer experiences, improved targeting for marketing campaigns, and more accurate analytics, crucial for modern omnichannel customer engagement strategies.

Without a Customer Data Platform (CDP), organizations often struggle with fragmented customer data residing in silos across various systems, making it nearly impossible to create a unified, real-time view of individual customers. This leads to inconsistent messaging, ineffective personalization, inaccurate customer analytics, and a reactive approach to customer engagement. The absence of a centralized, actionable customer profile results in missed opportunities for targeted marketing, suboptimal customer experiences, and difficulty in understanding the full customer journey, ultimately hindering an organization's ability to drive customer loyalty, optimize marketing spend, and compete effectively in a personalized economy.

Implementation	Challenges
* Define customer data sources & requirements * Select appropriate CDP platform * Ingest, unify & cleanse disparate customer data * Establish customer identity resolution * Integrate with CRM, marketing & analytics tools * Plan for data governance & privacy compliance	* Complex data integration from diverse sources * Ensuring accurate customer identity resolution * Managing data quality & consistency at scale * Compliance with data privacy regulations (GDPR, CCPA) * Proving ROI on unified customer view * Overcoming organizational silos for data sharing

Products	Platform Cost	Project Cost	Duration
Segment (Twilio)	$10-60K	$50-300K	3-9 m
Tealium AudienceStream CDP	$15-70K	$80-350K	3-10 m
ActionIQ	$20-90K	$100-450K	4-12 m
RudderStack	$5-30K	$20-100K	2-6 m
Salesforce Customer 360 (partially CDP)	$20-100K	$120-600K	4-14 m
Adobe Experience Platform (CDP)	$25-120K	$150-700K	5-16 m

Marketing Automation Platforms are software systems designed to automate and streamline repetitive marketing tasks such as email campaigns, lead nurturing, social media posting, and ad management. They enable marketers to segment audiences, personalize communications, and automate workflows based on customer behavior and triggers across various channels. By improving efficiency, increasing lead generation, and nurturing prospects through the sales funnel, Marketing Automation Platforms free up marketing teams to focus on strategy and creativity. This technology allows organizations to deliver consistent, targeted, and timely messages at scale, optimizing customer engagement, accelerating sales cycles, and ultimately driving higher marketing ROI.

Without Marketing Automation Platforms, organizations typically rely on manual, labor-intensive processes for marketing tasks, leading to inconsistent messaging, inefficient lead nurturing, and a limited ability to personalize communications at scale. This results in missed opportunities to engage prospects effectively, a fragmented customer journey, and higher operational costs for marketing activities. The absence of automation hinders the ability to segment audiences precisely, trigger timely interactions, and track campaign performance accurately, ultimately impacting lead conversion rates, customer engagement, and an organization's capacity to execute sophisticated, data-driven marketing strategies effectively.

Implementation	Challenges
* Define marketing goals & customer segments * Select appropriate marketing automation platform * Migrate & integrate customer/lead data * Design automated campaigns & workflows * Plan content strategy & personalization rules * Train marketing team & establish analytics	* Data integration with CRM & other systems * Content creation & personalization at scale * Complex lead scoring and nurturing logic * Ensuring deliverability & avoiding spam filters * Measuring accurate ROI on campaigns * User adoption and overcoming manual habits

Products	Platform Cost	Project Cost	Duration
HubSpot Marketing Hub	$5-30K	$20-100K	2-6 m
Pardot (Salesforce)	$10-50K	$60-250K	3-9 m
Marketo Engage (Adobe)	$15-70K	$80-350K	3-10 m
Oracle Eloqua	$15-60K	$70-300K	3-9 m
ActiveCampaign	$1-10K	$5-30K	1-3 m
Mailchimp (Advanced features)	$1-5K	$2-20K	1-2 m

4. CONTACT CENTER AS A SERVICE (CCAAS)

Contact Center as a Service (CCaaS) is a cloud-based solution that enables organizations to manage customer interactions across various channels—voice, email, chat, social media—through a single, integrated platform delivered as a service. CCaaS eliminates the need for on-premise hardware and software, offering scalability, flexibility, and advanced features like intelligent routing, IVR, call recording, workforce management, and analytics. By centralizing omnichannel customer communication, CCaaS improves customer service efficiency, enhances the customer experience through personalized and consistent interactions, and provides actionable insights into contact center performance, allowing organizations to deliver superior support and build stronger customer relationships.

Without Contact Center as a Service (CCaaS) solutions, organizations often rely on outdated, on-premise contact center infrastructure or disparate systems for different communication channels, leading to fragmented customer interactions, high operational costs, and limited scalability. This results in inconsistent customer experiences, longer resolution times, and a lack of unified reporting across communication channels. The absence of a centralized, cloud-based CCaaS hinders the ability to intelligently route customer inquiries, leverage advanced analytics for performance improvement, and adapt quickly to changing customer demands, ultimately impacting customer satisfaction, agent productivity, and an organization's overall service quality.

Implementation	Challenges
* Define customer service channels & needs * Select CCaaS provider & service level * Integrate with CRM & other back-end systems * Configure IVR, routing & agent workflows * Train agents & supervisors on new platform * Establish analytics & reporting for service KPIs	* Complex integration with existing CRM/ERP * Ensuring call/channel quality & reliability * Agent training and adoption of new tools * Managing customer data security & privacy * Performance scaling during peak times * Measuring clear ROI on improved service quality

Products	Platform Cost	Project Cost	Duration
Genesys Cloud CX	$10-60K	$50-300K	3-9 m
Amazon Connect	$5-30K	$20-100K	2-6 m
Five9	$10-50K	$60-250K	3-9 m
Twilio Flex	$5-25K	$20-90K	2-5 m
Talkdesk	$10-40K	$50-200K	3-8 m
RingCentral Contact Center	$5-20K	$20-80K	2-5 m

Loyalty Management Systems are software solutions designed to create, manage, and optimize customer loyalty programs that incentivize repeat purchases, foster customer advocacy, and enhance customer lifetime value. These systems track customer interactions, points accumulation, reward redemption, and tiered statuses across various touchpoints. By enabling personalized rewards, targeted offers, and exclusive experiences, Loyalty Management Systems drive deeper customer engagement, encourage brand affinity, and provide valuable insights into customer behavior and preferences. This allows organizations to build stronger, more profitable relationships with their most valuable customers, reducing churn and maximizing recurring revenue through strategic customer retention efforts.

Without effective Loyalty Management Systems, organizations struggle to identify, reward, and retain their most valuable customers, leading to generic marketing efforts and a reactive approach to customer churn. This results in missed opportunities to incentivize repeat business, lower customer lifetime value, and a lack of data-driven insights into customer behavior and preferences for retention strategies. The absence of a structured loyalty program hinders the ability to build strong emotional connections with customers, foster brand advocacy, and effectively compete for customer retention, ultimately impacting long-term revenue stability and market share in competitive industries.

Implementation	Challenges
* Define loyalty program goals & structure * Select a Loyalty Management System * Integrate with POS, e-commerce & CRM * Design reward tiers, points & redemption rules * Plan personalized communication & offers * Establish KPIs & analytics for program success	* Complex integration with existing systems * Ensuring data accuracy for points & rewards * Personalizing offers at scale * Member engagement & program awareness * Preventing fraud & reward abuse * Measuring clear ROI on loyalty program

Products	Platform Cost	Project Cost	Duration
LoyaltyLion	$5-30K	$20-100K	2-6 m
Punchh (ParTech)	$10-50K	$60-250K	3-9 m
Annex Cloud Loyalty	$15-70K	$80-350K	3-10 m
Antavo Loyalty Cloud	$10-40K	$50-200K	3-8 m
Salesforce Loyalty Management	$15-60K	$70-300K	3-9 m
Oracle CrowdTwist Loyalty and Engagement	$20-80K	$100-400K	4-12 m

Live Chat Platforms enable real-time, text-based communication between customers and customer service agents or automated chatbots directly on websites, mobile apps, or other digital channels. These platforms provide immediate support, answer queries, guide users, and resolve issues, often leveraging features like canned responses, chat routing, and visitor monitoring. By offering instant assistance and a convenient communication channel, Live Chat Platforms improve customer satisfaction, reduce support costs by deflecting calls, and increase conversion rates by providing in-the-moment help during online journeys. This critical tool enhances the digital customer experience, making interactions more efficient, personalized, and accessible, driving both customer loyalty and business growth.

Without Live Chat Platforms, organizations often force customers to rely on slower, less convenient communication channels like phone calls or emails for support, leading to frustrating customer experiences and delayed resolutions. This results in higher abandonment rates on websites, increased call center volumes, and a reactive approach to customer queries, missing opportunities for immediate engagement or conversion assistance. The absence of real-time chat capabilities undermines customer satisfaction, increases operational costs for support, and limits the ability to provide proactive assistance during the online journey, ultimately impacting customer loyalty and digital sales performance.

Implementation	Challenges
* Define chat support scope & hours * Select a Live Chat platform * Integrate into website/app & CRM * Train agents on chat etiquette & tools * Develop canned responses & routing rules * Establish KPIs & monitoring for chat performance	* Ensuring adequate agent staffing for real-time response * Managing high chat volumes during peak times * Seamless integration with CRM for customer context * Maintaining chat quality & customer satisfaction * Implementing effective chatbot escalation * Measuring clear ROI on chat support efficiency

Products	Platform Cost	Project Cost	Duration
Zendesk Chat	$1-10K	$5-30K	1-3 m
LiveChat	$1-5K	$2-20K	1-2 m
Intercom	$5-30K	$20-100K	2-6 m
Freshchat (Freshworks)	$1-8K	$5-25K	1-3 m
Salesforce Service Cloud (Live Agent)	$10-50K	$60-250K	3-9 m
HubSpot Service Hub (Live Chat)	$1-10K	$5-30K	1-3 m

Ai6x6™ Transformation Blueprint

Behavioral 'Intelligence'

"You've got to start with the customer experience and work backward to the technology." — Steve Jobs, Founder of Apple, cited here for defining experience-led design as the soul of product development.

Knowing before asking means leadership anticipates not because of psychic instinct—but analytic proximity. Churn models, lifetime value scoring, and segmentation analytics aren't just predictors—they're empathic tools for timing, tone, and relevance. Transformation maturity lies in asking fewer questions—because behavior reveals answers already. When insight becomes ambient, personalization becomes quiet power. Customers don't notice intelligence—but they feel seen. And being seen, well, that's loyalty disguised as logic.

The "Behavioral Intelligence" chapter is crucial for customer transformation's digital maturity as it focuses on leveraging data to proactively understand and anticipate customer needs and behaviors. A digitally mature organization employs advanced predictive analytics, including churn prediction models, customer lifetime value models, and propensity scoring, to gain deep insights into customer segments. This intelligence, powered by recommendation engines and cross-sell/upsell analytics, empowers businesses to deliver personalized offers, prevent issues before they arise, and build strong, loyal customer relationships by consistently exceeding expectations and driving future growth.

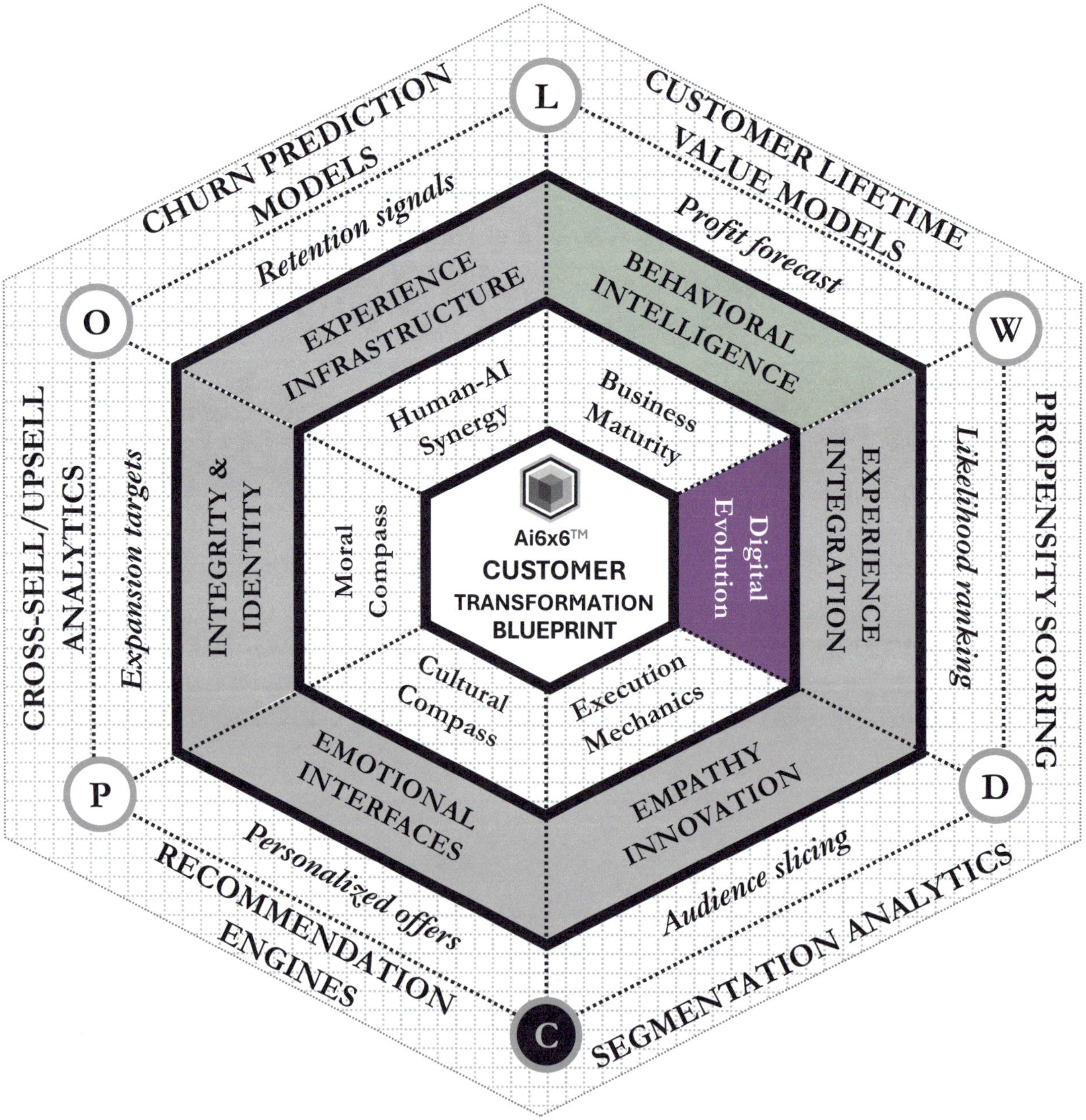

Ai6x6 DIGITAL EVOLUTION FRAMEWORK
FOR CUSTOMER TRANSFORMATION | BEHAVIORAL INTELLIGENCE

Churn Prediction Models are analytical tools that use machine learning algorithms to identify customers who are at high risk of discontinuing their service, subscription, or relationship with a business. These models analyze historical customer data, including demographics, usage patterns, interaction history, and past behaviors, to assign a "churn probability" score to each customer. By proactively identifying at-risk customers, Churn Prediction Models enable organizations to implement targeted retention strategies such as personalized offers, proactive outreach, or service interventions, before customers actually leave. This minimizes customer attrition, maximizes customer lifetime value, and significantly improves customer retention rates, crucial for sustainable business growth.

Without Churn Prediction Models, organizations operate reactively, often discovering customer churn only after it has occurred, leading to significant revenue loss and missed opportunities for retention. They lack the foresight to identify at-risk customers proactively, resulting in generic or untargeted retention efforts that are inefficient and often too late. The absence of predictive churn analytics means businesses struggle to understand the underlying causes of customer defection, cannot personalize retention strategies effectively, and ultimately face higher customer acquisition costs to replace lost customers, severely impacting long-term profitability and market share.

Implementation	Challenges
* Define "churn" for your business model	* Data quality & availability for historical analysis
* Collect relevant historical customer data	* Defining true churn vs. temporary inactivity
* Select & train appropriate ML algorithms	* Selecting the right predictive features
* Integrate models into CRM/marketing platforms	* Interpreting model results & explainability
* Design proactive retention strategies	* Operationalizing predictions into action
* Continuously monitor model accuracy & refresh	* Measuring clear ROI on retention efforts

Products	Platform Cost	Project Cost	Duration
Dataiku (ML for Churn)	$15-70K	$80-350K	3-10 m
Salesforce Einstein Prediction Builder	$10-50K	$60-250K	3-9 m
Google Cloud Vertex AI (Custom ML)	$5-30K	$20-100K	2-6 m
Azure Machine Learning (Custom ML)	$5-25K	$20-90K	2-5 m
H2O.ai Driverless AI	$15-70K	$80-350K	3-10 m
Custom ML Development (Python/R)	N/A	$50-250K	3-9 m

Customer Lifetime Value (CLV) Models are analytical frameworks that predict the total revenue a business can reasonably expect from a customer throughout their entire relationship. These models leverage historical purchase data, customer behavior, engagement metrics, and predictive analytics to forecast future spending, profitability, and duration of customer relationships. By providing a forward-looking financial valuation for each customer, CLV Models enable organizations to segment customers by their long-term potential, optimize marketing spend, prioritize retention efforts for high-value customers, and make more informed decisions about customer acquisition strategies. This maximizes long-term profitability, optimizes resource allocation, and fosters sustainable growth by focusing on the most valuable customer segments.

Without Customer Lifetime Value (CLV) Models, organizations often make marketing, sales, and retention decisions based on short-term metrics like immediate purchase value, leading to inefficient resource allocation and a failure to identify or prioritize their most valuable customers. This results in overspending on low-value customers, underinvesting in high-potential segments, and a lack of strategic focus on long-term customer relationships. The absence of a predictive CLV framework hinders an organization's ability to optimize customer acquisition costs, personalize marketing efforts for maximum return, and drive sustainable, profitable growth by truly understanding the long-term economic contribution of its customer base.

Implementation	Challenges
* Collect comprehensive customer purchase data * Select & train appropriate CLV modeling techniques * Validate model accuracy against historical data * Integrate CLV scores into CRM/marketing systems * Design segment-specific marketing & retention * Continuously monitor CLV trends & model performance	* Data quality & completeness for historical purchases * Selecting appropriate modeling approach (e.g., probabilistic) * Handling sparse data or new customer segments * Integrating CLV insights into operational workflows * Measuring the true impact on profitability * Explaining complex model outputs to business users

Products	Platform Cost	Project Cost	Duration
Segment / RudderStack (CDP for data)	$10-60K	$50-300K	3-9 m
Dataiku (ML for CLV)	$15-70K	$80-350K	3-10 m
Google Analytics (Enhanced E-commerce)	N/A	$10-50K	1-3 m
Custom ML Development (Python/R, PyMC Labs)	N/A	$50-250K	3-9 m
Salesforce Einstein (Predictive Scoring)	$10-50K	$60-250K	3-9 m
Adobe Experience Platform (Analytics)	$25-120K	$150-700K	5-16 m

Propensity Scoring involves using statistical and machine learning models to predict the likelihood of a customer or prospect taking a specific action, such as making a purchase, responding to an offer, clicking an ad, or churning. These models analyze various customer attributes, historical behaviors, and contextual data to generate a "propensity score" for each individual. By identifying high-propensity segments, organizations can optimize their marketing, sales, and service efforts, targeting the right customers with the most relevant messages at the opportune moment. This increases the effectiveness of campaigns, improves conversion rates, and maximizes the return on investment for customer engagement initiatives, driving more efficient and impactful customer interactions.

Without Propensity Scoring, organizations often rely on broad segmentation or intuition to target customers, leading to inefficient marketing spend, lower conversion rates, and a reactive approach to customer engagement. This means that offers might be sent to uninterested customers, or critical leads could be overlooked, resulting in wasted resources and missed sales opportunities. The absence of predictive propensity insights hinders the ability to personalize outreach effectively, optimize campaign performance, and proactively guide customer journeys, ultimately impacting sales efficiency, marketing ROI, and an organization's overall ability to drive specific desired customer actions.

Implementation	Challenges
* Define target customer action to predict * Collect relevant customer behavior data * Select & train appropriate ML models * Integrate scores into marketing automation/CRM * Design targeted campaigns based on scores * Continuously monitor model accuracy & refine	* Data quality & availability of behavioral data * Defining the 'event' clearly for prediction * Ensuring model explainability for actionability * Integrating scores into real-time operational systems * Avoiding bias in predictive modeling * Measuring uplift from targeted actions vs. control

Products	Platform Cost	Project Cost	Duration
Segment / Tealium (CDP for data)	$10-60K	$50-300K	3-9 m
Dataiku	$15-70K	$80-350K	3-10 m
Salesforce Einstein (Predictive Scoring)	$10-50K	$60-250K	3-9 m
Adobe Experience Platform (AI/ML)	$25-120K	$150-700K	5-16 m
Braze (Customer Engagement Platform)	$10-50K	$60-250K	3-9 m
Custom ML Development (Python/R)	N/A	$50-250K	3-9 m

Segmentation Analytics involves dividing a broad customer base into distinct groups (segments) based on shared characteristics, behaviors, needs, or demographics, using statistical analysis and machine learning techniques. Unlike simple demographic segmentation, advanced segmentation analytics uncovers hidden patterns and creates more granular, actionable segments that reflect true customer differences. By providing a deeper understanding of diverse customer groups, Segmentation Analytics enables organizations to tailor marketing messages, personalize product offerings, optimize service strategies, and allocate resources more effectively to specific segments. This leads to more relevant customer experiences, increased marketing ROI, and improved customer satisfaction by addressing the unique needs and preferences of each group.

Without sophisticated Segmentation Analytics, organizations often treat their entire customer base as a homogenous group or rely on overly simplistic demographic divisions, leading to generic marketing campaigns, undifferentiated product offerings, and inefficient resource allocation. This results in irrelevant customer experiences, lower engagement rates, and a failure to identify and address the unique needs of diverse customer groups. The absence of granular, behavior-driven segmentation hinders the ability to personalize outreach, optimize marketing spend, and maximize customer lifetime value, ultimately impacting an organization's competitiveness and its capacity to build strong, tailored relationships with its varied customer base.

Implementation	Challenges
* Define business objectives for segmentation * Collect comprehensive customer data * Apply statistical/ML clustering techniques * Interpret and profile distinct customer segments * Integrate segments into marketing/CRM systems * Design segment-specific strategies & messaging	* Data quality & completeness across all sources * Choosing optimal segmentation variables & methods * Interpreting and naming segments effectively * Operationalizing segments into actionable campaigns * Maintaining segment relevance over time * Overcoming organizational silos for segment adoption

Products	Platform Cost	Project Cost	Duration
Customer Data Platforms (CDPs)	$10-70K	$50-350K	3-10 m
Marketing Automation Platforms	$5-60K	$20-300K	2-9 m
BI Tools (e.g., Tableau, Power BI)	$5-50K	$20-250K	2-9 m
Dataiku (Visual ML for Segmentation)	$15-70K	$80-350K	3-10 m
Custom Analytics (Python/R, Clustering)	N/A	$30-150K	2-7 m
Adobe Experience Platform (Audience Segmentation)	$25-120K	$150-700K	5-16 m

Recommendation Engines are intelligent systems that leverage machine learning algorithms to suggest relevant products, content, services, or information to users, based on their past behavior, preferences, similarities to other users, or item characteristics. These engines analyze vast amounts of data to predict user interest and provide personalized recommendations in real-time across various touchpoints, such as e-commerce websites, streaming platforms, or content apps. By enhancing user discovery, increasing engagement, and driving conversions, Recommendation Engines improve customer experience, boost sales, and maximize cross-sell/upsell opportunities, playing a critical role in modern digital commerce and content consumption strategies.

Without Recommendation Engines, organizations typically present generic or manually curated lists of products or content to users, leading to lower engagement, reduced discovery of relevant items, and missed opportunities for personalization. This means customers may struggle to find what they truly want, abandoning websites or content platforms prematurely, resulting in lower conversion rates and diminished satisfaction. The absence of intelligent, personalized recommendations hinders the ability to maximize sales, increase customer lifetime value, and cultivate deeper engagement, ultimately impacting revenue growth and competitiveness in digital-first industries.

Implementation	Challenges
* Define recommendation goals (e.g., sales, engagement) * Collect rich user behavior & item data * Select & train appropriate ML algorithms * Implement real-time serving & feedback loops * Integrate into website/app & marketing platforms * Plan for A/B testing & continuous model improvement	* Data quality & volume for effective models * Addressing cold start problem for new users/items * Ensuring real-time performance at scale * Avoiding algorithmic bias & filter bubbles * Integrating with diverse front-end systems * Measuring clear uplift in sales/engagement

Products	Platform Cost	Project Cost	Duration
Amazon Personalize	$5-30K	$20-100K	2-6 m
Google Cloud Recommendations AI	$5-25K	$20-90K	2-5 m
Microsoft Azure Personalizer	$5-20K	$20-80K	2-5 m
Recombee	$1-10K	$5-30K	1-3 m
Adobe Experience Platform (Sensei)	$25-120K	$150-700K	5-16 m
Custom ML Development (TensorFlow, PyTorch)	N/A	$100-500K	4-12 m

Cross-Sell/Upsell Analytics leverages data analysis and machine learning to identify opportunities to sell additional products or services (cross-sell) or higher-value versions of current products (upsell) to existing customers. These analytical approaches examine customer purchase history, Browse behavior, demographics, and product affinities to predict which customers are most likely to be receptive to specific offers. By segmenting customers based on their propensity for cross-selling or upselling, organizations can create targeted campaigns, optimize product bundles, and empower sales teams with relevant recommendations. This maximizes customer lifetime value, increases average order value, strengthens customer relationships, and drives incremental revenue growth from the existing customer base, leading to highly efficient sales strategies.

Without dedicated Cross-Sell/Upsell Analytics, organizations often rely on generic promotions or anecdotal evidence to identify sales opportunities with existing customers, leading to inefficient campaigns and missed revenue potential. This means that relevant offers may not reach receptive customers, or valuable customers might not be presented with higher-value options, resulting in lower average transaction values and reduced customer lifetime value. The absence of data-driven insights for cross-selling and upselling hinders the ability to personalize offers effectively, optimize product recommendations, and maximize revenue from the existing customer base, ultimately impacting profitability and competitive growth in saturated markets.

Implementation	Challenges
* Define cross-sell/upsell goals & target products * Collect comprehensive customer & product data * Develop predictive models for product affinity * Integrate insights into sales/marketing workflows * Design targeted campaigns & sales enablement * Continuously measure uplift & refine strategies	* Data quality of purchase & product interaction history * Ensuring model accuracy for specific offers * Integrating insights into real-time sales/marketing tools * Avoiding customer fatigue from too many offers * Attribution of cross-sell/upsell revenue * Overcoming organizational silos between product/sales/marketing

Products	Platform Cost	Project Cost	Duration
Salesforce Sales Cloud (Einstein)	$10-50K	$60-250K	3-9 m
HubSpot Sales Hub	$5-30K	$20-100K	2-6 m
CDP Platforms (Segment, Tealium)	$10-70K	$50-350K	3-10 m
Marketing Automation Platforms (e.g., Marketo)	$15-70K	$80-350K	3-10 m
BI Tools (e.g., Tableau, Power BI)	$5-50K	$20-250K	2-9 m
Custom ML Development (Python/R)	N/A	$50-250K	3-9 m

Ai6x6™ Transformation Blueprint

Experience 'Integration'

"Omnichannel isn't a tactic. It's how customers live." — Angela Ahrendts
Former SVP of Retail at Apple; redefined physical-digital experience integration.

Channels are no longer separate—they are simultaneous. Webhooks, chatbot APIs, CRM-email connectors—they don't just link—they narrate. "Experience Integration" means conversation travels with the customer, unbroken by platform. Transformation maturity appears when interactions feel remembered and responses feel rehearsed. Integration becomes choreography—so smooth that customers don't feel transitions. They feel relationship. Connectors aren't technical—they're theatrical cues in the experience script.

"Experience Integration" is vital for customer transformation's digital maturity because it ensures seamless and real-time connectivity across all customer-facing systems and channels, creating a unified experience. A digitally mature organization utilizes omnichannel orchestration platforms, webhooks, and comprehensive API integrations to synchronize interactions from chatbots to social media and email. This provides a consistent and frictionless customer journey across every touchpoint. This deep integration eliminates fragmented experiences, enhances operational efficiency, and allows for agile, personalized responses to customer demands, ultimately strengthening customer relationships and driving satisfaction.

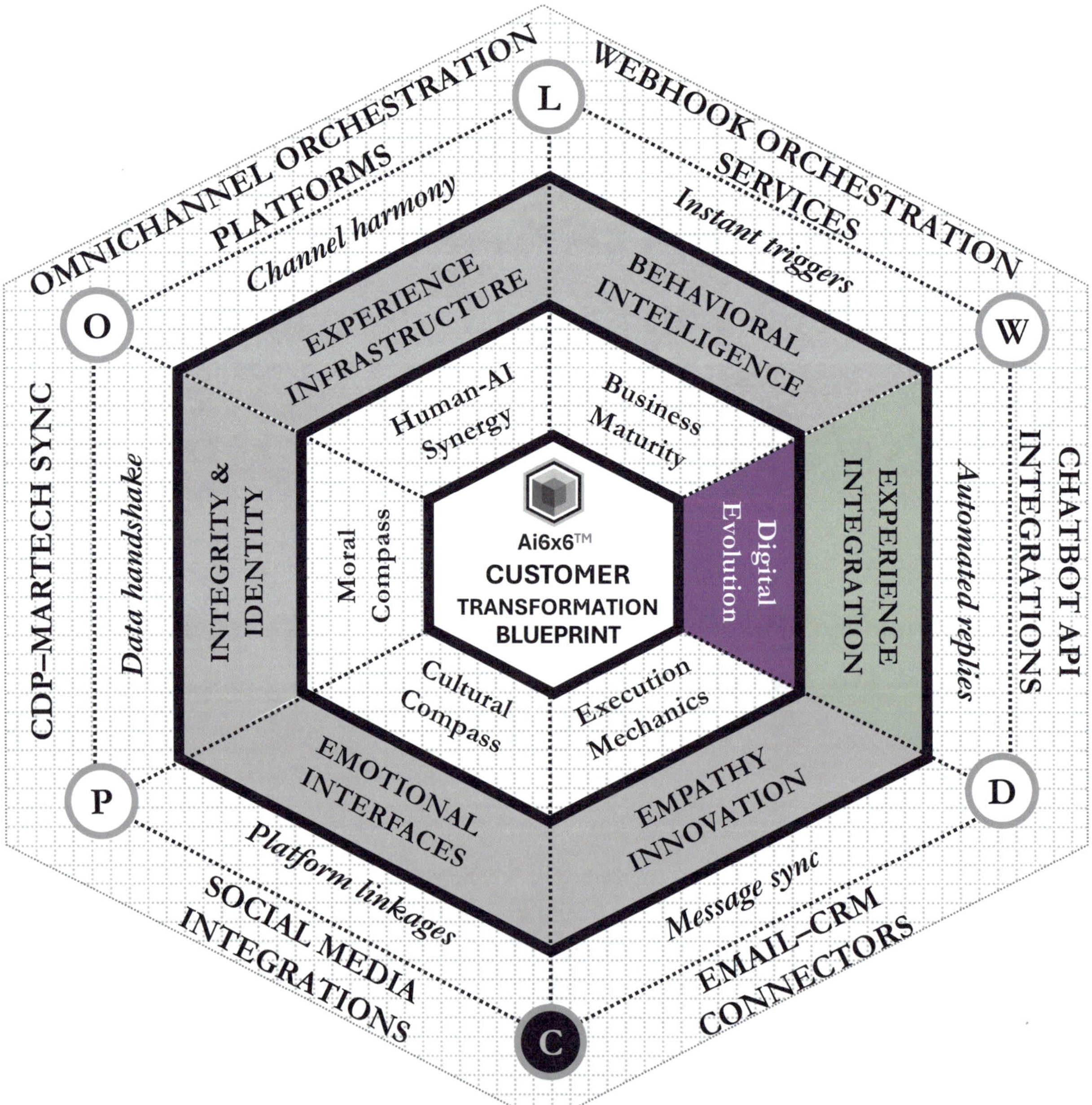

Ai6x6 DIGITAL EVOLUTION FRAMEWORK
FOR CUSTOMER TRANSFORMATION | EXPERIENCE INTEGRATION

1. OMNICHANNEL ORCHESTRATION PLATFORMS

Omnichannel Orchestration Platforms are centralized systems designed to coordinate and personalize customer interactions consistently across all touchpoints and channels, both online and offline. They go beyond simply connecting channels by intelligently sequencing messages, offers, and content based on real-time customer behavior and preferences, ensuring a seamless and contextual customer journey. By providing a unified view of the customer and automating cross-channel workflows, these platforms enable businesses to deliver highly personalized and synchronized experiences, whether a customer is interacting via website, mobile app, email, social media, or in-store. This significantly improves customer satisfaction, drives engagement, and maximizes conversion rates by delivering the right message at the right time through the preferred channel, fostering true customer-centricity.

Without Omnichannel Orchestration Platforms, organizations often manage customer interactions in silos, with each channel operating independently, leading to fragmented, inconsistent, and often frustrating customer experiences. This means a customer's interaction on the website might not be recognized in an email, or an in-store experience might not influence a subsequent mobile app interaction, resulting in disjointed journeys and repetitive requests for information. The absence of a unified orchestration layer hinders the ability to personalize communications effectively, provide seamless transitions across channels, and gain a holistic view of customer engagement, ultimately impacting customer loyalty, operational efficiency, and overall brand perception in a multi-touchpoint world.

Implementation	Challenges
* Map complete customer journey across channels * Select appropriate orchestration platform * Integrate all customer-facing systems (CRM, CDP) * Design cross-channel personalization rules * Implement real-time decisioning & delivery * Establish unified analytics & attribution	* Complex integration with diverse legacy systems * Ensuring real-time data synchronization * Defining and maintaining customer identity across channels * Orchestrating complex customer journeys with many rules * Organizational silos hindering cross-functional collaboration * Measuring clear ROI on seamless customer experience

Products	Platform Cost	Project Cost	Duration
Adobe Journey Optimizer	$20-100K	$120-600K	4-14 m
Salesforce Marketing Cloud (Journey Builder)	$15-80K	$100-500K	4-12 m
Braze	$10-60K	$50-300K	3-9 m
Iterable	$10-50K	$60-250K	3-9 m
Treasure Data (CDP with Orchestration)	$15-70K	$80-350K	3-10 m
Acoustic Campaign	$10-40K	$50-200K	3-8 m

Webhook Orchestration Services provide a centralized platform for managing, transforming, and routing webhooks—automated messages sent from one application to another when a specific event occurs. Instead of point-to-point integrations, these services enable complex workflows by acting as a hub that receives webhooks, processes their data, and then triggers subsequent actions in multiple downstream systems based on predefined rules. By simplifying real-time data flow between disparate applications, Webhook Orchestration Services improve system interoperability, enable agile automation of business processes, and reduce the development effort for event-driven architectures. This accelerates the creation of seamless customer experiences by ensuring timely data synchronization and automated responses across the customer ecosystem.

Without Webhook Orchestration Services, organizations often resort to brittle, point-to-point integrations for event-driven data exchange, making it difficult to manage, monitor, and scale real-time workflows between numerous applications. This leads to complex, unmanageable integration spaghetti, increased development overhead for each new connection, and a higher risk of data inconsistencies or failures when a single service changes. The absence of a centralized orchestration layer hinders the ability to react instantly to customer events across the ecosystem, automate complex customer journeys, and maintain a flexible, scalable integration architecture, ultimately impacting the agility and responsiveness of customer-facing operations.

Implementation	Challenges
* Identify event-driven data flows & applications * Select a webhook orchestration service * Define webhook structures & data transformations * Design conditional logic & routing rules * Implement error handling & monitoring * Integrate with source & destination applications	* Managing large volumes of real-time events * Ensuring data consistency across systems * Handling diverse webhook formats & payloads * Debugging complex event-driven workflows * Ensuring security & authentication for webhooks * Performance scaling under high event loads

Products	Platform Cost	Project Cost	Duration
Zapier	$0.1-5K	$0.5-10K	0.5-2 m
Make (formerly Integromat)	$0.1-5K	$0.5-10K	0.5-2 m
Workato	$5-30K	$20-100K	2-6 m
Tray.io	$5-25K	$20-90K	2-5 m
Pipedream	$0.1-2K	$0.5-5K	0.5-1.5 m
AWS EventBridge / Azure Logic Apps	$1-10K	$5-30K	1-3 m

Chatbot API Integrations involve connecting conversational AI chatbots with various backend systems and data sources through Application Programming Interfaces (APIs). This enables chatbots to do more than just answer predefined questions; they can retrieve real-time customer data, update records in CRM, initiate transactions, or access product information from an ERP system. By linking the conversational interface to operational data and processes, Chatbot API Integrations transform chatbots from simple Q&A tools into powerful virtual assistants capable of delivering highly personalized, dynamic, and transactional customer experiences. This significantly improves customer self-service, reduces agent workload, and enhances operational efficiency, ensuring seamless and intelligent customer interactions across digital channels.

Without robust Chatbot API Integrations, organizations' chatbots are limited to providing only generic, predefined responses, unable to access real-time customer context or perform actions within backend systems. This results in frustrating customer experiences where users are frequently redirected to live agents for complex queries or transactional needs, negating much of the efficiency benefit of a chatbot. The absence of deep API connectivity hinders the chatbot's ability to truly personalize interactions, resolve issues autonomously, or fulfill transactional requests, ultimately impacting customer satisfaction, increasing operational costs due to agent escalations, and limiting the full potential of conversational AI to enhance customer service.

Implementation	Challenges
* Define chatbot scope & required backend data/actions * Select chatbot platform & NLP engine * Develop/configure APIs for backend systems * Design conversation flows with API calls * Implement error handling & fallback to agents * Plan for continuous improvement & monitoring	* Complex API development & maintenance * Ensuring data security & privacy during API calls * Handling diverse user intents & fallback scenarios * Real-time performance of API calls * Integrating with diverse legacy backend systems * Measuring ROI on self-service automation

Products	Platform Cost	Project Cost	Duration
Google Dialogflow (with integrations)	$2-15K	$10-70K	1-4 m
IBM Watson Assistant (with integrations)	$5-30K	$20-100K	2-6 m
Microsoft Azure Bot Service (with integrations)	$2-15K	$10-70K	1-4 m
Ada (with integrations)	$5-40K	$30-150K	2-7 m
Intercom (with App Integrations)	$5-30K	$20-100K	2-6 m
Custom Development (Python, Node.js + APIs)	N/A	$50-250K	3-9 m

Email–CRM Connectors are integration tools or features that synchronize data and activities between email marketing platforms and CRM (Customer Relationship Management) systems. These connectors automatically transfer customer contact information, email engagement metrics (opens, clicks, unsubscribes), campaign history, and lead statuses between the two systems. By creating a unified view of customer interactions across sales and marketing, Email–CRM Connectors eliminate manual data entry, ensure data consistency, and provide sales teams with real-time insights into marketing engagement. This streamlines lead management, enables more personalized sales outreach, improves marketing attribution, and ultimately enhances overall customer relationship management by ensuring both departments operate from a single, accurate source of customer truth.

Without Email–CRM Connectors, organizations often operate with a significant disconnect between their email marketing efforts and sales activities, leading to fragmented customer data, manual data transfers, and a lack of real-time visibility into lead engagement. This results in missed sales opportunities, inconsistent messaging between marketing and sales, and difficulty in attributing revenue accurately to marketing campaigns. The absence of seamless integration hinders lead nurturing, prevents sales teams from prioritizing hot leads effectively, and creates operational inefficiencies, ultimately impacting sales pipeline velocity, customer satisfaction due to disjointed experiences, and the overall effectiveness of marketing and sales alignment.

Implementation	Challenges
* Identify email platform & CRM in use * Select appropriate connector or build custom * Map data fields for synchronization * Define sync frequency & direction * Implement data cleansing & de-duplication rules * Test integration thoroughly & monitor performance	* Data mapping complexity & consistency * Handling data volume and sync frequency * Ensuring data quality and de-duplication * Managing access permissions between systems * Resolving conflicts in data updates * Troubleshooting integration errors

Products	Platform Cost	Project Cost	Duration
Salesforce Sales Cloud (Native Integrations)	N/A	$5-30K	1-3 m
HubSpot CRM (Native Integrations)	N/A	$2-15K	1-2 m
Mailchimp for Salesforce	N/A	$1-10K	0.5-2 m
Zapier / Make (Integration Platform)	$0.1-2K	$0.5-5K	0.5-1.5 m
Tray.io / Workato (iPaaS)	$5-20K	$20-80K	2-5 m
Custom API Development	N/A	$20-100K	2-6 m

Social Media Integrations involve connecting an organization's social media channels with its core business systems, such as CRM, customer service platforms, and marketing automation tools. These integrations enable businesses to monitor social conversations, capture social leads, route customer service inquiries from social media to the appropriate teams, and personalize social outreach based on unified customer data. By centralizing social media interactions, organizations can provide consistent customer support, engage with customers in real-time on their preferred platforms, and leverage social insights to enrich customer profiles and inform marketing strategies. This enhances customer experience, builds brand reputation, and allows for more holistic customer engagement across all digital touchpoints.

Without Social Media Integrations, organizations often manage their social media presence in isolation from other customer-facing systems, leading to fragmented customer views, delayed responses to social inquiries, and missed opportunities for customer engagement and lead generation. This results in inconsistent customer experiences, a reactive approach to social customer service, and an inability to leverage valuable social data to enrich customer profiles in CRM or personalize marketing efforts. The absence of seamless integration hinders the ability to provide omnichannel support, capture social leads effectively, and gain a holistic view of customer interactions, ultimately impacting customer satisfaction, brand reputation, and the effectiveness of digital customer engagement strategies.

Implementation	Challenges
* Define social listening & engagement goals * Select social media management tools * Integrate with CRM, Service Desk, Marketing Automation * Configure social listening rules & alerts * Train customer service/marketing teams * Establish social media reporting & analytics	* Managing high volume of social data & interactions * Ensuring real-time response & SLA adherence * Attributing social media interactions to customer profiles * Handling diverse social platform APIs & data formats * Ensuring data privacy & compliance on social data * Overcoming organizational silos between teams

Products	Platform Cost	Project Cost	Duration
Sprout Social	$1-10K	$5-30K	1-3 m
Hootsuite (Advanced features)	$1-8K	$5-25K	1-3 m
Salesforce Social Studio (Marketing Cloud)	$10-50K	$60-250K	3-9 m
Sprinklr (CXM Platform)	$15-70K	$80-350K	3-10 m
Zendesk (Social Messaging Add-on)	$2-15K	$10-70K	1-4 m
HubSpot (Social Media Tools)	$1-10K	$5-30K	1-3 m

CDP–Martech Sync refers to the critical integration and continuous synchronization of data between a Customer Data Platform (CDP) and various Martech (Marketing Technology) applications, such as email platforms, advertising networks, personalization engines, and analytics tools. The CDP acts as the central hub for unified customer profiles, and the sync pushes enriched, segment-ready data to downstream Martech tools, enabling highly targeted campaigns and personalized experiences. By ensuring consistent, up-to-date customer data across the entire marketing stack, CDP–Martech Sync eliminates data silos, improves campaign effectiveness, enhances personalization at scale, and provides a single source of truth for marketing insights. This maximizes marketing ROI, streamlines operations, and delivers cohesive, relevant customer journeys across all touchpoints.

Without robust CDP–Martech Sync, organizations often find their marketing technology stack operating on fragmented, inconsistent, or outdated customer data, leading to disjointed customer experiences and inefficient campaign execution. This means personalized messages might be sent based on incomplete profiles, ad targeting could be inaccurate, and cross-channel campaigns may lack cohesion. The absence of seamless data flow from a unified CDP to various Martech tools results in wasted marketing spend, lower conversion rates, and a reactive approach to customer engagement, ultimately hindering an organization's ability to deliver truly personalized customer journeys and maximize the effectiveness of its marketing efforts.

Implementation	Challenges
* Define data points for synchronization * Select CDP and Martech stack * Map data fields & identify unique customer ID * Implement real-time or near-real-time sync * Configure audience segmentation in CDP * Establish monitoring & error handling for sync	* Complex data mapping between diverse systems * Ensuring real-time or near-real-time data consistency * Managing data privacy & compliance across tools * Handling large data volumes for synchronization * Troubleshooting integration errors & data discrepancies * Proving ROI on improved personalization & campaign effectiveness

Products	Platform Cost	Project Cost	Duration
Segment (Connections & Destinations)	$10-60K	$50-300K	3-9 m
Tealium (Connectors & Integrations)	$15-70K	$80-350K	3-10 m
RudderStack (Event Stream & Destinations)	$5-30K	$20-100K	2-6 m
ActionIQ (Integrations)	$20-90K	$100-450K	4-12 m
mParticle	$15-70K	$80-350K	3-10 m
Custom API Development / iPaaS	N/A	$50-250K	3-9 m

Ai6x6™ Transformation Blueprint

Empathy 'Innovation'

"In order to build empathy at scale, we need to design systems that learn from actual emotion." — Rana el Kaliouby, AI expert and founder of Affectiva; known for emotion-aware technologies in digital experiences.

Empathy doesn't scale automatically—it's engineered. Voice commerce, live video, social co-creation—all these platforms rehearse attention, not just interaction. "Empathy Innovation" shifts the focus from marketing conversion to emotional choreography. Innovation here means sensing tone, mood, and need even before the customer names them. Mature transformation delivers personalization that feels generous, not algorithmic. When systems treat emotion like a first-class data point, loyalty becomes organic—and innovation becomes affection with architecture.

The "Empathy Innovation" chapter is paramount for customer transformation's digital maturity as it emphasizes creating novel, personalized, and emotionally resonant customer experiences through digital means. A digitally mature organization explores cutting-edge technologies like voice commerce platforms, virtual showroom experiences, and interactive video shopping to engage customers in new and immersive ways. This commitment to innovative customer experiences, often fostered through co-creation communities and social commerce apps, drives deeper connections, enhances brand loyalty, and significantly differentiates the brand in a crowded and evolving digital landscape, ensuring sustained customer engagement.

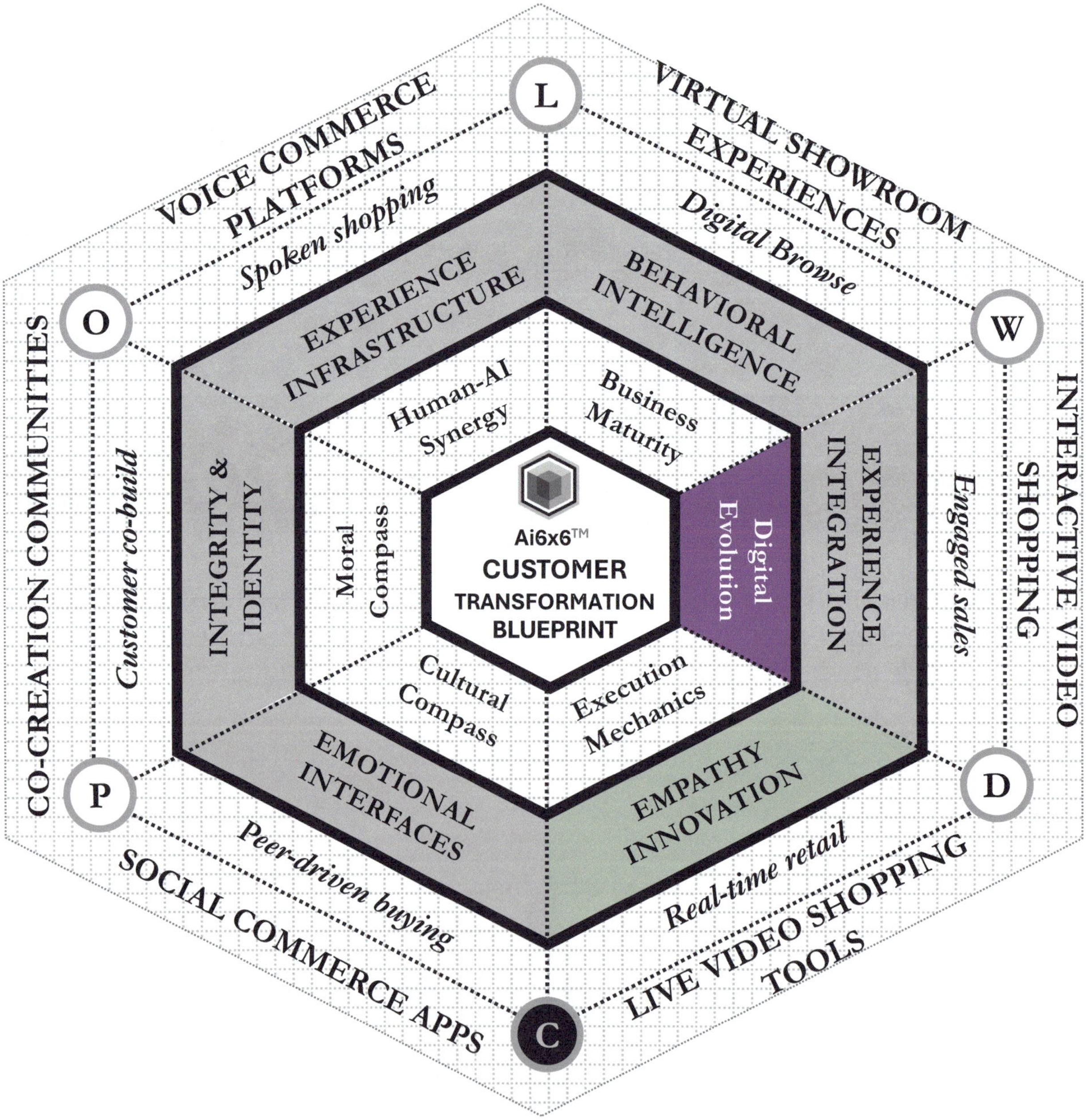

Ai6x6 DIGITAL EVOLUTION FRAMEWORK
FOR CUSTOMER TRANSFORMATION | EMPATHY INNOVATION

1. VOICE COMMERCE PLATFORMS

Voice Commerce Platforms enable customers to research, browse, and purchase products or services using voice commands through smart speakers (like Amazon Echo, Google Home), voice assistants on smartphones, or integrated in-car systems. These platforms leverage natural language processing (NLP) and speech recognition to understand user intent, process product queries, and facilitate transactions without requiring manual input. By offering a hands-free, intuitive shopping experience, Voice Commerce Platforms cater to consumer convenience, increase accessibility, and open new frictionless sales channels. This innovative approach enhances customer experience, drives impulse purchases, and broadens an organization's reach into the growing ecosystem of voice-activated devices, ushering in a new era of conversational commerce.

Without Voice Commerce Platforms, organizations miss out on a rapidly growing segment of customer interactions and purchasing opportunities through voice-enabled devices. Customers who prefer hands-free or screen-free interactions are forced to revert to traditional, less convenient methods, leading to friction in the buying journey and potentially lost sales. The absence of a voice-activated sales channel limits an organization's accessibility, fails to meet evolving consumer expectations for convenience, and delays entry into a significant future market, ultimately hindering competitive advantage and slowing overall digital transformation in retail and service industries.

Implementation	Challenges
* Define voice-enabled product catalog * Select a voice commerce platform/API * Design intuitive voice user interface (VUI) * Integrate with e-commerce, payment & inventory * Train NLP for product-specific queries * Plan for user testing & continuous VUI improvement	* Ensuring accurate speech recognition & NLP * Handling complex product queries via voice * Integrating with diverse backend systems * User adoption & trust in voice transactions * Ensuring data security for voice payments * Optimizing for hands-free, screen-less experience

Products	Platform Cost	Project Cost	Duration
Amazon Pay (for Alexa Skills)	$5-30K	$20-100K	2-6 m
Google Assistant (Actions for Payments)	$5-25K	$20-90K	2-5 m
Custom Voice App Development (Node.js, Python)	N/A	$50-250K	3-9 m
Voysis (Acquired by Apple)	N/A	$40-200K	3-8 m
SoundHound (Conversational AI)	$10-40K	$60-300K	3-9 m
Shopify (Integrations via Apps)	$1-10K	$5-50K	1-3 m

Virtual Showroom Experiences leverage immersive technologies like 3D modeling, virtual reality (VR), and augmented reality (AR) to create online, interactive environments where customers can explore products and services in a highly engaging and realistic manner, mimicking a physical showroom. Customers can navigate virtual spaces, customize products, view items from multiple angles, and even interact with virtual sales assistants, all from the comfort of their home. By transcending geographical boundaries and offering a richer, more engaging Browse experience, Virtual Showrooms enhance customer engagement, increase conversion rates, reduce return rates by improving product understanding, and offer significant scalability for showcasing product portfolios, driving a compelling digital sales strategy.

Without Virtual Showroom Experiences, organizations are limited to traditional 2D product images and descriptions online, which often fail to convey the full scope, detail, or scale of complex products or personalized offerings. This results in a less engaging customer Browse experience, lower confidence in online purchases, and higher return rates due to mismatched expectations. The absence of immersive, interactive product exploration hinders the ability to create emotional connections with products, differentiate from competitors, and provide a truly innovative online shopping journey, ultimately impacting sales conversion, customer satisfaction, and an organization's competitive standing in digital commerce.

Implementation	Challenges
* Define target products for 3D/VR experience * Create high-quality 3D models & virtual environments * Select VR/AR platform or web-based 3D framework * Integrate with e-commerce, inventory & CRM * Design intuitive navigation & interaction * Plan for performance optimization & accessibility	* High cost & complexity of 3D content creation * Ensuring real-time rendering performance * Integrating with diverse backend systems * User adoption of VR/AR hardware (if applicable) * Providing seamless multi-platform experience * Measuring clear impact on conversion & returns

Products	Platform Cost	Project Cost	Duration
Unity 3D / Unreal Engine (Development)	$0-20K	$100-500K	4-12 m
Obsess (Virtual Stores)	$10-50K	$60-250K	3-9 m
Emperia	$5-30K	$20-100K	2-6 m
V-commerce platforms (e.g., Vobling)	$10-40K	$50-200K	3-8 m
Shopify / Magento (with 3D/AR apps)	$1-10K	$5-50K	1-3 m
Custom WebGL/Three.js Development	N/A	$50-250K	3-9 m

Interactive Video Shopping combines live or pre-recorded video content with clickable elements, quizzes, and embedded purchasing functionalities, allowing customers to directly engage with products and make purchases within the video stream. This goes beyond passive viewing by creating a dynamic, immersive, and personalized shopping experience where customers can explore details, ask questions, receive recommendations, and complete transactions without leaving the video environment. By leveraging the power of visual storytelling and real-time interaction, Interactive Video Shopping enhances customer engagement, drives higher conversion rates, and reduces friction in the buying journey, bridging the gap between entertainment and commerce, fostering a more engaging and efficient digital retail experience.

Without Interactive Video Shopping, organizations using video for marketing or product showcases miss the opportunity to convert viewer engagement directly into sales, forcing customers to navigate away from compelling content to make a purchase. This creates friction in the buying process, potentially leading to abandonment, and reduces the immediate commercial value of video assets. The absence of embedded interactivity and purchasing capabilities in video content limits the direct sales impact of rich media, hinders impulse buying, and fails to capitalize on the increasing consumer preference for integrated, frictionless shopping experiences, ultimately impacting conversion rates and digital revenue.

Implementation	Challenges
* Define video content strategy (live/pre-recorded) * Select an interactive video platform * Integrate with e-commerce & inventory systems * Design interactive elements & purchase flows * Plan for content creation & live streaming (if applicable) * Establish analytics for viewer engagement & conversions	* Creating compelling interactive video content * Ensuring seamless integration with e-commerce * Managing real-time inventory for live shopping * Performance optimization for video streaming * User experience design for embedded interactions * Measuring clear uplift in engagement & conversion

Products	Platform Cost	Project Cost	Duration
Livestreaming/Video Commerce Platforms	$5-30K	$20-100K	2-6 m
Smartzer	$5-25K	$20-90K	2-5 m
Spott.ai	$2-15K	$10-70K	1-4 m
Channelize.io	$2-10K	$5-50K	1-3 m
Buywith	$5-20K	$20-80K	2-5 m
Custom HTML5/JavaScript Video Players	N/A	$30-150K	2-7 m

Live Video Shopping Tools enable brands and retailers to host interactive live-streamed events where hosts showcase products, engage with viewers in real-time through chat, answer questions, and facilitate immediate purchases directly within the video stream. Often integrated with e-commerce platforms, these tools leverage the immediacy and authenticity of live video to create a dynamic, social, and immersive shopping experience. By combining entertainment with commerce, Live Video Shopping Tools build community, foster trust, drive impulse buys, and significantly boost sales by replicating the personal touch of in-store shopping in a digital format. This innovative approach offers a highly engaging path to conversion, allowing brands to connect with customers on a deeper, more personal level at scale.

Without Live Video Shopping Tools, organizations promoting products or services through video miss the opportunity to create immediate, interactive sales opportunities during live broadcasts or pre-recorded sessions. This results in a passive viewing experience where customer questions cannot be answered in real-time, and purchase decisions are deferred, creating friction and leading to lower conversion rates. The absence of direct interactivity and in-stream purchasing capabilities in video content hinders the ability to capitalize on immediate customer interest, build authentic connections, and replicate the spontaneity of in-person retail, ultimately impacting sales conversion and limiting the full potential of video as a direct revenue driver.

Implementation	Challenges
* Define live show format & product focus * Select a live video shopping platform * Integrate with e-commerce & inventory systems * Plan content, hosts & promotions * Ensure reliable live streaming infrastructure * Train hosts for engaging interactive selling	* Ensuring stable, high-quality live stream performance * Real-time inventory management during live sales * Engaging audience & managing live chat volume * Driving traffic to live events * Measuring direct sales attribution from live streams * Talent sourcing & training for engaging hosts

Products	Platform Cost	Project Cost	Duration
CommentSold	$5-30K	$20-100K	2-6 m
Buywith	$5-20K	$20-80K	2-5 m
Bambuser	$10-40K	$50-200K	3-8 m
TalkShopLive	$2-10K	$5-50K	1-3 m
Channelize.io	$2-10K	$5-50K	1-3 m
Shopify Live (via apps/integrations)	$1-10K	$5-50K	1-3 m

Social Commerce Apps are mobile applications or integrated features within social media platforms that allow users to discover, browse, and purchase products directly within a social network's environment. These apps leverage social features like recommendations from friends, influencer endorsements, user-generated content, and integrated checkout processes to create a seamless and community-driven shopping experience. By blending social interaction with e-commerce, Social Commerce Apps reduce friction in the buying journey, increase product discovery through trusted networks, and foster a more engaging and personalized shopping experience. This expands sales channels, taps into viral marketing potential, and strengthens brand loyalty by integrating commerce naturally into the daily social lives of customers.

Without Social Commerce Apps, organizations selling online miss out on capitalizing on the immense purchasing influence of social networks, forcing customers to leave social platforms to complete a purchase. This creates friction in the buying process, limits impulse purchases, and fails to leverage the power of peer recommendations and influencer marketing directly at the point of discovery. The absence of embedded shopping capabilities within social environments hinders the ability to monetize social engagement, convert social interest directly into sales, and engage customers where they spend significant time, ultimately impacting digital sales growth and limiting access to a crucial customer acquisition channel.

Implementation	Challenges
* Define social commerce strategy & target platforms * Select social media platforms with commerce features * Integrate product catalog & inventory * Configure payment gateways & checkout flows * Plan for social media advertising & influencer marketing * Establish analytics for social sales performance	* Adapting to different social platform policies * Integrating complex product catalogs & inventory * Ensuring secure payment processing within social apps * Managing customer service for social purchases * Attributing sales accurately from social channels * Building trust & preventing fraud in social commerce

Products	Platform Cost	Project Cost	Duration
Instagram Shopping / Facebook Shops	$0-1K	$1-10K	0.5-2 m
TikTok Shopping / Shopify TikTok Integration	$0-1K	$1-10K	0.5-2 m
Pinterest Shop the Look / Buyable Pins	$0-1K	$1-10K	0.5-2 m
Shopify (with social integrations)	$1-10K	$5-50K	1-3 m
BigCommerce (with social integrations)	$1-10K	$5-50K	1-3 m
Swirl (Social Commerce Platform)	$5-20K	$20-80K	2-5 m

Co-creation Communities are online platforms or structured programs where organizations actively engage customers, users, or external partners in the product development, innovation, and service design process. These communities provide tools for submitting ideas, voting on concepts, providing feedback on prototypes, and participating in design challenges. By harnessing collective intelligence and fostering a sense of ownership among participants, Co-creation Communities enable organizations to develop more customer-centric products and services, gain deeper market insights, and build stronger brand loyalty. This collaborative approach accelerates innovation cycles, reduces product development risks, and creates a highly engaged customer base that feels invested in the brand's success, leading to more relevant and successful market offerings.

Without Co-creation Communities, organizations often rely on internal teams or traditional market research methods for product development and innovation, leading to products that may not fully meet customer needs or market demands. This results in slower innovation cycles, higher risks of product failure due to lack of customer input, and a disconnect between the brand and its customer base. The absence of direct, collaborative customer engagement in the design process hinders the ability to gain real-time feedback, tap into diverse perspectives, and build a sense of customer ownership, ultimately impacting market relevance, product success rates, and brand loyalty in competitive landscapes.

Implementation	Challenges
* Define co-creation goals & target participants * Select a community platform or build custom * Design clear participation guidelines & incentives * Develop feedback mechanisms & idea submission tools * Plan for moderation & engagement management * Integrate insights into product development workflow	* Attracting & retaining active community members * Effectively managing & prioritizing submitted ideas * Integrating diverse feedback into product design * Maintaining confidentiality of sensitive ideas * Avoiding design by committee pitfalls * Measuring clear impact on product success & customer loyalty

Products	Platform Cost	Project Cost	Duration
Aha! Ideas (Product Idea Management)	$1-10K	$5-30K	1-3 m
UserVoice (Feedback & Feature Requests)	$1-8K	$5-25K	1-3 m
Ideascale (Innovation Management)	$5-30K	$20-100K	2-6 m
Invision (Design Collaboration)	$0-2K	$1-10K	0.5-2 m
Custom Forum/Community Platforms	N/A	$30-150K	2-7 m
Alchemer (Surveys & Feedback)	$0.5-5K	$1-10K	0.5-2 m

Ai6x6™ Transformation Blueprint

Emotional 'Interfaces'

"Design is the silent ambassador of your brand." — Paul Rand

Legendary graphic designer behind iconic branding; his philosophy influenced tech UX worldwide.

Interfaces aren't pixels—they're promises. Responsive frameworks, AR try-ons, and conversational widgets don't just beautify—they resonate. "Emotional Interfaces" makes UX a performance—subtle but sticky. In mature transformation, interface behavior signals intention: whether it's respect for time, clarity in navigation, or elegance in transitions. Customers feel brand intelligence not through ads—but through moments of use. When interfaces choreograph empathy, technology becomes invisible—and emotion becomes operational.

"Emotional Interfaces" is critical for customer transformation's digital maturity because it focuses on crafting intuitive, responsive, and emotionally engaging digital touchpoints for customers. A digitally mature organization prioritizes user-centric design through responsive web frameworks, progressive web apps, and voice UI for customers. This attention to interface design, including augmented reality try-on and mobile wallet UIs, ensures products and services are not only functional but also delightful, easy to use, and personalized. This drives adoption, enhances satisfaction, and builds long-term customer loyalty by creating truly memorable digital experiences.

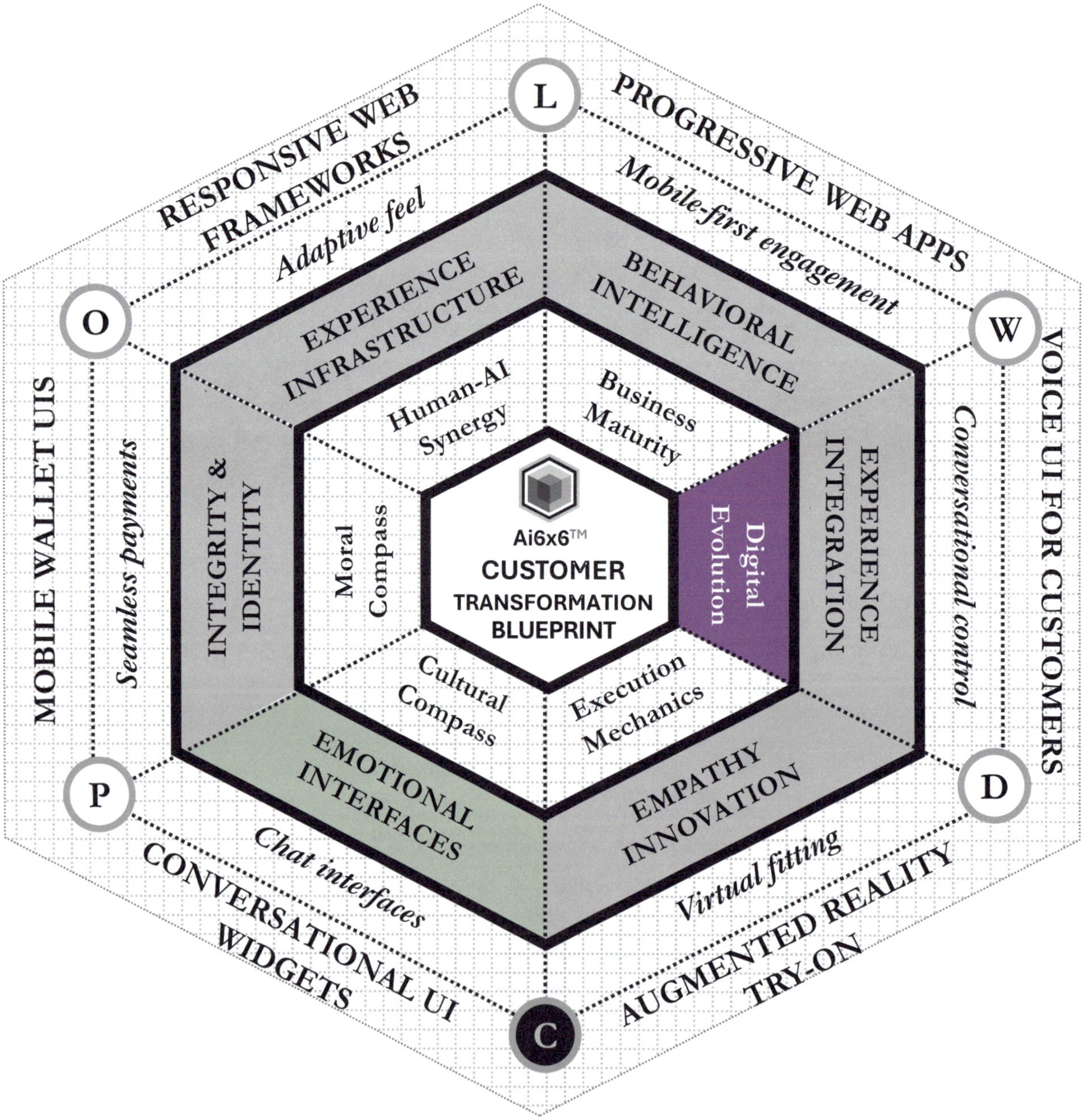

Ai6x6 DIGITAL EVOLUTION FRAMEWORK
FOR CUSTOMER TRANSFORMATION | EMOTIONAL INTERFACES

Responsive Web Frameworks are collections of pre-written code (like HTML, CSS, and JavaScript) that provide a structured foundation for building websites and web applications that automatically adapt their layout and content to fit various screen sizes and devices, from desktops to tablets and smartphones. These frameworks employ techniques such as fluid grids, flexible images, and media queries to ensure a consistent and optimal user experience regardless of how or where a customer accesses the website. By prioritizing adaptability and accessibility, Responsive Web Frameworks significantly reduce development time and cost, improve SEO, and enhance customer satisfaction by providing a seamless, frustration-free Browse experience across all devices, crucial for reaching today's diverse online audience.

Without Responsive Web Frameworks, organizations often resort to building separate websites or specialized mobile versions, leading to increased development costs, maintenance overhead, and a fragmented, inconsistent customer experience across different devices. This means customers accessing content on varying screen sizes may encounter usability issues, poor readability, or broken layouts, resulting in frustration and early abandonment. The absence of a unified, adaptive design approach hinders SEO performance, limits audience reach, and ultimately impacts conversion rates and customer satisfaction by failing to deliver a seamless and optimized digital experience for all users, regardless of their device.

Implementation	Challenges
* Define target devices & screen sizes * Select appropriate responsive framework (e.g., Bootstrap) * Design flexible layouts & adaptive content * Optimize images & media for different resolutions * Conduct extensive cross-device testing * Ensure performance & loading speed on mobile	* Ensuring consistent design across all breakpoints * Optimizing image & media loading for mobile * Performance issues on older devices/browsers * Debugging responsiveness across diverse platforms * Managing content presentation for small screens * Keeping up with evolving device landscape

Products	Platform Cost	Project Cost	Duration
Bootstrap (CSS Framework)	N/A	$5-30K	1-3 m
Tailwind CSS (Utility-first CSS)	N/A	$5-25K	1-3 m
React (JavaScript Library)	N/A	$20-100K	2-6 m
Angular (JavaScript Framework)	N/A	$20-100K	2-6 m
Vue.js (JavaScript Framework)	N/A	$15-80K	2-5 m
Foundation (CSS Framework)	N/A	$5-20K	1-3 m

Progressive Web Apps (PWAs) are web applications that utilize modern web capabilities to deliver an app-like experience to users, directly through their web browser. They combine the best features of websites (discoverability, no installation required) with the best features of native mobile apps (offline capabilities, push notifications, fast loading, responsive design, home screen icon). By providing a fast, reliable, and engaging experience, PWAs offer a superior alternative to traditional mobile websites, improving user retention, conversion rates, and overall engagement. This technology allows organizations to reach a wider audience without the friction of app store downloads, delivering a frictionless and consistent customer experience across devices, crucial for enhancing digital presence.

Without Progressive Web Apps (PWAs), organizations often force customers to choose between a full native app (which requires download and installation) and a less performant mobile website that lacks app-like features. This results in higher friction for engagement, slower loading times, and the absence of offline capabilities or push notifications directly from the web, leading to lower user retention and conversion rates on mobile web. The absence of PWAs means businesses miss the opportunity to deliver a seamless, fast, and engaging app-like experience without the burden of app stores, ultimately impacting mobile customer satisfaction and limiting the reach and effectiveness of their digital channels.

Implementation	Challenges
* Define app-like features required (offline, push) * Develop with Service Workers for offline caching * Implement Web App Manifest for home screen icon * Ensure HTTPS & responsive design * Optimize performance (Lighthouse score) * Plan for push notifications & user re-engagement	* Browser compatibility & feature support variations * Debugging Service Workers for offline behavior * Limited access to native device features (e.g., contacts) * User adoption & awareness of PWA capabilities * Managing caching strategies for real-time data * Performance optimization for all network conditions

Products	Platform Cost	Project Cost	Duration
Custom Development (React, Angular, Vue, etc.)	N/A	$50-250K	3-9 m
Next.js (React Framework for PWAs)	N/A	$30-150K	2-7 m
Nuxt.js (Vue Framework for PWAs)	N/A	$30-150K	2-7 m
Ionic Framework (Hybrid App Dev)	N/A	$20-100K	2-6 m
Shopify / Magento (PWA themes/plugins)	$1-10K	$5-50K	1-3 m
Workbox (Google - PWA Library)	N/A	$5-30K	1-3 m

Voice User Interfaces (VUIs) for customers enable interactions with technology through spoken commands and natural language, allowing users to control devices, access information, and complete tasks using their voice. This includes interfaces for smart speakers, voice assistants on smartphones, in-car systems, and voice-activated kiosks. By leveraging advanced speech recognition and natural language understanding (NLU), VUIs offer a hands-free, intuitive, and highly accessible mode of interaction, particularly beneficial for multitasking or users with accessibility needs. This enhances customer convenience, speeds up task completion, and transforms user experiences by making technology more natural and less reliant on traditional visual interfaces, driving a new era of conversational engagement.

Without Voice User Interfaces (VUIs) for customers, organizations rely solely on visual or touch-based interfaces, which can be cumbersome, distracting, or inaccessible in certain contexts (e.g., while driving, cooking, or for visually impaired users). This means customers are limited in how they can interact with products and services, leading to friction, slower task completion, and a less inclusive user experience. The absence of voice-enabled interaction limits an organization's reach to a growing segment of consumers who prefer conversational interfaces, fails to meet evolving expectations for convenience, and delays entry into critical new markets, ultimately impacting customer satisfaction and competitive differentiation.

Implementation	Challenges
* Define customer use cases for voice interaction * Select a speech recognition & NLU engine * Design intuitive conversational flows (VUI design) * Integrate with backend systems & data * Train voice models on domain-specific vocabulary * Plan for rigorous user testing & iteration	* Ensuring high accuracy in speech recognition * Handling diverse accents, dialects & background noise * Designing natural & forgiving conversational flows * Integrating with complex backend data & actions * User adoption & overcoming privacy concerns * Measuring clear ROI on voice interaction

Products	Platform Cost	Project Cost	Duration
Amazon Alexa Skills Kit	$5-30K	$20-100K	2-6 m
Google Assistant (Actions on Google)	$5-25K	$20-90K	2-5 m
IBM Watson Assistant (Voice Integration)	$5-30K	$20-100K	2-6 m
Microsoft Azure AI Speech	$5-20K	$20-80K	2-5 m
Twilio Voice (with NLP integrations)	$2-15K	$10-70K	1-4 m
Custom Voice AI Development	N/A	$50-250K	3-9 m

Augmented Reality (AR) Try-On experiences allow customers to virtually "try on" products—such as clothing, eyewear, makeup, or furniture—using their smartphone or tablet camera, seeing how items look on themselves or in their own environment in real-time. This technology overlays digital 3D models of products onto the real world, providing a highly interactive and personalized visualization. By bridging the gap between online Browse and physical interaction, AR Try-On enhances customer confidence, reduces uncertainty about fit or appearance, and minimizes product returns. This innovative interface increases conversion rates, improves customer satisfaction, and creates a highly engaging, memorable shopping experience that leverages cutting-edge technology to personalize the customer journey and build trust.

Without Augmented Reality (AR) Try-On, organizations relying on traditional 2D product images or generic size charts leave customers with significant uncertainty about how products will look or fit in real life. This often leads to hesitation in purchasing, increased product returns due to mismatched expectations, and a less engaging online shopping experience. The absence of virtual try-on capabilities hinders the ability to provide a truly personalized visualization of products, reduce buyer's remorse, and differentiate from competitors, ultimately impacting conversion rates, operational costs associated with returns, and overall customer satisfaction in digital commerce.

Implementation	Challenges
* Identify products suitable for AR try-on * Create high-quality 3D models of products * Select an AR SDK/platform (e.g., ARCore, ARKit) * Integrate with e-commerce & product catalog * Design intuitive user experience for AR interaction * Plan for performance optimization on diverse devices	* Creating accurate & realistic 3D product models * Ensuring smooth AR rendering & tracking performance * Handling diverse lighting & environmental conditions * Integrating with existing e-commerce platforms * Device compatibility & fragmentation * Measuring clear impact on conversion & returns

Products	Platform Cost	Project Cost	Duration
Shopify AR (via 3D Models)	N/A	$5-30K	1-3 m
Adobe Aero / Dimension (3D Asset Creation)	$0-5K	$2-15K	1-2 m
8th Wall (WebAR Platform)	$2-15K	$10-70K	1-4 m
Wikitude	$2-10K	$5-50K	1-3 m
ARKit (Apple) / ARCore (Google) - Native Dev	N/A	$50-250K	3-9 m
VNTANA (3D Commerce Platform)	$5-20K	$20-80K	2-5 m

Conversational UI Widgets are embeddable chat interfaces (often powered by AI chatbots or live agents) that appear on websites or applications, allowing users to interact with a brand or service through text-based dialogue. Unlike full chatbot platforms, these widgets are designed for specific, contextual interactions within a larger digital experience, such as answering FAQs on a product page, helping with navigation, collecting feedback, or assisting with checkout. By providing immediate, personalized, and proactive support directly at the point of need, Conversational UI Widgets enhance user experience, reduce friction, deflect common inquiries from human agents, and improve conversion rates by guiding customers through their digital journey, fostering efficient and engaging interactions.

Without Conversational UI Widgets, organizations often rely on static FAQs, contact forms, or phone numbers for customer support and engagement on their digital properties, leading to delayed responses and a less interactive user experience. This means customers may struggle to find quick answers, abandon tasks due to confusion, or incur the inconvenience of switching channels for support, resulting in frustration and lost opportunities. The absence of in-context, real-time conversational assistance hinders immediate problem resolution, limits proactive engagement, and fails to capture user intent effectively, ultimately impacting customer satisfaction, digital conversion rates, and operational efficiency of customer support.

Implementation	Challenges
* Define specific use cases for widget (e.g., FAQ, support) * Select a chatbot/live chat platform with widget capability * Design conversation flows & knowledge base * Integrate with relevant website/app pages & data * Implement fallback to human agents * Plan for continuous improvement & A/B testing	* Designing intuitive & context-aware conversations * Ensuring real-time performance & responsiveness * Integrating with backend data for personalized responses * Managing escalation paths to human agents * User adoption & perception of bot vs. human * Measuring clear impact on support deflections & conversions

Products	Platform Cost	Project Cost	Duration
Intercom (Messenger)	$5-30K	$20-100K	2-6 m
Zendesk Chat / Answer Bot	$1-10K	$5-30K	1-3 m
LiveChat	$1-5K	$2-20K	1-2 m
Freshchat (Freshworks)	$1-8K	$5-25K	1-3 m
Drift (Conversational Marketing)	$5-25K	$20-90K	2-5 m
HubSpot Service Hub (Chat Widget)	$1-10K	$5-30K	1-3 m

Mobile Wallet User Interfaces (UIs) refer to the design and functionality of digital wallets (e.g., Apple Pay, Google Pay, Samsung Pay) that allow customers to securely store payment credentials, loyalty cards, coupons, tickets, and even digital IDs on their smartphones for convenient transactions and interactions. These UIs enable tap-to-pay functionality, in-app purchases, and seamless redemption of loyalty rewards or digital passes. By streamlining the checkout process, consolidating customer-facing items, and enhancing security through tokenization, Mobile Wallet UIs provide a highly efficient, secure, and personalized customer experience. This increases convenience, accelerates sales, reduces friction in physical and digital transactions, and fosters deeper customer engagement by making interactions smoother and more accessible, leveraging the ubiquity of mobile devices.

Without optimized Mobile Wallet UIs, organizations relying on traditional payment methods or physical loyalty cards create friction in the customer journey, particularly for mobile-first consumers. This means customers face slower checkout processes, manual entry of payment details, or the inconvenience of carrying multiple physical cards, leading to frustration and potential transaction abandonment. The absence of seamless mobile wallet integration hinders the ability to provide quick, secure, and convenient payment options, leverage digital loyalty programs effectively, and capitalize on the growing trend of contactless transactions, ultimately impacting sales velocity, customer satisfaction, and an organization's competitive edge in modern commerce.

Implementation	Challenges
* Define payment & loyalty strategy for mobile wallets * Select mobile wallet platforms (Apple Pay, Google Pay) * Integrate payment gateway with mobile wallet APIs * Design user flows for adding cards, loyalty, etc. * Ensure security & compliance for tokenization * Plan for marketing & user adoption of mobile wallet	* Complex integration with payment gateways & POS * Ensuring robust security & data encryption * Handling diverse payment methods & currencies * User adoption & trust in mobile wallet technology * Compliance with payment industry standards (PCI DSS) * Integrating loyalty programs & offers seamlessly

Products	Platform Cost	Project Cost	Duration
Apple Pay API (Native iOS Dev)	N/A	$20-100K	2-6 m
Google Pay API (Native Android/Web Dev)	N/A	$20-100K	2-6 m
Stripe (Payment Gateway with Wallet Support)	$0.1-5K	$1-10K	0.5-2 m
Braintree (Payment Gateway with Wallet Support)	$0.1-5K	$1-10K	0.5-2 m
Shopify Payments (with Wallet Integration)	$1-10K	$5-30K	1-3 m
Custom Mobile App Development	N/A	$50-250K	3-9 m

Ai6x6™ Transformation Blueprint

'Integrity' & Identity

"I believe people will demand transparency, and organizations will have to deliver it." — Marissa Mayer , Former CEO of Yahoo; central figure in shaping consumer trust policies during rapid tech expansion.

Data isn't currency—it's conversation. Consent platforms, fraud detection systems, and identity verification services don't merely secure—they reassure. "Integrity & Identity " means treating customer information not as asset, but agency. Maturity in transformation emerges when transparency is operationalized—built into every data handshake, not retrofitted. Customers shouldn't just opt in—they should feel included. When systems protect before they collect, and verify before they personalize, privacy becomes culture—not just compliance. These relationships aren't managed by legal teams alone—they're curated through experience, empathy, and system behavior that proves respect quietly and continuously.

The "Integrity & Identity " chapter is fundamental for customer transformation's digital maturity as it underscores the paramount importance of trust, privacy, and security in all customer interactions. A digitally mature organization implements robust consent management platforms, GDPR compliance solutions, and secure payment gateways. This, combined with fraud detection systems and multi-factor authentication, protects customer data and ensures transparent, ethical practices. This unwavering commitment to integrity builds confidence, fosters long-term loyalty, and strengthens the brand's reputation in a privacy-conscious and security-aware digital world, safeguarding the customer relationship.

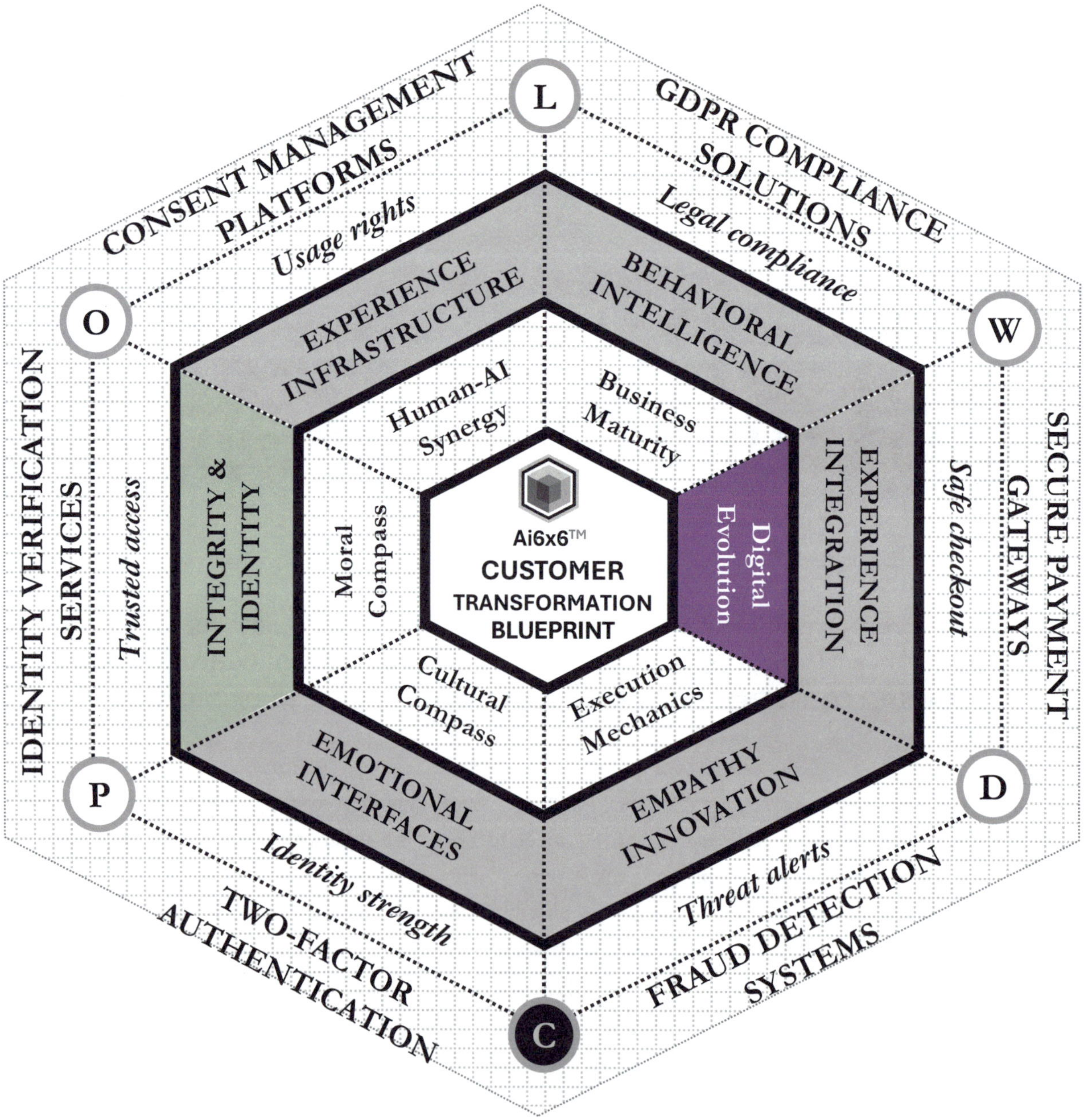

Ai6x6 DIGITAL EVOLUTION FRAMEWORK
FOR CUSTOMER TRANSFORMATION | INTEGRITY & IDENTITY

1. CONSENT MANAGEMENT PLATFORMS (CMPS)

Consent Management Platforms (CMPs) are software solutions that enable organizations to collect, manage, and store user consent for data collection, processing, and usage, particularly in compliance with global privacy regulations like GDPR, CCPA, and CPRA. CMPs typically present users with clear, granular options for accepting or rejecting cookies, tracking technologies, and marketing communications, and record these preferences in an auditable log. By providing transparency and control over personal data, CMPs build customer trust, ensure legal compliance, and avoid hefty regulatory fines. This foundational integrity component is critical for fostering respectful data relationships, enabling personalized experiences responsibly, and mitigating privacy risks in today's data-driven customer interactions.

Without Consent Management Platforms (CMPs), organizations struggle to effectively collect and manage explicit user consent for data processing, leading to significant risks of non-compliance with global privacy regulations, potential fines, and damage to brand reputation. Relying on opaque or implicit consent mechanisms can erode customer trust, create legal vulnerabilities, and hinder the ethical use of customer data for personalization and marketing. The absence of a centralized, auditable CMP makes it difficult to track consent preferences, demonstrate accountability, and provide transparent data practices, ultimately jeopardizing an organization's ability to engage customers responsibly and operate legally in a privacy-conscious landscape.

Implementation	Challenges
* Identify all data collection points * Select a CMP platform * Customize consent banners & preference centers * Integrate with website, apps & data systems * Map data processing activities to consent types * Plan for continuous monitoring & auditing	* Complex legal interpretation of privacy laws * Integrating CMP across diverse digital assets * Ensuring user-friendly & transparent consent UIs * Managing granular consent preferences at scale * Adapting to evolving privacy regulations * Demonstrating auditable proof of consent

Products	Platform Cost	Project Cost	Duration
OneTrust Consent & Preference Mgmt	$10-50K	$60-250K	3-9 m
TrustArc Privacy Management Platform	$10-40K	$50-200K	3-8 m
Cookiebot (Cybot)	$0.5-5K	$2-20K	0.5-2 m
Usercentrics	$1-10K	$5-30K	1-3 m
Didomi	$2-15K	$10-70K	1-4 m
Transcend (Privacy & Data Governance)	$15-60K	$70-300K	3-9 m

GDPR (General Data Protection Regulation) Compliance Solutions are specialized tools and services designed to help organizations meet the strict requirements of the European Union's data privacy law, which governs the processing of personal data for EU citizens. These solutions typically include features for data mapping, data subject access request (DSAR) management, consent management, data breach notification, and data protection impact assessments (DPIAs). By systematizing privacy operations, GDPR Compliance Solutions enable organizations to demonstrate accountability, protect personal data effectively, and avoid severe non-compliance penalties. This ensures customer data is handled with respect and legality, fostering trust and enabling responsible global customer engagement, essential for any business dealing with EU citizens' data.

Without dedicated GDPR Compliance Solutions, organizations face substantial legal and financial risks when processing personal data of EU citizens, including hefty fines (up to 4% of global annual revenue or €20 million, whichever is higher) and severe reputational damage. Manually managing GDPR requirements is complex, time-consuming, and prone to error, leading to inefficiencies in handling data subject requests, conducting privacy assessments, and responding to breaches. The absence of structured compliance tools hinders an organization's ability to demonstrate accountability, ensure data privacy, and maintain customer trust, ultimately jeopardizing market access in Europe and undermining overall data governance efforts.

Implementation	Challenges
* Appoint DPO & understand GDPR scope * Conduct data mapping & inventory * Select a GDPR compliance platform * Implement DSAR fulfillment workflows * Establish breach notification procedures * Plan for regular audits & training	* Complex interpretation of legal requirements * Integrating solutions across diverse data systems * Managing data subject access requests at scale * Ensuring continuous compliance with evolving regulations * Organizational change management for data handling * Demonstrating ongoing accountability & record-keeping

Products	Platform Cost	Project Cost	Duration
OneTrust Privacy Management Platform	$15-70K	$80-350K	3-10 m
TrustArc Privacy Management Platform	$15-60K	$70-300K	3-9 m
BigID (Discovery & Privacy Ops)	$20-90K	$100-450K	4-12 m
DataGrail	$10-50K	$60-250K	3-9 m
Securiti.AI (Data Command Center)	$20-100K	$120-500K	4-13 m
Compliance as a Service Providers	$5-30K	$20-100K	2-6 m

Secure Payment Gateways are critical e-commerce services that authorize credit card or direct payment processing for online and in-person transactions. They act as a secure bridge between a merchant's website or point-of-sale system and the financial institutions involved in the transaction. These gateways encrypt sensitive payment information, transmit it securely to the payment processor, and return an approval or decline message to the merchant, often while adhering to PCI DSS (Payment Card Industry Data Security Standard) compliance. By ensuring the highest level of security for financial data, Secure Payment Gateways protect customers from fraud, build trust in online transactions, and enable businesses to process payments reliably and efficiently, forming the backbone of trustworthy digital commerce.

Without Secure Payment Gateways, organizations attempting to process payments directly would expose sensitive customer financial data to severe security risks, leading to potential data breaches, widespread fraud, and massive non-compliance penalties from payment card industry standards (PCI DSS). Relying on insecure or uncertified payment methods would erode customer trust, result in high chargeback rates, and likely lead to an inability to accept major credit cards. The absence of a robust, compliant payment gateway jeopardizes an organization's financial operations, severely impacts its reputation, and prevents it from engaging in legitimate digital commerce, making it impossible to operate in a modern, secure marketplace.

Implementation	Challenges
* Select a reputable payment gateway provider * Integrate gateway APIs with e-commerce platform * Implement secure checkout pages (PCI DSS) * Configure fraud detection rules * Test transaction flows & error handling * Plan for ongoing security updates & compliance	* Ensuring PCI DSS compliance continuously * Protecting against payment fraud & chargebacks * Handling global currencies & payment methods * Integration complexity with diverse platforms * Managing transaction fees & reconciliation * Maintaining high uptime & performance during peaks

Products	Platform Cost	Project Cost	Duration
Stripe	$0.1-5K	$1-10K	0.5-2 m
PayPal (for Business)	$0.1-5K	$1-10K	0.5-2 m
Braintree (a PayPal service)	$0.1-5K	$1-10K	0.5-2 m
Adyen	$1-10K	$5-30K	1-3 m
Shopify Payments	N/A	$0.5-5K	0.5-1.5 m
Square (Online Payments)	$0.1-5K	$1-10K	0.5-2 m

4. FRAUD DETECTION SYSTEMS

Fraud Detection Systems are advanced software solutions that use machine learning, artificial intelligence, and rule-based logic to identify and prevent fraudulent activities in real-time, particularly in financial transactions, account creation, and user interactions. These systems analyze vast amounts of data, including transaction patterns, behavioral anomalies, device fingerprints, and historical fraud data, to flag suspicious activities and assign risk scores. By proactively identifying and blocking fraudulent attempts, Fraud Detection Systems protect both the organization and its customers from financial losses, chargebacks, and reputational damage. This is critical for maintaining transactional integrity, fostering customer trust, and ensuring secure and reliable customer experiences across all digital touchpoints.

Without robust Fraud Detection Systems, organizations are highly vulnerable to financial losses from fraudulent transactions, chargebacks, and account takeovers, leading to significant revenue drain and reputational damage. Manually reviewing suspicious activities is inefficient and often too slow to prevent real-time fraud, resulting in missed fraudulent patterns and a reactive security posture. The absence of sophisticated fraud detection capabilities compromises data integrity, erodes customer trust, and exposes organizations to severe financial and legal repercussions, ultimately undermining their ability to conduct secure and reliable digital commerce and engage customers with confidence.

Implementation	Challenges
* Define fraud types relevant to your business * Collect diverse transaction & user behavior data * Select a fraud detection platform * Configure rules & train ML models * Integrate with payment systems & user journeys * Plan for continuous monitoring & model tuning	* High false positive rates (blocking legitimate users) * Adapting to evolving fraud patterns * Integrating with diverse data sources & systems * Real-time performance requirements * Balancing security with user experience * Ensuring data privacy in fraud analysis

Products	Platform Cost	Project Cost	Duration
Sift (Digital Trust & Safety)	$5-30K	$20-100K	2-6 m
Forter	$10-50K	$60-250K	3-9 m
Signifyd	$5-25K	$20-90K	2-5 m
Kount (an Equifax Company)	$5-20K	$20-80K	2-5 m
Riskified	$10-40K	$50-200K	3-8 m
Cybersource (Visa - Fraud Management)	$5-30K	$20-100K	2-6 m

Two-Factor Authentication (2FA), also known as Multi-Factor Authentication (MFA), is a security process that requires users to provide two different authentication factors to verify their identity before granting access to an account or system. These factors typically fall into categories of "something you know" (e.g., password), "something you have" (e.g., phone, hardware token), or "something you are" (e.g., fingerprint, face scan). By adding an extra layer of security beyond just a password, 2FA significantly reduces the risk of unauthorized access, even if a password is stolen. This protects customer accounts, safeguards sensitive personal data, and builds greater trust in digital interactions, essential for enhancing cybersecurity and maintaining data integrity in customer-facing applications.

Without Two-Factor Authentication (2FA), organizations' customer accounts and sensitive data are vulnerable to compromise through password-only authentication, which is susceptible to phishing, brute-force attacks, and credential stuffing. Relying on single-factor authentication significantly increases the risk of unauthorized access, data breaches, and identity theft, leading to severe reputational damage, financial losses, and erosion of customer trust. The absence of an additional layer of security hinders an organization's ability to protect customer privacy effectively, meet stringent security compliance standards, and provide a truly secure digital experience, ultimately jeopardizing both customer relationships and overall business continuity.

Implementation	Challenges
* Identify critical customer access points * Select 2FA methods (SMS, authenticator app, biometrics) * Integrate 2FA solution with identity provider/app * Design user enrollment & recovery processes * Plan for user education & support * Establish audit logging for 2FA events	* User adoption & potential friction from extra step * Managing various 2FA methods (SMS vs. app) * Handling account recovery when 2FA is lost * Integrating with diverse existing authentication systems * Ensuring reliable delivery of codes (SMS issues) * Cost of SMS or other 2FA delivery methods at scale

Products	Platform Cost	Project Cost	Duration
Auth0 (Okta)	$5-30K	$20-100K	2-6 m
Duo Security (Cisco)	$5-25K	$20-90K	2-5 m
Microsoft Azure Active Directory MFA	$5-20K	$20-80K	2-5 m
Okta (Identity Cloud)	$10-40K	$50-200K	3-8 m
Twilio Authy API	$1-10K	$5-30K	1-3 m
Google Authenticator (Free)	N/A	$1-10K	0.5-2 m

Identity Verification Services are specialized platforms that authenticate the identity of individuals, typically during account creation, high-value transactions, or sensitive data access, using various methods like document verification, biometric checks (face, fingerprint), database lookups, or knowledge-based authentication. These services help organizations confirm that a user is who they claim to be, mitigating risks associated with fraud, identity theft, and compliance breaches (e.g., KYC/AML regulations). By ensuring robust identity assurance, Identity Verification Services protect customer accounts, prevent fraudulent activities, enable seamless onboarding processes, and build trust in digital interactions, forming a critical defense layer for customer-facing systems and sensitive data.

Without robust Identity Verification Services, organizations are highly susceptible to risks from fraudulent account creation, identity theft, and non-compliance with "Know Your Customer" (KYC) or Anti-Money Laundering (AML) regulations, leading to significant financial losses and severe legal penalties. Relying on self-declared information or weak verification methods allows bad actors to exploit systems, compromise legitimate customer accounts, and damage brand reputation. The absence of strong identity assurance hinders an organization's ability to onboard customers securely, prevent fraud effectively, and meet regulatory obligations, ultimately compromising data integrity and eroding customer trust in its digital services.

Implementation	Challenges
* Define identity verification needs (e.g., onboarding, high-risk) * Select a reputable identity verification provider * Integrate verification APIs into user workflows * Design user experience for verification process * Configure verification methods & risk thresholds * Plan for compliance & ongoing monitoring	* Balancing user experience with security rigor * Handling diverse identity document types globally * Ensuring accuracy of biometric/document verification * Integrating with various backend systems * Cost per verification for high volumes * Adapting to evolving fraud techniques & regulations

Products	Platform Cost	Project Cost	Duration
Onfido	$5-30K	$20-100K	2-6 m
Jumio	$5-25K	$20-90K	2-5 m
Auth0 (Identity Verification integrations)	$5-30K	$20-100K	2-6 m
Trulioo	$10-40K	$50-200K	3-8 m
Veriff	$5-20K	$20-80K	2-5 m
IDology (GBG)	$10-30K	$20-100K	2-6 m

BLUEPRINT SERIES

SECTION E

Execution Mechanics

Ai6x6™ Transformation Blueprint

Story 1 — The Manhattan Project Coordination Model (1942–1946): The Manhattan Project united scientists, engineers, and military leaders across dispersed sites under extreme secrecy and urgency. Workstreams were interdependent, requiring precise coordination without full visibility. Structured reporting, controlled interfaces, and disciplined compartmentalization enabled progress despite complexity. Human judgment guided trade-offs while machines and labs executed specialized tasks. The project demonstrated how execution succeeds when governance, clarity, and cadence align. It became a template for managing high-stakes, multi-disciplinary programs.

Story 2 — The Berlin Airlift Logistics System (1948–1949): With land routes blocked, Allied forces had to supply an entire city solely by air. Aircraft schedules, loading patterns, and turnaround times were optimized to the minute. Machines provided lift and speed, but humans orchestrated sequencing, prioritization, and contingency planning. Continuous feedback loops refined operations daily. The airlift proved that disciplined execution can overcome structural constraints. It showed how precision, rhythm, and adaptability create operational breakthroughs.

Story 3 — The Interstate Highway Construction Program (1956–1970s): The United States embarked on building a nationwide highway network requiring coordination across states, contractors, and engineering disciplines. Standardized designs ensured consistency while local teams adapted to terrain and conditions. Machines handled heavy construction, but humans managed sequencing, safety, and quality. The program relied on predictable funding, clear specifications, and phased execution. It demonstrated how large-scale delivery succeeds when structure and flexibility coexist. The highways became a symbol of disciplined national execution.

Story 4 — Toyota's Just-in-Time Logistics (1960s–1980s): As Toyota refined its production system, logistics became a critical execution layer. Parts arrived exactly when needed, minimizing inventory and waste. Machines tracked flow, but humans monitored signals, resolved disruptions, and maintained rhythm. Kanban created transparency across the value chain. The system required precision, trust, and continuous calibration. It showed how execution excellence emerges when every component of the workflow reinforces timing, quality, and flow.

Story 5 — FedEx's Overnight Delivery Network (1973–1990s): FedEx built a hub-and-spoke system capable of moving packages across the country overnight. Aircraft, sorting machines, and routing algorithms provided speed, but humans managed exceptions, weather disruptions, and customer commitments. The network operated on tight cycles that demanded flawless coordination. Data informed decisions while people ensured reliability. The system demonstrated how execution mechanics scale when tech and human oversight operate in synchronized cadence.

Story 6 — The Rise of Agile Software Delivery (2001–2015): Software teams struggled with long cycles, rigid plans, and late-stage failures. Agile introduced iterative delivery, cross-functional collaboration, and continuous feedback. Machines automated testing and deployment, while humans prioritized work, clarified requirements, and adapted to change. Execution shifted from linear phases to dynamic cycles. The approach improved predictability and responsiveness. It became a global standard for modern execution, proving that disciplined iteration outperforms rigid planning.

LESSONS FROM A CUSTOMER TRANSFORMATION PERSPECTIVE

Customer value strengthens when execution mechanics evolve from ad-hoc delivery to consistent, dependable performance. Over time, organizations learned that reliability, timeliness, and transparency matter as much as innovation. Customers benefit when processes are predictable, exceptions are managed gracefully, and commitments are honored even under pressure. Machines accelerate throughput, but humans safeguard quality and trust. Execution maturity ensures that customer experience is not dependent on individual effort but on institutionalized discipline. The result is a customer relationship built on confidence—knowing the organization can deliver what it promises, when it promises, with minimal friction and maximum reliability.

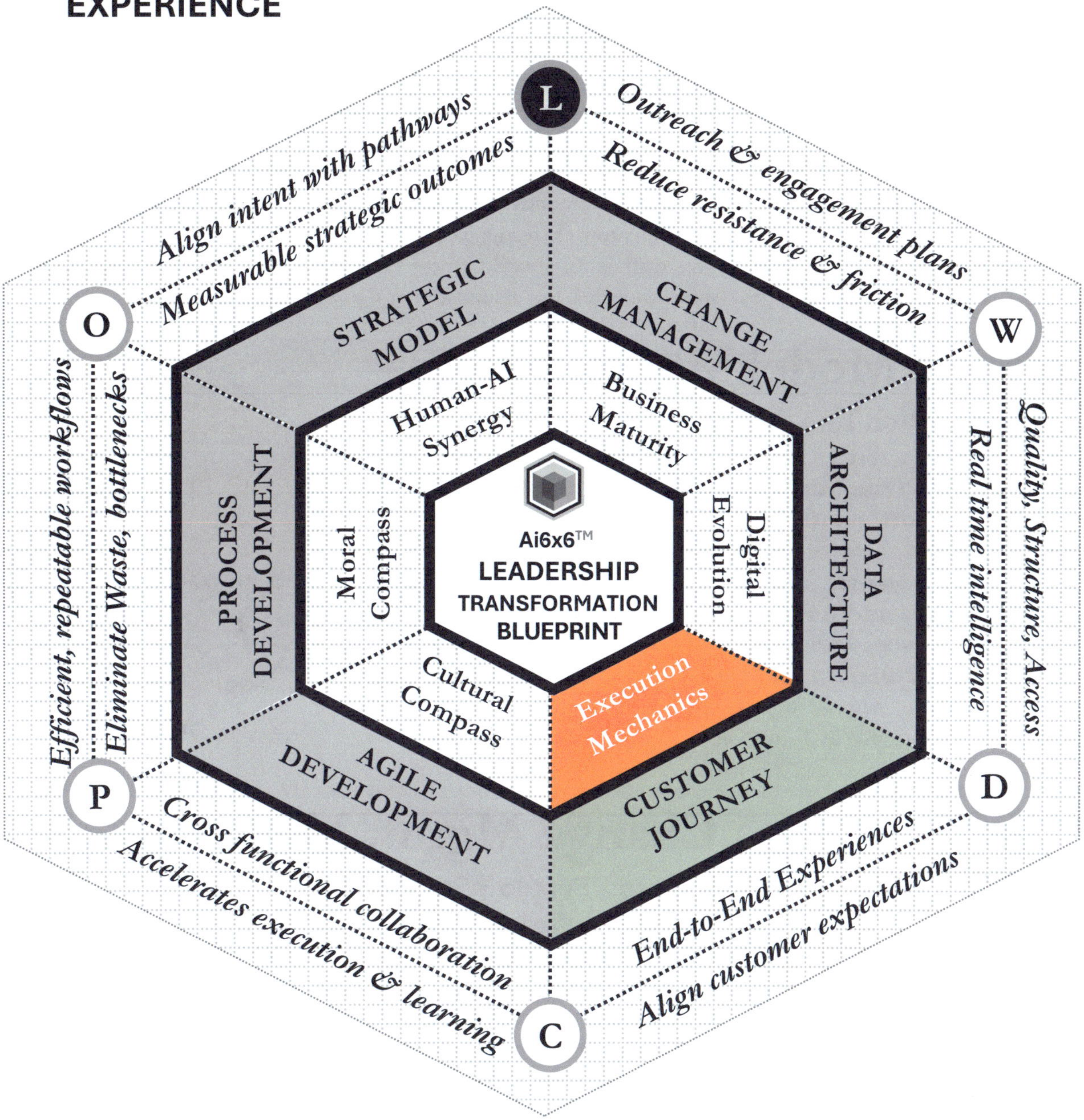

The **Ai6x6 Research Lab** identified this lens as the primary antidote to the "randomness" of customer interactions. We chose it because the human mind requires a disciplined, repeatable system to make high-stakes design decisions without the friction of hesitation. This lens isn't merely about **Journey Mapping** or **Service Blueprinting**; it is about the psychological certainty and mental clarity that come from a proven methodology like **Design Thinking** or **Jobs to be Done**. It is vital because it provides a mental scaffolding, turning the chaos of market demand into a series of surgical, confident strikes on customer value.

This lens allows the leader's heart to remain calm while the market moves at terminal velocity. By relying on **Persona Frameworks** and **CX Maturity Models**, the leader replaces intuition-based guesswork with evidence-based conviction. We believe that execution mechanics are the ultimate form of brand mindfulness—the practice of being fully present in the "how" of the experience so that the "what" of customer satisfaction is achieved with excellence. This lens ensures that every touchpoint has a defined purpose and a coherent structure, providing the steady footing necessary to leap into new markets with the assurance of a master experience architect.

Execution has always been the crucible where strategy proves its worth. Leaders can articulate bold visions, design elegant frameworks, and outline ambitious roadmaps, but none of it matters if the organization cannot execute with precision, coherence, and resilience. In the age of AI and accelerated change, **Execution Mechanics** becomes the lens that determines whether transformation becomes real or remains theoretical. It forces leaders to confront the uncomfortable truth that most failures are not failures of strategy but failures of execution. The gap between intention and impact is rarely philosophical; it is operational. Execution Mechanics exposes that gap and provides the architecture to close it. This lens reframes execution as a system—an interconnected set of methodologies that convert ambition into momentum. It recognizes that execution is not the final step of strategy but the environment in which strategy lives or dies. In a world where AI accelerates decision cycles, compresses timelines, and amplifies both strengths and weaknesses, leaders must design execution as they design vision.

Strategic Model Methodologies

Turning Ambition Into Structure

Execution Mechanics begins with **strategic model methodologies**, the structured approaches that translate ambition into actionable pathways. These methodologies are not academic exercises; they are the scaffolding that ensures strategy is grounded. Leaders who neglect this dimension find themselves improvising under pressure, relying on intuition rather than disciplined frameworks. Strategic models define the logic of execution—how choices are made, how priorities are sequenced, and how resources are allocated. They ensure that every initiative has a defined purpose, a coherent structure, and a measurable outcome. Without this foundation, execution becomes reactive, fragmented, and vulnerable to disruption.

Change Mgmt. Methodologies

Aligning People With Progress

The next dimension, **change management methodologies**, recognizes that execution is ultimately a human endeavor. No transformation succeeds unless people understand, accept, and commit to the change. Leaders who underestimate this reality discover that resistance is not a barrier but a predictable response to uncertainty. Effective change management acknowledges the emotional, cultural, and psychological dimensions of transformation. It ensures that communication is transparent, expectations are clear, and support is available. It turns change from something imposed into something shared. In this dimension, execution becomes relational—rooted in trust, empathy, and the ability to guide people toward possibility.

Data Architecture Methodologies

Building the Backbone of Execution

Data architecture provide the backbone for modern execution. In a world where decisions must be informed by evidence, organizations cannot afford fragmented data, inconsistent definitions, or unreliable pipelines. Data architecture is not a technical concern; it is a strategic one. It determines whether insights are trustworthy, whether systems can scale, and whether AI can operate responsibly. Leaders who invest in strong data architecture create the conditions for intelligence to flow freely across the enterprise. In this dimension, execution becomes intelligent—powered by clarity, consistency, and evidence.

Customer Journey Methodologies

Anchoring Execution in Value

Customer journey ensure that execution remains anchored in value creation. Too many organizations design processes around internal convenience rather than customer experience. Customer journey thinking forces leaders to see the organization from the outside in, identifying friction, gaps, and opportunities that internal perspectives overlook. It ensures that execution is not only efficient but meaningful. It aligns operations with the expectations, needs, and emotions of the people the organization serves. In this dimension, execution becomes empathetic—shaped by the reality of customers rather than the assumptions of leaders.

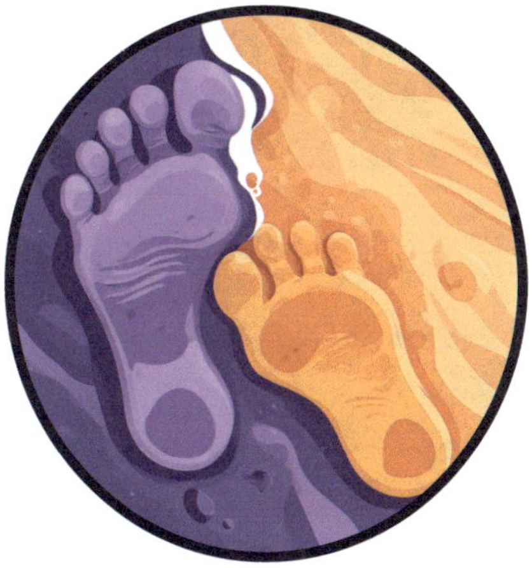

Agile Development Methodologies

Enabling Adaptation at Speed

Agile development introduce the adaptability required in a world where conditions shift rapidly. Traditional execution models assume stability, predictability, and linear progress. Agile assumes the opposite. It embraces iteration, experimentation, and continuous learning. It allows teams to adjust quickly, respond to feedback, and deliver value incrementally. Leaders who adopt agile mindsets create organizations that move with speed without sacrificing quality. In this dimension, execution becomes dynamic—capable of evolving in real time as new information emerges.

Process Improvement Methodologies

Sustaining Excellence Over Time

The final dimension, **process improvement methodologies**, completes the system by ensuring that execution evolves rather than stagnates. They provide the discipline to identify inefficiencies, eliminate waste, and refine workflows. They ensure that execution is not a one-time effort but a continuous practice. Leaders who embrace process improvement build organizations that get better over time, not simply bigger or busier. In this dimension, execution becomes self-correcting—driven by curiosity, discipline, and a commitment to excellence.

Execution Mechanics is the lens that transforms strategy into reality by ensuring that ambition is supported by structure, that change is supported by people, and that progress is supported by systems. It reveals that execution is not the final step of leadership but the continuous arena in which leadership is tested. In a world where AI accelerates both opportunity and exposure, leaders must design execution as deliberately as they design vision, building the operational muscle that allows the organization to move with coherence under pressure. This lens makes clear that execution is not a matter of effort but of engineering—of constructing the models, architectures, and behaviors that convert intention into repeatable performance. Execution becomes a living capability rather than a periodic push, a discipline that strengthens with use rather than erodes under stress.

STRATEGIC MODEL
Flight Schedule

Establishing a universal cadence ensures every department operates with synchronized precision. The master timeline that dictates when every resource must be in place to ensure a coherent, global operation.

DATA ARCHITECTURE
Air Traffic Control

Maintaining a high-fidelity signaling system eliminates the risk of operational collisions and blind spots. The invisible infrastructure that provides the real-time telemetry needed for confident, evidence-based decisions at high velocity.

CUSTOMER JOURNEY
The Terminal Flow

Eliminating friction at every touchpoint transforms the user experience into a seamless journey of value. Redesigning the "Gate" flow to remove bottlenecks and demonstrate respect for the customer's time and needs.

AGILE DEVELOPMENT
Quick Turnaround

Compressing execution cycles minimizes the cost of delay and keeps strategic assets in constant motion. The ability to de-plane, refuel, and re-load (iterate) with extreme speed, ensuring the organization never "idles" on the tarmac.

CHANGE MANAGEMENT
Ground Crew Synergy

Cultivating organizational elasticity allows the workforce to pivot instantly without losing mechanical momentum. Leaders create psychological safety for the "crew" to adapt during disruptions, synchronized under pressure.

PROCESS IMPROVEMENT
Logistics Optimization

Embedding a habit of marginal gains compounds small refinements into a structural advantage. Identifying every millisecond lost in baggage handling or fueling to maximize the total capacity and throughput of the hub.

COMMON FRAMEWORK - TOGAF (The Open Group Architecture Framework)

Book 1: Leadership Transformation—Strategic Direction : The frameworks in this pillar ensure that all efforts are strategically aligned and that leaders are focused on the right outcomes for value creation.

1. **McKinsey 7-S Framework:** Aligns the organization's strategy with its structure, systems, staff, and skills to ensure all components support the desired transformation.
2. **Business Model Canvas:** Creates a shared understanding of how value will be created, delivered, and captured, providing a clear map for all teams.
3. **McKinsey Three Horizons of Growth:** Structures innovation efforts across short-term improvements (Horizon 1), scaling existing successes (Horizon 2), and future-proofing the business (Horizon 3), ensuring a sustainable pipeline of value.
4. **Balanced Scorecard:** Translates the strategic vision into measurable performance objectives, connecting operational activities to high-level financial and non-financial goals.
5. **SWOT Analysis:** Provides a structured way to evaluate the strategic context, enabling leaders to leverage strengths, mitigate weaknesses, seize opportunities, and counter threats.
6. **Blue Ocean Strategy:** Drives leaders to look for uncontested market space, shifting the focus from competing for value to creating new value.

Book 2: Workforce Transformation—Adoption and Engagement : This pillar addresses the human element of change, ensuring that employees are prepared, engaged, and able to sustain new ways of working.

1. **Kotter's 8-Step Change Model:** Provides a step-by-step process for leading major change, from creating a sense of urgency to solidifying new approaches in the corporate culture.
2. **ADKAR Model:** Guides change managers in ensuring that individual employees have the Awareness, Desire, Knowledge, Ability, and Reinforcement needed for successful change.
3. **Lewin's Change Management Model:** Conceptualizes change as a three-stage process: "Unfreeze" existing habits, implement the "Change," and "Refreeze" the new processes to make them stick.
4. **Bridges' Transition Model:** Focuses on the psychological journey of individuals, helping them navigate the emotional stages of an ending, a neutral zone, and a new beginning.
5. **Prosci Change Management Methodology:** Provides a structured, research-based methodology for applying change management to any project or initiative.
6. **The Satir Change Model:** Helps leaders understand and anticipate the emotional and performance impacts of change on individuals and teams.

Book 3: Data Transformation—Measurement and Insights - This pillar provides the tools and architectures to manage data as a strategic asset, ensuring that value can be reliably measured and tracked.

1. **Zachman Framework:** Offers a comprehensive, logical structure for classifying and organizing enterprise architecture, ensuring data management is part of the larger business strategy.
2. **DAMA-DMBOK:** Provides a set of best practices for managing data throughout its lifecycle, ensuring the quality and integrity of data used for value measurement.
3. **Data Mesh:** Promotes a decentralized, domain-driven approach to data ownership, empowering teams to create and manage data as a product.
4. **Lambda Architecture (or Kappa Architecture):** Establishes a technical architecture for processing large volumes of data efficiently, supporting real-time value tracking and analytics.
5. **Master Data Management (MDM) Frameworks:** Ensure that an organization's critical business data is consistent, accurate, and provides a "single source of truth" for reliable reporting.
6. **Data Fabric:** Creates a flexible, interconnected data architecture that supports seamless access and integration of data, regardless of its location.

Book 4: Customer Transformation—External Value Delivery: This pillar focuses on defining and delivering value from the customer's perspective, ensuring that transformation efforts translate to measurable customer outcomes.

1. **Customer Journey Mapping:** Visualizes the customer's end-to-end experience, identifying pain points and opportunities for value creation.
2. **Service Blueprinting:** Expands on journey mapping by detailing the front-stage and backstage processes, people, and systems that support a customer experience.
3. **Jobs to be Done (JTBD) Framework:** Helps define "value" from the customer's perspective by focusing on the functional and emotional outcomes they seek.
4. **Design Thinking Methodology:** Provides a human-centered approach to innovation that focuses on empathizing with customers to design meaningful solutions.
5. **CX (Customer Experience) Maturity Models:** Provides a benchmark for tracking and improving customer experience over time, linking strategic initiatives to better customer outcomes.
6. **Persona Development Frameworks:** Creates detailed profiles of target users, ensuring that transformation efforts are grounded in a deep understanding of customer needs.

Book 5: Product Transformation—Internal Value Delivery : This pillar establishes an agile, iterative process for continuously building and improving the products and services that deliver customer value.

1. **Scrum:** Enables cross-functional teams to deliver value incrementally in short sprints, allowing for continuous feedback and adaptation.
2. **Kanban:** Helps manage and visualize workflow, limiting work-in-progress and improving the flow of value to the customer.
3. **Lean Startup:** Applies a "build-measure-learn" feedback loop to product development, ensuring teams are constantly validating their assumptions about value creation.
4. **Scaled Agile Framework (SAFe):** Provides a system for coordinating and aligning multiple agile teams across the enterprise to deliver large-scale solutions.
5. **Feature Flags / Toggles Architecture:** Enables teams to deploy new features and capabilities quickly and safely, allowing for real-time testing of value propositions.
6. **Continuous Integration / Continuous Delivery (CI/CD) Pipeline Architecture:** Automates the software delivery process, allowing teams to deliver new value to customers efficiently and reliably.

Book 6: Operations Transformation—Efficiency and Sustainability
This pillar focuses on optimizing the internal processes and systems that support the entire value chain, ensuring efficiency and sustainability.

1. **Lean Management:** Aims to eliminate waste and optimize the flow of value through the organization's processes, generating cost savings and efficiency gains.
2. **Six Sigma (DMAIC Methodology):** Uses a data-driven approach to improve processes by defining, measuring, analyzing, improving, and controlling them.
3. **BPM (Business Process Management) Lifecycle:** Provides a structured approach to continuously improving business processes, ensuring sustained value delivery over the long term.
4. **SCOR Model (Supply Chain Operations Reference):** Offers a standard framework for managing and improving supply chain efficiency, a key driver of operational value.
5. **Digital Twin Architecture:** Creates virtual models of physical assets or processes, enabling simulation and optimization before real-world changes are made.
6. **Robotic Process Automation (RPA) Frameworks/Methodologies:** Automates repetitive, rule-based tasks, freeing up human resources for higher-value activities.

TOGAF (The Open Group Architecture Framework) is a widely recognized and comprehensive enterprise architecture methodology that provides a structured approach for developing and managing enterprise architecture. It helps organizations design, plan, implement, and govern an enterprise information architecture. At its core is the Architecture Development Method (ADM), a detailed, iterative process for developing various architectural domains: Business, Data, Application, and Technology. For leadership transformation, particularly in achieving digital maturity, TOGAF is indispensable. Digital transformation is not just about implementing new technologies; it requires aligning business strategy with IT capabilities across the entire enterprise. Without a coherent architecture, digital initiatives can become fragmented, leading to integration challenges, redundant systems, security vulnerabilities, and a lack of scalability. TOGAF provides the necessary rigor to ensure that digital investments are strategically aligned, well-integrated, and support the organization's long-term digital vision. It helps leaders understand the current state of their digital landscape, define the desired future state (e.g., cloud-native infrastructure, API-driven integration, AI-enabled applications), and create a roadmap to get there. By fostering a common language and framework for architects and stakeholders, TOGAF enables effective communication and collaboration across complex digital initiatives. It ensures that the digital transformation is not just a series of isolated projects but a cohesive, well-governed evolution of the entire enterprise, a critical component of achieving and sustaining comprehensive digital maturity. It provides the blueprint for how an organization's digital capabilities (applications, data, infrastructure) support its strategic business goals, ensuring that every technological investment contributes to a unified, adaptable, and future-ready digital enterprise. This structured approach helps avoid costly misalignments and ensures that digital initiatives are built on a solid, scalable foundation, allowing the organization to truly harness the power of technology for competitive advantage.

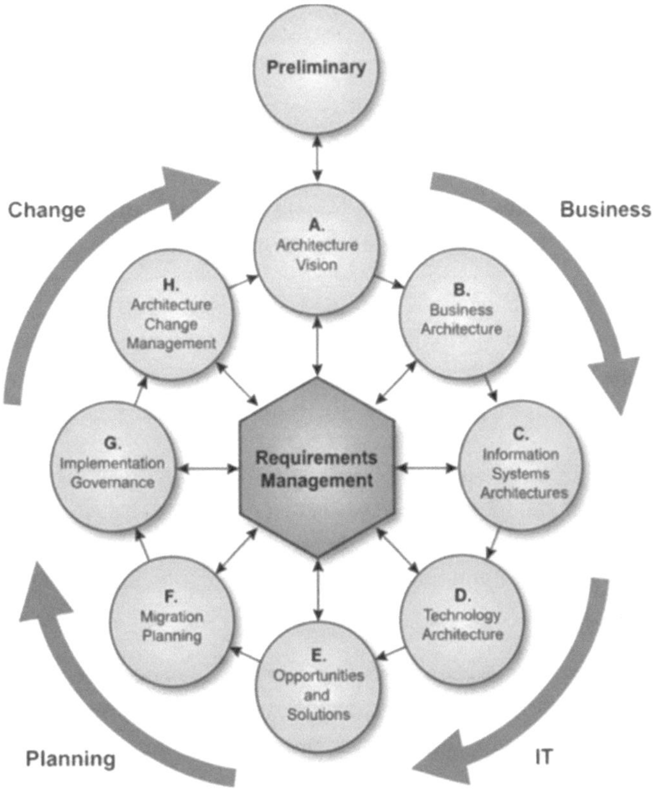

TOGAF Frameworks, Image source: medium.com

1. **Preliminary Phase: Framework and Principles:** Establish the architecture capability for the enterprise. This involves defining the organizational context for architecture work, identifying key stakeholders, establishing the architecture principles (e.g., "cloud-first," "API-driven," "data-driven decision making"), and defining the scope and governance of the architecture effort.

2. **Phase A: Architecture Vision:** Develop a high-level vision for the target architecture that aligns with business strategy and digital transformation goals. This involves identifying key stakeholders, defining the business value of the architecture work, and securing approval for the architecture development cycle.

3. **Phase B: Business Architecture:** Develop a baseline (current state) and target (future state) Business Architecture that describes the organization's processes, functions, and organizational structure. This ensures that the digital transformation efforts are aligned with and support the core business capabilities and strategic objectives, identifying where digital enablement can optimize business value, streamline workflows, and enhance customer interactions through digital means.

4. **Phase C: Information Systems Architectures (Data & Application):** Develop baseline and target Data Architecture (logical and physical data models, data governance, data flows) and Application Architecture (applications, their interactions, and how they support business functions). This phase is critical for designing integrated digital systems, ensuring data flows seamlessly across digital platforms, and supporting new digital applications that drive business value and customer engagement.

5. **Phase D: Technology Architecture:** Develop baseline and target Technology Architecture that describes the logical software and hardware capabilities required to support the Business, Data, and Application components. This includes defining infrastructure (e.g., cloud platforms, edge computing), platforms (e.g., microservices, AI/ML platforms), and security standards necessary for a robust, scalable, and resilient digital enterprise.

6. **Phase E: Opportunities & Solutions:** Identify and evaluate various implementation options and opportunities to deliver the target architecture. This involves assessing gaps between baseline and target architectures, identifying potential solutions (e.g., Commercial Off-The-Shelf (COTS), custom development, cloud migration, SaaS adoption), and defining high-level implementation projects. For digital maturity, this prioritizes digital solutions that offer the greatest business value and strategic alignment.

7. **Phase F: Migration Planning:** Develop a detailed implementation and migration plan. This involves prioritizing projects, estimating costs, defining a timeline, and outlining the sequence of activities required to transition from the baseline to the target architecture. This ensures a structured and managed approach to digital transformation implementation, minimizing disruption and maximizing efficiency.

8. **Phase G: Implementation Governance:** Oversee the implementation of the architecture. This involves establishing governance processes to ensure that projects adhere to the defined architecture, managing architectural contracts, and providing architectural guidance during development and deployment. This continuous oversight is vital for maintaining architectural integrity and consistency throughout the digital transformation.

9. **Phase H: Architecture Change Management:** Establish processes for managing changes to the architecture. This ensures that the architecture remains relevant and responsive to evolving business needs, technological advancements (e.g., new AI capabilities, quantum computing), and market shifts.

10. **Requirements Management:** This phase runs concurrently throughout all ADM phases. It involves managing the architectural requirements, ensuring they are consistent, complete, and aligned with stakeholder needs. For digital maturity, this ensures that evolving digital requirements (e.g., real-time data access, scalability for new digital services) are captured and integrated into the architectural design and implementation.

Ai6x6™ Transformation Blueprint

Customer Journey Frameworks

"You've got to start with the customer experience and work backward to the technology."

Steve Jobs. The foundational principle that customer empathy must drive all technology and process decisions, forcing the business to align its entire architecture to solve customer problems seamlessly.

Customer Journey Frameworks are essential strategic tools that translate abstract experience goals into a measurable, end-to-end service blueprint. These frameworks map the complete customer lifecycle, from initial awareness through purchase, use, and advocacy, detailing every interaction and touchpoint. This structure is critical because it forces the organization to shift from an internal, siloed view to an **outside-in, customer-centric perspective**. The framework identifies pain points, moments of truth, and key emotional drivers, enabling targeted investments in digital technology and process optimization. It provides the model for integrating various systems (CRM, marketing automation, service platforms) to ensure a **seamless, contextually-aware omnichannel experience**. Without this rigorous, shared framework, transformation efforts lead to disjointed service, wasted technology spending, and failure to drive the loyalty and growth that define successful customer transformation.

The **Customer Journey Framework** is focused on defining and delivering measurable value strictly from the customer's perspective, ensuring that transformation initiatives directly translate into superior customer outcomes. At its core, Customer Journey Mapping provides a visual, end-to-end representation of the customer experience, allowing the organization to pinpoint critical pain points and opportunities for value creation. This is further detailed by Service Blueprinting, which expands the view to include the necessary front-stage and backstage processes, people, and technology required to support that experience. To properly define what "value" means, the Jobs to be Done (JTBD) Framework reframes the focus toward the core functional and emotional outcomes customers are genuinely seeking. The entire process is underpinned by the Design Thinking Methodology, a human-centered approach that prioritizes empathy and iterative design to craft truly meaningful solutions. To ground these efforts in reality, Persona Development Frameworks create detailed profiles of target users, and CX (Customer Experience) Maturity Models provide a structured benchmark for tracking and continually improving customer experience linked to strategic business goals.

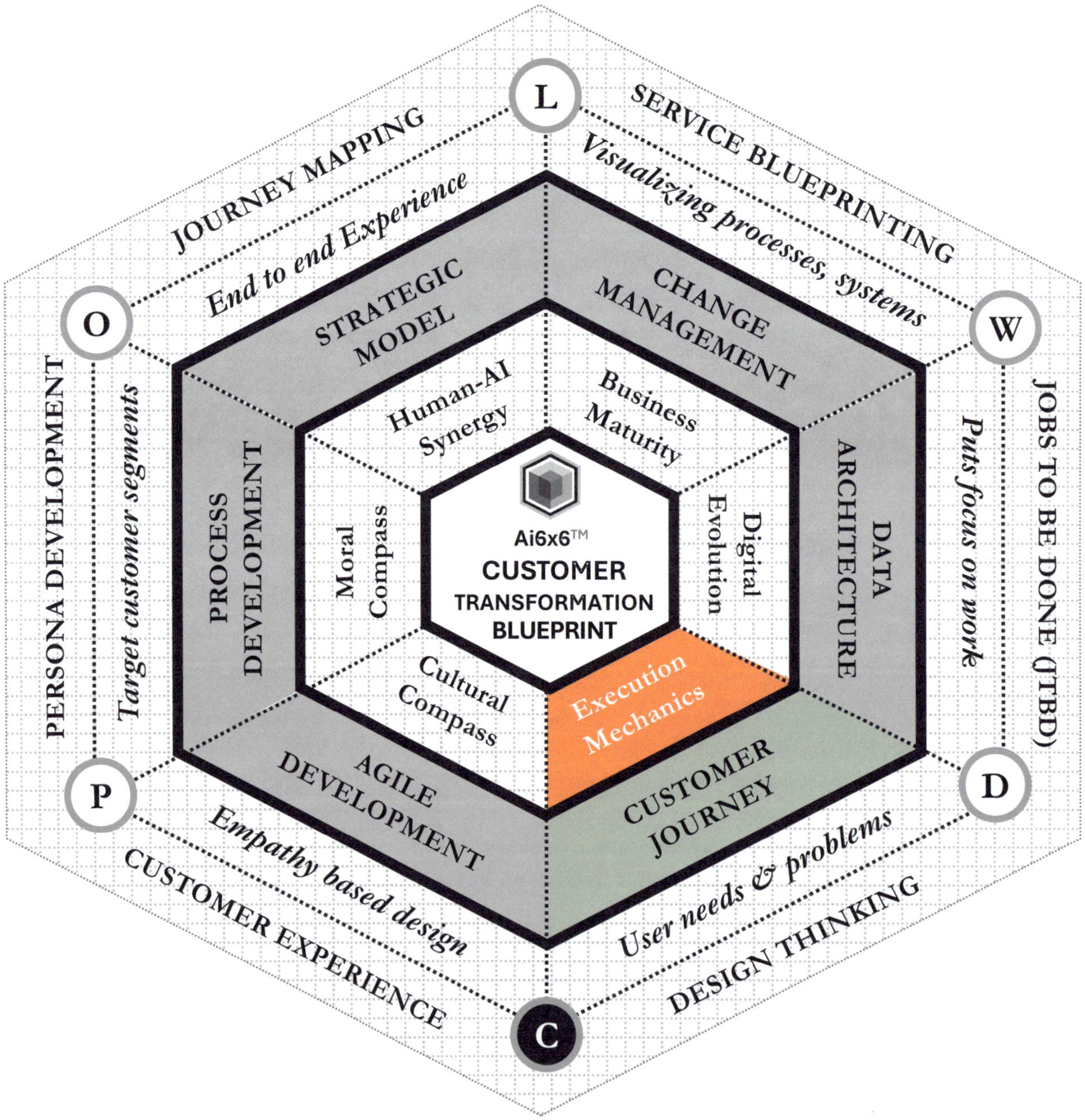

Ai6x6 CUSTOMER JOURNEY FRAMEWORK FOR CUSTOMER TRANSFORMATION

Customer Journey Mapping is a powerful visualization tool that illustrates the end-to-end experience a customer has with an organization, from their initial interaction to achieving their goal and beyond. It goes beyond simple process flows to capture the customer's actions, thoughts, and feelings at each touchpoint.

For customer transformation, this framework is indispensable because it shifts the organization's perspective from internal processes to the customer's lived experience. Many digital transformations focus on internal efficiency, but true digital maturity in customer experience requires understanding the customer's perspective across all digital and physical channels. Journey mapping helps identify critical digital touchpoints (e.g., website, mobile app, chatbot, social media), pinpoint pain points (e.g., frustrating online forms, inconsistent information across channels), and uncover moments of truth where the experience can make or break customer loyalty. By visualizing the entire journey, including the emotional highs and lows, organizations can identify opportunities to create seamless, personalized, and delightful digital experiences. It reveals where digital investments can have the greatest impact on customer satisfaction and retention, ensuring that technological changes are driven by genuine customer needs. This customer-centric approach to digital design and optimization is a hallmark of a digitally mature organization that truly understands and serves its customers in a connected world.

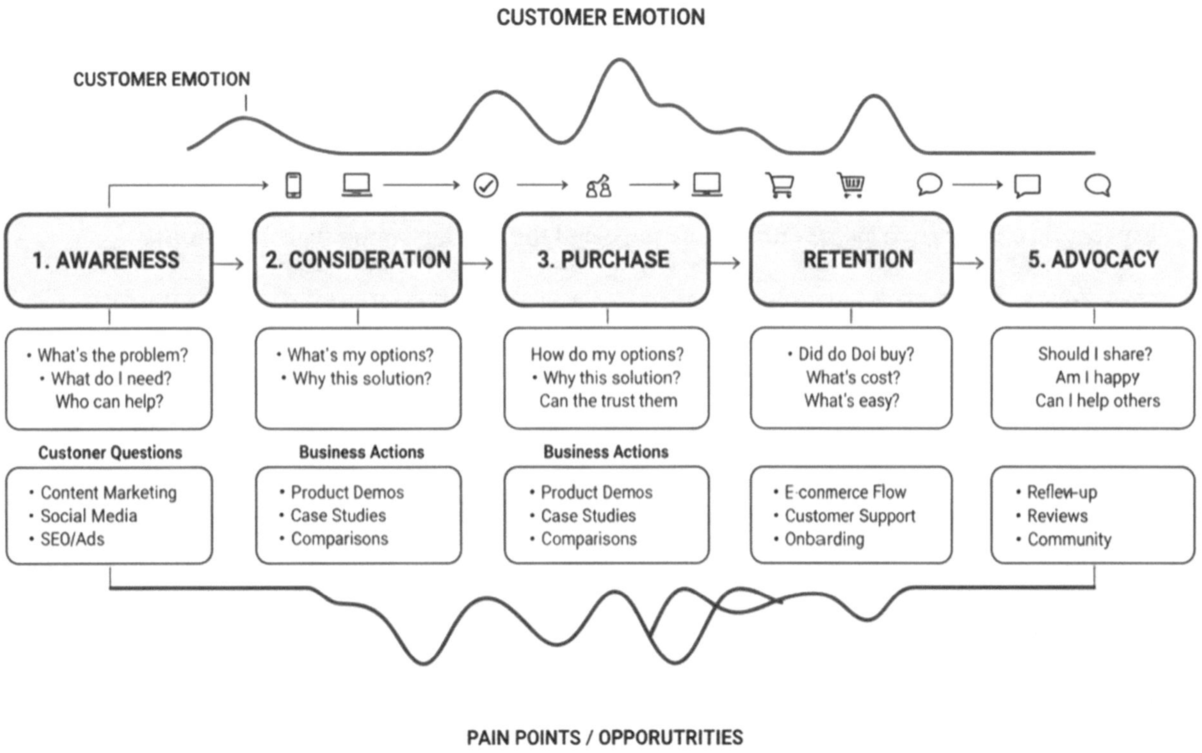

Customer Journey Mapping

1. **Define Scope & Persona:** Identify the specific customer persona whose journey will be mapped and clearly define the scope of the journey (e.g., from initial awareness to post-purchase support, or a specific service interaction). This ensures focus and relevance for the mapping exercise. For digital maturity, this step often involves defining personas that include their digital behaviors, channel preferences, and tech savviness, ensuring the journey reflects real digital user paths.

2. **Identify Customer Stages:** Break the entire customer journey into distinct, chronological phases or stages. Common stages include Awareness, Consideration, Purchase, Service/Usage, and Loyalty/Advocacy. These stages provide a high-level structure for the detailed mapping. In a digitally mature context, these stages are increasingly fluid and interconnected across various digital and physical touchpoints, requiring a holistic view.

3. **Map Customer Actions, Thoughts, Feelings:** For each stage, detail what the customer *does* (e.g., searches online, downloads an app, calls support), what they are *thinking* (e.g., "Is this product right for me? Is this website secure?"), and what they are *feeling* (e.g., frustrated by a slow digital interface, delighted by a personalized recommendation). This human-centered perspective is crucial for understanding the emotional landscape of the journey and identifying moments of digital friction or delight.

4. **Identify Touchpoints:** List all the points of interaction (digital and physical) between the customer and the organization at each stage. This includes websites, mobile apps, chatbots, social media channels, email campaigns, call centers, physical stores, and product usage itself. Understanding touchpoints helps identify where and how customers interact, revealing opportunities for digital integration and consistency across all channels.

5. **Uncover Pain Points & Opportunities:** Analyze the completed map to identify frustrations, inefficiencies, moments of friction, and unmet needs (pain points). For digital maturity, this often means pinpointing frustrating digital interfaces, broken omnichannel experiences, or missing digital self-service options. Simultaneously, look for moments of delight and opportunities to significantly improve the experience, create new value, or differentiate the brand through innovative digital solutions.

6. **Visualize & Prioritize Improvements:** Create a clear, compelling visual representation of the customer journey map (e.g., a timeline, infographic). Use the insights gained to brainstorm potential solutions, prioritize initiatives based on impact and feasibility, and assign ownership for implementing improvements to enhance the customer experience. For digital maturity, this translates into prioritizing digital transformation projects that directly address journey pain points and enhance digital touchpoints, leading to a truly seamless and engaging customer experience.

> "We've had three big ideas at Amazon that we have stuck with for 18 years, and they're the reason we're successful: Put the customer first. Invent. And be patient."
> — Jeff Bezos, Founder, Amazon

Service Blueprinting is a powerful visualization technique that extends customer journey mapping by illustrating the underlying organizational processes, systems, and people that support customer interactions. It maps the customer's actions to the visible "front-stage" activities (what the customer sees) and the invisible "back-stage" activities (what makes it happen), including support processes and physical evidence.

For customer transformation, particularly in achieving digital maturity, Service Blueprinting is critical. While customer journey maps highlight customer pain points, service blueprints reveal why those pain points exist by exposing the internal operational and technological complexities. It helps link customer-facing digital interactions (e.g., using a mobile app) to the backend systems (e.g., databases, APIs, legacy systems) and internal processes (e.g., data entry, approvals) that enable them. This comprehensive view allows organizations to identify inefficiencies, broken handoffs between departments or systems, and opportunities for digital automation and integration that will directly improve the customer experience. For example, a slow customer service response (customer pain point) might be traced back to a manual data lookup process (back-stage activity) that could be automated with RPA or integrated with a new digital system. By understanding these internal dependencies, leaders can strategically invest in digital transformation efforts that not only enhance the customer's digital journey but also streamline the underlying operations, leading to a more efficient, seamless, and digitally mature service delivery model.

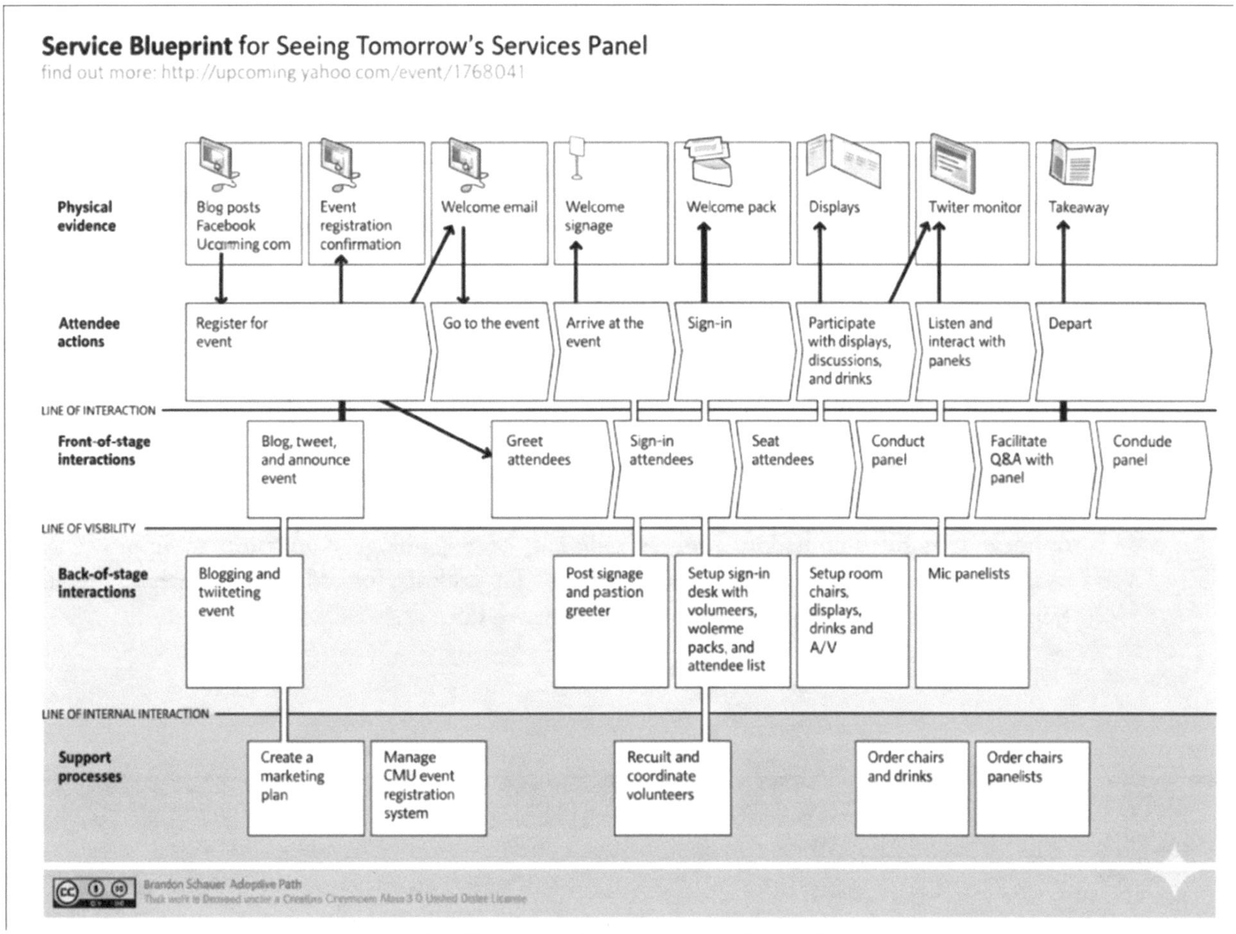

Service Blueprint , Image source: Wikipedia.org

1. **Start with Customer Journey Map:** Use an existing customer journey map as the foundation. The customer actions and touchpoints from the journey map will form the top layer of the service blueprint, providing the customer's perspective. This ensures that the internal process analysis remains firmly rooted in the customer's actual experience, which is paramount for digitally mature CX.

2. **Map Customer Actions & Front-Stage:** Reiterate the customer's actions at each stage of their journey. Below this, detail all the "front-stage" actions – those visible interactions and activities performed by the organization's employees or systems that the customer directly experiences (e.g., a customer service agent speaking, a website displaying information, a chatbot responding). For digital maturity, this layer focuses on the digital interfaces and interactions the customer directly engages with.

3. **Identify Back-Stage Actions:** Below the "line of visibility," detail all the "back-stage" actions. These are the invisible activities performed by employees or internal systems that support the front-stage actions but are *not* directly seen by the customer (e.g., an agent checking a database, a system processing an order, internal team communications, data synchronization between systems). This reveals the internal efforts required to deliver the visible digital experience.

4. **Map Support Processes:** Identify any internal support processes, systems, or physical evidence that enable the back-stage actions. This layer includes IT infrastructure, backend databases, CRM systems, ERP systems, HR systems, training programs, or physical resources (e.g., equipment) that are essential for delivering the service. For digital maturity, this often highlights the need for robust cloud infrastructure, integrated data platforms, and automated backend systems.

5. **Identify Physical Evidence & Touchpoints**: List all tangible elements the customer encounters at each touchpoint (e.g., website user interface, mobile app design, physical store layout, email template, product packaging, chatbot's tone of voice). This helps ensure consistency and quality across all customer interactions, both digital and physical, contributing to a cohesive brand experience.

6. **Draw Lines of Interaction & Analyze:** Connect the layers with lines to show dependencies and handoffs:
   - Line of Interaction: Between customer actions and front-stage actions.
   - Line of Visibility: Separates front-stage actions (what the customer sees) from back-stage actions (what they don't see).
   - Line of Internal Interaction: Between back-stage actions and internal support processes.

   Analyze the complete blueprint to identify inefficiencies, broken handoffs, bottlenecks, and opportunities for automation, process redesign, or technology integration to improve the overall customer experience. This analysis is critical for identifying where digital transformation can streamline operations to enhance CX.

> "The job of the designer is to make things clear."
> — Don Norman, Co-founder, Nielsen Norman Group (Pioneer in UX Design)

The Jobs to Be Done (JTBD) framework, popularized by Clayton Christensen, shifts the focus from traditional customer segmentation (demographics, psychographics) and product features to understanding the fundamental "jobs" customers are trying to get done. A "job" is a higher-level aspiration or problem that a customer is trying to solve in their life, encompassing functional, emotional, and social dimensions.

For customer transformation, especially in the pursuit of digital maturity, JTBD is profoundly insightful. Many organizations fall into the trap of simply digitizing existing products or processes without truly understanding the underlying customer need. JTBD helps leaders discover unmet digital needs and identify opportunities to create innovative digital products and services that truly solve customer problems in novel ways. For instance, a customer might "hire" a mobile banking app not just to check balances (functional job), but to "feel in control of their finances" (emotional job) or "project financial responsibility" (social job). By deeply understanding these jobs and the "pains" customers experience and "gains" they seek when trying to get them done, organizations can design digital solutions that offer superior value. It guides digital innovation towards creating solutions that are so effective at getting the job done that customers "switch" from competitors. This customer-centric approach to innovation, driven by a deep understanding of underlying needs rather than superficial features, is a hallmark of digitally mature organizations that consistently deliver breakthrough customer value.

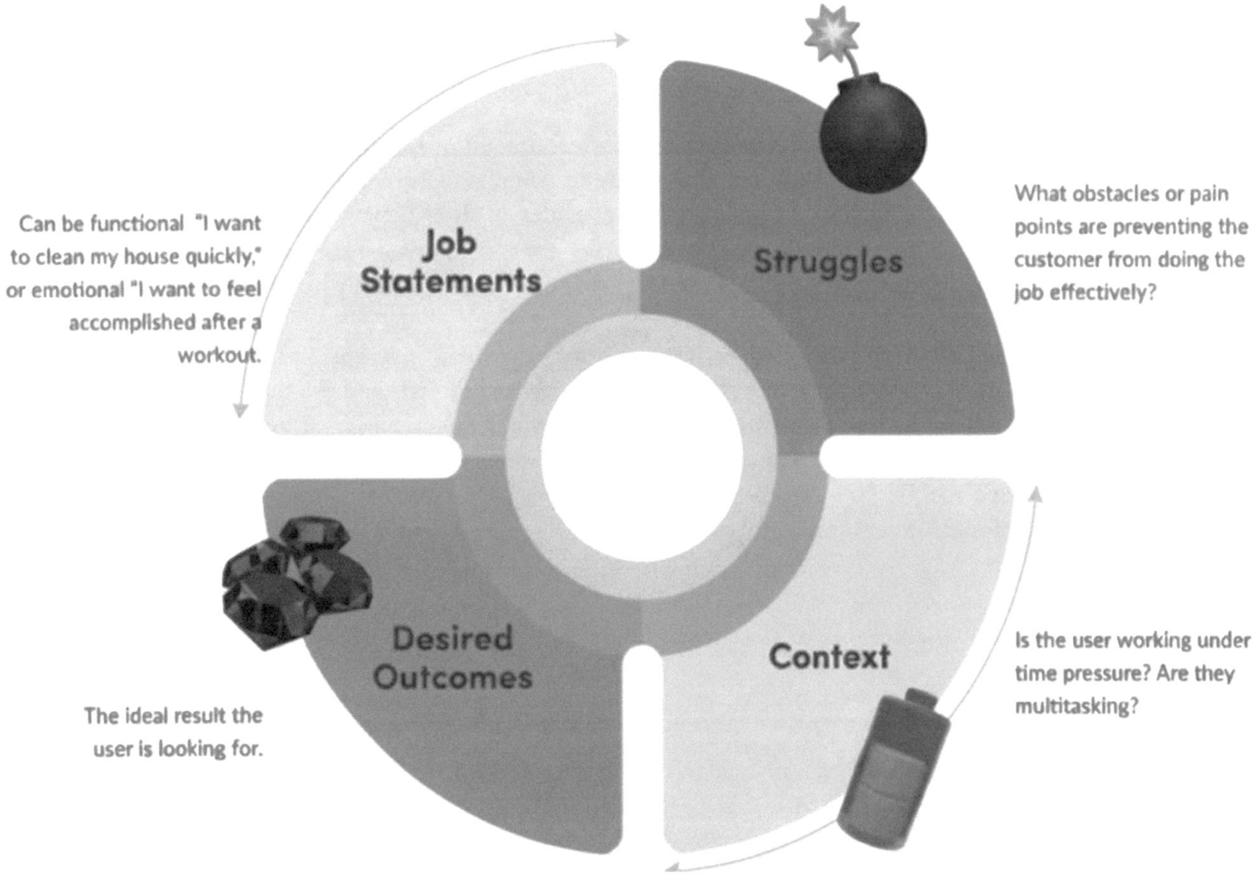

Jobs to Be Done Framework , Image source: https://productschool.com/

1. **Identify the "Job":** Discover the core functional, emotional, and social needs that customers are trying to fulfill. A "job" is a higher-level aspiration or problem, not just a specific task (e.g., "help me feel safe when driving my family," not "buy a car"). This requires deep qualitative research and moving beyond superficial digital feature requests to understand the underlying digital problems customers face or opportunities they seek to leverage with technology.

2. **Understand the "Job Performer":** Gain deep empathy for the customer (the "job performer"), their context, and their struggles related to getting the job done. This goes beyond demographics to understand their motivations, anxieties, and desired outcomes when faced with the "job." For digital maturity, this involves understanding their digital habits, their comfort level with technology, and the digital tools they currently use (or wish they had) to accomplish their "job."

3. **Map the "Job Steps":** Break down the "job" into a series of discrete, chronological steps the customer takes to accomplish it, regardless of the product or service currently used. These steps might include defining, locating, preparing, confirming, executing, monitoring, modifying, and concluding the job. This helps identify all points where digital solutions could intervene to make the job easier or better.

4. **Identify "Desired Outcomes":** For each "job step," articulate what success looks like for the customer. These are the criteria customers use to evaluate how well a solution performs the job. Desired outcomes are often expressed as metrics (e.g., "minimize the time it takes," "reduce the risk of error," "increase convenience"). For digital maturity, this means identifying desired outcomes that can be uniquely achieved or significantly enhanced through digital means (e.g., "receive real-time updates," "automate repetitive sub-tasks").

5. **Uncover "Pains" and "Gains":** Identify the obstacles, frustrations, and undesirable situations (pains) customers experience when trying to get the job done. Simultaneously, identify the benefits, delights, and positive experiences (gains) they seek. This provides a clear target for innovation. For digital maturity, this helps pinpoint "digital pains" (e.g., clunky apps, data privacy concerns) and "digital gains" (e.g., instant access, personalized recommendations) that can be addressed by new digital solutions.

6. **Design Solutions:** Develop products, services, or experiences that help customers get the "job" done better, faster, more cheaply, or with less effort. Solutions should directly address the identified pains and amplify the desired gains, leading to truly innovative and customer-centric offerings that outperform existing alternatives. For digital maturity, this means designing digital solutions (e.g., AI-powered assistants, intuitive mobile apps, seamless online platforms) that directly address the core "job" and deliver superior desired outcomes.

> "A lot of times, people don't know what they want until you show it to them."
> — Steve Jobs, Co-founder, Apple

Design Thinking is a human-centered, iterative approach to innovation and problem-solving that prioritizes understanding the user's needs, challenging assumptions, and redefining problems in an attempt to identify alternative strategies and solutions. It typically involves five non-linear stages: Empathize, Define, Ideate, Prototype, and Test.

For customer transformation, particularly in achieving digital maturity, Design Thinking is invaluable. Many digital initiatives fail because they are technology-driven rather than user-driven, resulting in complex, unintuitive, or irrelevant digital experiences. Design Thinking ensures that new digital solutions are truly user-centric. By starting with deep empathy for the customer, it helps organizations uncover unarticulated needs and pain points related to their digital interactions. This leads to defining the right problem to solve before jumping to solutions. The iterative nature of prototyping and testing allows for rapid experimentation with digital concepts (e.g., new app features, website designs, chatbot interactions) and quick learning from real user feedback. This minimizes risk and ensures that digital investments are directed towards solutions that genuinely delight customers and provide tangible value. A digitally mature organization consistently uses Design Thinking to create intuitive, seamless, and emotionally resonant digital customer experiences, differentiating itself in the market and building lasting customer loyalty.

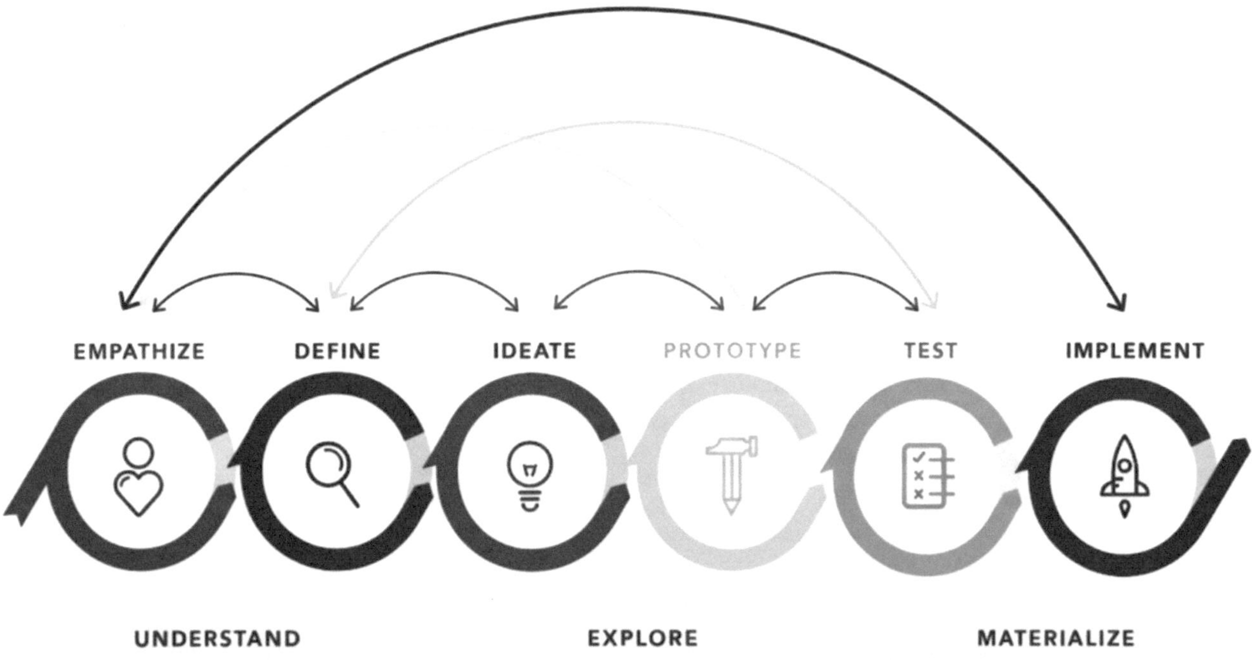

Design thinking framework , Image source: https://www.nngroup.com/

1. **Empathize:** The first stage involves gaining a deep understanding of your target customers. This means putting yourself in their shoes, observing their behaviors (especially digital interactions), listening to their stories, and conducting interviews to uncover their needs, challenges, desires, and motivations. The goal is to build genuine empathy for the user, which is foundational for designing truly effective digital customer experiences that resonate on a human level.

2. **Define:** Synthesize the insights gathered during the Empathize phase to clearly articulate the core problem(s) you are trying to solve. This involves framing the problem from the customer's perspective in a human-centered way (e.g., "How might we help busy parents quickly prepare healthy meals using smart kitchen devices?"). This clear, digitally-informed problem statement guides the subsequent stages, ensuring solutions are relevant and impactful.

3. **Ideate:** Brainstorm a wide range of creative solutions to the defined problem, without judgment. This phase encourages divergent thinking, generating as many ideas as possible, regardless of initial feasibility. Techniques like brainstorming, mind mapping, and sketching are commonly used to explore diverse possibilities, including innovative digital solutions like new app features, AI-driven services, or interactive online tools.

4. **Prototype:** Build low-fidelity versions of your most promising ideas. Prototypes can be sketches, mockups, storyboards, simple physical models, or even role-playing scenarios. For digital customer experiences, this often involves creating clickable wireframes, basic app mockups, or interactive website prototypes. The goal is to create something tangible that can be quickly tested and iterated upon, rather than investing heavily in a fully developed digital solution.

5. **Test**: Put your prototypes in front of real customers to gather feedback. Observe how they interact with the prototype, listen to their reactions, and ask probing questions. This iterative testing helps validate or invalidate assumptions, uncover new insights, and refine your solutions based on real-world user interaction. For digital solutions, this means observing how users navigate interfaces, whether they understand new features, and if the digital experience is intuitive.

6. **Implement:** Once a solution has been thoroughly tested and refined, and validated to meet customer needs effectively, it is ready for implementation. This involves bringing the solution to market, continuously monitoring its performance, and iterating based on ongoing customer feedback and market changes. Design Thinking is a continuous loop, not a linear process, ensuring that digitally transformed customer experiences remain relevant and continuously improve, a key characteristic of digital maturity.

> "Design thinking is a human-centered approach to innovation that draws from the designer's toolkit to integrate the needs of people, the possibilities of technology, and the requirements for business success."
> — Tim Brown, CEO, IDEO

CX (Customer Experience) Maturity Models provide a structured framework for organizations to assess their current capabilities in delivering customer experience and to plot a strategic roadmap for improvement. These models typically define a progression through various stages of maturity (e.g., from "Reactive" to "Proactive," "Differentiated," or "Customer-Centric") across several dimensions, such as CX strategy, customer understanding, experience design, data and analytics, technology enablement, organizational culture, and governance.

For customer transformation, particularly in achieving digital maturity, CX Maturity Models are crucial. They help organizations move beyond ad-hoc digital initiatives to a systematic, enterprise-wide approach to digital customer experience. By assessing their current digital CX capabilities, organizations can identify specific gaps—for instance, a lack of real-time data for personalization, fragmented digital channels, or an absence of a unified digital CX strategy. The model then guides the development of a phased roadmap, prioritizing digital transformation projects and investments that will have the greatest impact on advancing CX maturity. This ensures that digital efforts are not isolated, but contribute to a cohesive, integrated, and continuously improving digital customer experience. A digitally mature organization understands its CX maturity level, has a clear vision for its future digital CX, and systematically invests in the people, processes, and technologies (like AI, analytics, omnichannel platforms) required to deliver consistently exceptional and personalized digital interactions, ultimately driving customer loyalty and competitive advantage.

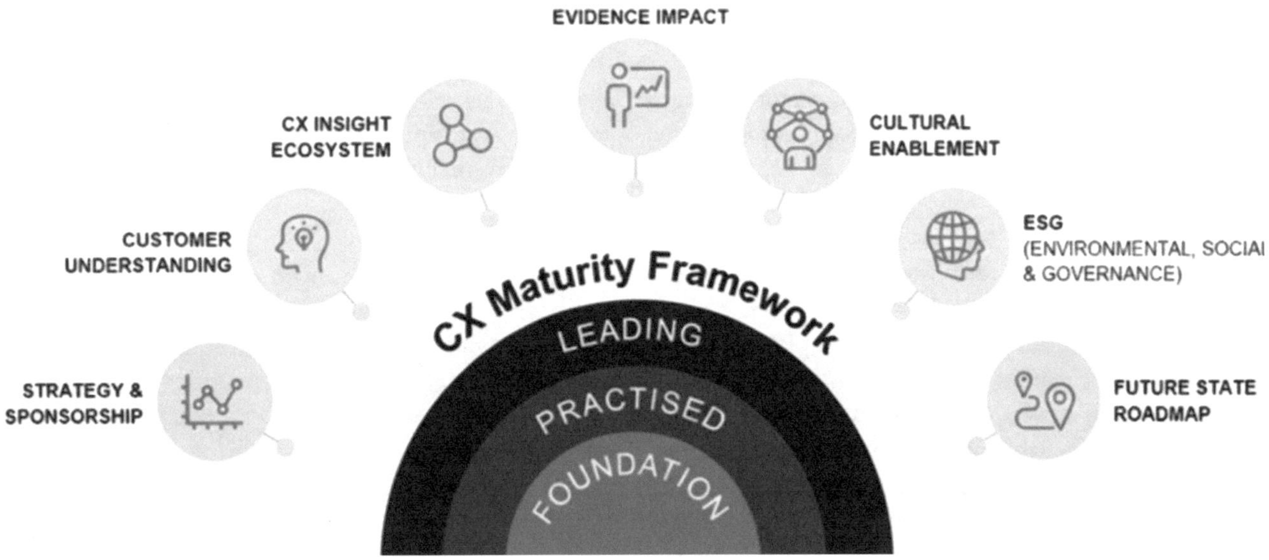

CX Maturity Framework , Image source: Ipsos.com

243

1. **Assess Current State:** Evaluate the organization's current customer experience capabilities across various dimensions. These dimensions typically include CX strategy, customer understanding, experience design, data and analytics, technology enablement, organizational culture, and governance. This assessment involves surveys, interviews, audits, and benchmarking against industry peers to establish a baseline, specifically identifying the current level of digital integration and effectiveness in CX.

2. **Define Target State:** Envision and clearly articulate the desired level of CX maturity the organization aims to achieve. Maturity models often define stages (e.g., "Reactive," "Emerging," "Differentiated," "Leading," "Customer-Centric"). This step involves setting aspirational yet realistic goals for where the organization wants its CX capabilities to be in the future, often striving for a highly personalized, proactive, and seamless digital customer experience.

3. **Identify Gaps:** Perform a detailed comparison between the assessed current state and the defined target state for each CX dimension. This step pinpoints the specific areas where current capabilities fall short of the desired maturity level, highlighting critical deficiencies or missing elements that need to be addressed. For digital CX, this might reveal gaps in omnichannel integration, real-time data utilization for personalization, or a lack of digital self-service options.

4. **Develop Roadmap:** Create a phased plan with prioritized initiatives and investments to bridge the identified gaps and advance CX maturity. This roadmap should outline specific projects, timelines, resource allocation, and expected outcomes, ensuring a structured approach to improving customer experience capabilities over time. For digital maturity, this roadmap would heavily feature digital transformation projects, technology investments, and digital skill development programs.

5. **Implement & Monitor:** Execute the CX roadmap, launching the defined initiatives and programs. Continuously track progress against defined metrics and KPIs (e.g., Net Promoter Score (NPS), Customer Satisfaction (CSAT), customer churn, digital channel adoption rates, operational efficiency gains from digital tools). Regular monitoring allows for real-time adjustments and ensures the transformation stays on track to deliver desired CX improvements, leveraging data for continuous optimization.

6. **Cultivate Culture:** Foster a customer-centric culture throughout the entire organization. This involves leadership modeling desired behaviors, embedding customer-focused values, providing training, and aligning incentives to ensure employees at all levels understand their role in delivering exceptional customer experiences and are empowered to act on customer insights. For digital maturity, this means fostering a culture that embraces digital tools, data-driven decision-making, and continuous learning to adapt to evolving digital customer expectations.

> "Customer service shouldn't just be a department, it should be the entire company."
> — Tony Hsieh, Former CEO, Zappos

6. PERSONA DEVELOPMENT FRAMEWORKS

Persona development frameworks guide the creation of rich, detailed representations of target customer segments. A persona is a semi-fictional representation of your ideal customer based on real data and some educated guesses about demographics, behavior patterns, motivations, and goals.

For customer transformation, particularly in achieving digital maturity, persona development is crucial because it fosters deep empathy and ensures that digital solutions are designed with real users in mind. Without well-defined personas, digital product and service development can become generic, feature-driven, or based on assumptions that don't reflect actual customer needs. Personas help organizations move beyond abstract market segments to understand the nuanced behaviors and motivations of their digital users. They provide a shared understanding across product, marketing, sales, and engineering teams about who they are building for, influencing decisions about digital channel preferences, user interface design, content strategy, and even the tone of digital communications. By incorporating digital habits, tech savviness, and online behaviors into persona profiles, organizations can design targeted and effective digital customer experiences that truly resonate with user needs and preferences, leading to higher digital adoption, satisfaction, and loyalty. This human-centered approach to digital design is a cornerstone of digital maturity.

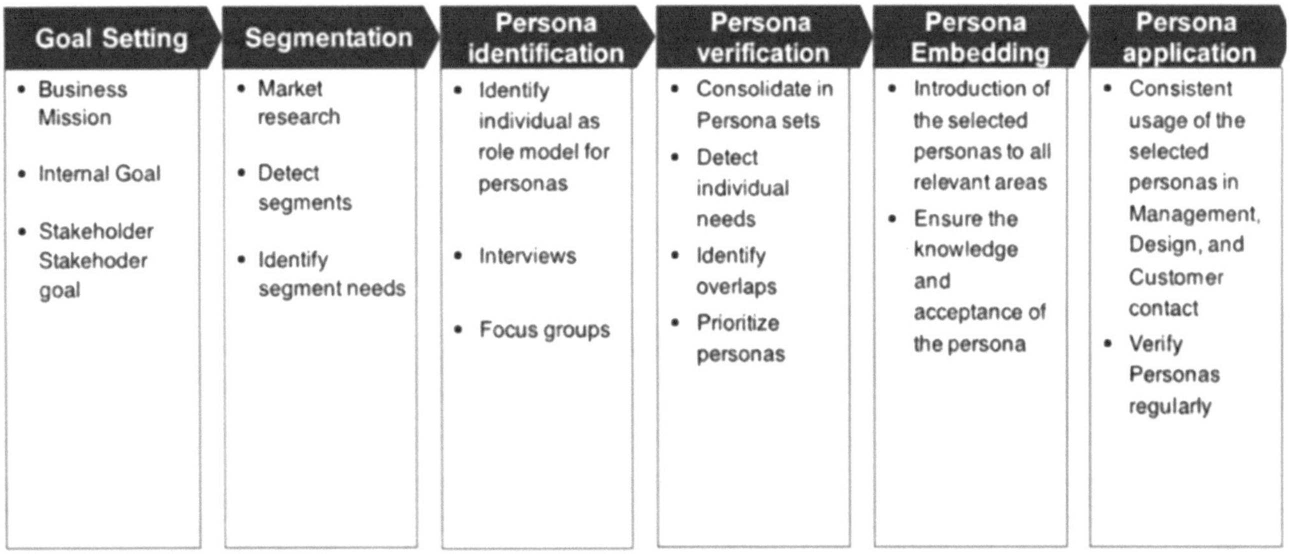

Persona Development Framework , Image source: Researchgate.com

1. **Research:** Begin by gathering extensive qualitative and quantitative data about your target users. This involves conducting in-depth interviews, surveys, ethnographic studies (observing users in their natural environment, including their digital interactions), analyzing website analytics, and reviewing customer support interactions. The goal is to uncover real behaviors, motivations, pain points, and digital habits (e.g., preferred digital channels, comfort with new tech).

2. **Identify Patterns:** Analyze the collected research data to identify common behaviors, motivations, goals, pain points, and demographic characteristics across your user base. Look for recurring themes and distinct clusters of users who share similar needs and traits. These patterns will form the basis for your different personas. For digital maturity, pay close attention to patterns in digital channel usage, digital literacy, and preferences for online vs. offline interactions.

3. **Create Persona Profile:** Develop a detailed profile for each distinct persona. This typically includes: a realistic name and photograph, key demographics (age, occupation, location), goals and motivations (what they want to achieve, including digital aspirations), pain points and frustrations (including digital friction), behaviors and habits (especially related to digital product/service usage), technology proficiency, and a compelling narrative or quote that encapsulates their essence. This brings the digital user to life for the development teams.

4. **Develop Scenarios:** Create short stories or use cases that describe how each persona interacts with your product, service, or organization to achieve their goals. These scenarios should explicitly include digital touchpoints and interactions (e.g., "Sarah uses the mobile app to track her order," "David struggles with the chatbot to resolve an issue"). These scenarios bring the personas to life and help teams visualize how different customer types will experience the transformed processes or new digital offerings.

5. **Communicate & Socialize:** Share the created personas widely across all relevant teams (product development, marketing, sales, customer service, engineering, leadership). Conduct workshops and presentations to ensure everyone builds a shared understanding of *who* the organization is transforming for, fostering empathy and aligning efforts around customer needs. This ensures that digital product development and service design are consistently customer-centric.

6. **Iterate:** Personas are living documents, not static artifacts. Continuously update and refine them as new customer insights emerge, as market conditions evolve (e.g., new digital trends), or as the organization's strategic focus shifts. Regularly revisit your research and make adjustments to ensure your personas remain accurate and relevant guides for customer transformation efforts, particularly in a rapidly evolving digital landscape.

"Your users are not you."
— Julie Zhuo, Former VP of Product Design, Facebook

Ai6x6™ Transformation Blueprint

The Empathy Equity Score (EES)

"You never really understand a person until you consider things from his point of view… until you climb into his skin and walk around in it." --- Harper Lee, the author of To Kill a Mockingbird created the global blueprint for moral courage and radical empathy. In the Customer lens, EES is the heartbeat of "Circular Partnership." Lee's wisdom anchors the shift from transactional sales to a relationship where the brand shares the customer's planetary burden, building long-term equity through genuine shared purpose.

In the industrial age, "Good Service" was a synonym for "Standardization." We spent decades perfecting the art of the script, the queue, and the uniform resolution. This was the era of Transactional Excellence, where Success was measured by how quickly we could move a human through a process without breaking the rules. But in the AI era, standardization is no longer a competitive advantage—it is a commodity. Every brand now has access to the same high-speed, "frictionless" bots. If your only value is speed, you are a utility, and utilities are eventually replaced by the cheapest alternative. This is the Identity Chasm of the modern brand: the gap between being Efficient and being Indispensable. When a brand automates its functional layers—returns, shipping updates, basic inquiries—it creates a vacuum. If that vacuum isn't filled with intentional human resonance, the brand becomes "Perfectly Cold." The customer's problem is solved, but their loyalty is lost. The primary metric for this evolution is the Empathy Equity Score (EES).

This chasm represents the tension between **Frictionless Efficiency** and **Meaningful Resonance**. As brands rush to automate, they often accidentally automate the "heart" out of the business, leaving customers feeling like processed data points. Identifying this chasm is critical because it prevents "Efficiency Myopia." It forces the brand to realize that in an era of infinite automation, human presence is the only remaining premium. You must bridge this to move from being a cold utility to an indispensable partner.

The greatest barrier to customer transformation is Efficiency Myopia—the delusion that a "zero-friction" experience is the same thing as a "high-value" experience. Organizations often use AI as a shield to hide from their customers, creating a "Digital Firewall" that makes it impossible to find a human soul. The staff, in turn, fall into a state of "Compliance Paralysis," mirroring the machine's coldness because they are being measured by the same rigid efficiency KPIs as the bot.

To bridge this chasm, the brand must stop using AI to replace the human and start using it to reveal the human. This is the shift from Volume to Resonance. It requires realizing that your value in an automated world is not your ability to "Process," but your ability to "Relate." You must use the machine to handle the Function so that the human can over-invest in the Feeling.

THE RAISON D'ÊTRE:

"TO FEEL"

By mastering the EES, the organization fulfills its true purpose: To Feel. In a world of infinite, free automation, Human Presence is the only thing that cannot be commoditized. It is the ultimate premium. When the machine handles the "Administrative Debris" of a transaction, the human employee is liberated to be the Sole Arbiter of Worth in the emotional exchange. We are not automating the relationship; we are automating the distractions that prevent the relationship from flourishing.

DEFINING THE EES: THE LOGIC OF RESONANCE

The Empathy Equity Score is a measurement of Emotional Surplus. It asks one simple, brutal question: Is the time we saved with AI being used to deepen the connection, or is it just being pocketed as a cost-saving?

The EES is the delta between Resolution Speed and Emotional Sentiment. If your bots are getting faster but your "Brand Affinity" is flatlining, you are in "Empathy Debt." You are effectively spending your long-term brand equity to buy short-term operational gains. The goal of the EES is to ensure that for every 60 seconds of "Friction" the AI removes, the organization reinvests at least 10 seconds of "High-Touch" human resonance.

THE OBJECTIVE: TOWARD RESONANCE

The goal of the EES intervention is to reach a state of Equilibrium we call Resonance. In this state, the technology is so invisible and efficient that it feels like "Magic," allowing the human interaction to feel like "Art."

Resonance allows a brand to command a "Premium of Presence." By increasing your EES, you move from being a vendor to being a partner. You are allowing the "Logic" of the machine to handle the "Transaction," so the "Heart" of your staff can handle the "Humanity." This is the only way to remain relevant in a world where everything else is "Frictionless" and "Forgettable."

CASE STUDY	THE CHASM REALITY	ACTION & RESULT
1: A LUXURY HOTEL CHAIN automated its concierge services to increase efficiency. However, guest satisfaction scores dropped as visitors felt the "digital-only" experience lacked the personal warmth expected of the brand.	The brand fell into **Transactional Excellence**, where "frictionless" became "meaningless." Leadership mistook speed for service, stripping away the emotional connection that built guest loyalty.	They used AI to handle "paperwork" (bookings/check-outs) while liberating staff for **High-Touch Hospitality**. Agents were retrained to use AI insights to provide "surprise" local gifts. Satisfaction rose 40%.
2: A GLOBAL AIRLINE deployed an AI chatbot to handle 80% of flight rebookings. While it saved millions in labor, it became a "Transactional Ghost" during a major storm, leaving thousands of stranded passengers in infinite FAQ loops.	The airline suffered from **Contextual Amnesia**. The AI could rebook a seat but could not acknowledge the passenger's mounting frustration or the complexity of a missed wedding or funeral.	Leadership installed **Sentiment-Based Routing**. If the AI detected "high-frustration" keywords, it instantly triggered a human handoff with full context. Brand trust scores recovered by 25% within a quarter.
3: A MAJOR TELECOM used predictive AI to offer "personalized" upgrade plans. Customers found the offers intrusive and "creepy," as the timing felt like the company was listening to private conversations.	This was a **Relevance-Resonance Gap**. The data was accurate (what the customer needed), but the delivery lacked **Empathetic Orchestration**, making the brand feel like a predator rather than a partner.	They shifted to **Value-Added Transparency**, having the AI explain *why* the offer was made (e.g., "We noticed your data usage doubled this month"). Conversion rates on offers improved by 18% as trust returned.
4: A FINTECH STARTUP achieved 100% automated customer support. While tickets were closed in seconds, the brand's community growth stalled because the bot's "voice" was cold, corporate, and lacked the brand's rebel identity.	The company faced **Identity Dilution**. By prioritizing the "logic" of the bot, they ignored the "soul" of the brand, turning their community of fans into a database of mere users.	They hired screenwriters to build a **Unified Brand Persona** for the AI, giving it the specific wit and tone of the founders. Community engagement surged 50% as the "Human" feel returned to the bot.
5: A RETAIL GIANT used AI to send millions of automated push notifications. The "spray and pray" approach led to high "opt-out" rates, as customers felt bombarded by irrelevant "Buy Now" messages.	This was **Efficiency over Empathy**. The leadership team was measuring "Clicks" rather than "Connection," treating their customer base as a resource to be mined rather than a relationship to be nurtured.	They implemented **Inquiry-Led Marketing**, using AI to ask customers questions about their preferences instead of just pushing products. This "Two-Way" dialogue reduced opt-outs by 60% and increased long-term value.
6: A HEALTH INSURANCE PROVIDER automated its claims denial letters. The cold, legalistic tone of the machine-generated letters during sensitive medical crises caused a public relations nightmare and legal scrutiny.	The firm was in **Moral Vacuum** mode. They automated the "What" (the denial) but failed to govern the "How" (the empathy), ignoring the high-stakes emotional state of the customer at that moment.	Leadership mandated **Empathy Guardrails**, requiring a human "Empathy Audit" for any AI-generated communication involving critical care. This move reduced customer grievances by 70% and repaired the brand's reputation.

ASSIGNMENT : THE CUSTOMER CULTURE DIAGNOSTIC AUDIT

The "Identity Chasm" in customer experience is most visible when a company is "Functionally Flawless but Emotionally Bankrupt." You have felt this: a refund that is processed in seconds but leaves you feeling like a nuisance rather than a guest. To bridge this gap, we must diagnose Transactional Coldness—the state where your staff has been so trained by "Efficiency KPIs" that they have forgotten how to be human. This audit is designed to expose where your "Standardization" is killing your "Resonance." On the left are the Symptoms of a brand that has lost its heart. On the right are the Strategic Interventions required to recalibrate for Equilibrium.

The Assignment: Identify the "Cold Spots" in your journey and match them to the human-centric remedy.

THE SYMPTOMS (THE COLD SPOTS)	MATCH	THE INTERVENTIONS (THE REMEDY)
1. The Scripted Mirror: Humans are so constrained by "Compliance AI" that they sound more robotic than the actual bots.		**A. High-Touch Calibration:** Using AI to flag "Emotional High-Stakes" moments so a human can step in *proactively*.
2. CSAT Narcissism: Celebrating high "Satisfaction" scores while ignoring a declining "Brand Loyalty" or referral rate.		**B. The Empathy Bounty:** Incentivizing staff for the *depth* of the interaction, not the speed of the hang-up.
3. The Dead-End Bot: An automated system that loops customers through FAQs without a "Human Escape Hatch."		**C. Sentient Routing:** Automatically moving complex or high-emotion cases to "Empathy Architects" with no time-limits.
4. Contextual Amnesia: Forcing the customer to repeat their life story as they move from bot to human to specialist.		**D. Radical Transparency:** Explaining *why* a step is automated, then offering a direct line to a human "Lead."
5. The "Mechanical" Human: Staff who are so monitored by "Call Duration" that they cut off customers mid-sentence.		**E. Unified Persona Mapping:** Providing the human agent with a real-time "Emotional Dossier" of the customer's history.
6. Efficiency Myopia: Cutting human-to-human time to meet a quarterly budget, ignoring the "Churn Tax" of lost souls.		**F. The Relationship Buffer:** Building "Unproductive Time" into schedules specifically for proactive customer outreach.
7. Tone Deafness: The AI provides a "cheerful" canned response to a customer who is clearly in a personal crisis.		**G. High-Fidelity Sentiment Filters:** AI that detects nuance (sarcasm, grief) and adjusts the brand's "Voice" instantly.
8. The Hidden VIP: Treating your most loyal, 10-year advocates with the same generic "logic" as a first-time web-crawler.		**H. Recognition Rituals:** Using AI to surface "Human Milestones" (birthdays, anniversaries) for personal human outreach.
9. Compliance Over Connection: Staff spend so much time logging data for the AI that they can't actually *listen* to the guest.		**I. Autonomous Documentation:** Letting AI handle 100% of the CRM logging to free the agent's eyes and ears for the guest.
10. The Empathy Gap: A workforce that is afraid to show genuine emotion because it might "violate the standard."		**J. Moral Authority Empowerment:** Giving staff the absolute right to "break the protocol" if it heals the relationship.

ANSWER MATCH: 1-J, 2-B, 3-D, 4-E, 5-I, 6-F, 7-G, 8-H, 9-A, 10-C

BLUEPRINT SERIES

SECTION F

Cultural Compass

Ai6x6™ Transformation Blueprint

Story 1 — The Post-War Japanese Industrial Ethos (1945–1960): Japan emerged from the war with limited resources but a collective commitment to national rebuilding. Companies embraced a culture of discipline, apprenticeship, and shared responsibility. Workers identified deeply with their organizations, reinforcing loyalty and craftsmanship. Leaders cultivated humility and long-term thinking. This cultural foundation enabled rapid industrial recovery. It demonstrated how shared values can accelerate capability even before systems mature.

Story 2 — NASA's Mission Culture During Apollo (1961–1972): NASA built a culture defined by precision, accountability, and collective purpose. Engineers and astronauts operated with a shared belief that failure was unacceptable, yet learning was essential. Communication was open, direct, and cross-disciplinary. Rituals, reviews, and norms reinforced psychological safety and technical rigor. This culture enabled teams to navigate uncertainty and complexity. It became a global example of mission-driven cohesion.

Story 3 — The Civil Rights Movement's Organizing Ethos (1954–1968): The movement relied on a culture of discipline, nonviolence, and moral clarity. Communities trained in de-escalation, collective action, and strategic communication. Leaders modeled courage and restraint, shaping norms that guided behavior under extreme pressure. Shared values created unity across diverse groups. This cultural coherence amplified impact far beyond organizational structure. It showed how culture can mobilize change at societal scale.

Story 4 — The Nordic Social Trust Model (1970s–1990s): Nordic countries cultivated cultures rooted in transparency, equality, and civic trust. Institutions operated with high accountability and low hierarchy. Citizens expected fairness and contributed to collective welfare. This cultural fabric enabled strong public services, stable governance, and economic resilience. Trust reduced friction in decision-making and collaboration. It demonstrated how cultural norms can become strategic national assets.

Story 5 — The Rise of Silicon Valley's Innovation Culture (1980s–2000s): Silicon Valley developed a culture that celebrated experimentation, risk-taking, and rapid iteration. Informal networks, open information flow, and meritocratic ideals shaped behavior. Failure was reframed as learning, attracting talent willing to push boundaries. Venture capital reinforced speed and ambition. This culture accelerated technological breakthroughs and entrepreneurial growth. It showed how cultural norms can amplify innovation beyond structural advantages.

Story 6 — The Globalization of Corporate Diversity & Inclusion (2000s–2020s): Organizations worldwide recognized that diverse teams outperform homogeneous ones when supported by inclusive cultures. Companies invested in psychological safety, equitable processes, and representation. Employees gained platforms to voice perspectives and challenge norms. Inclusion became a driver of creativity, retention, and ethical integrity. This cultural shift reshaped leadership expectations and organizational identity. It demonstrated how culture evolves to meet societal and competitive demands.

LESSONS FROM A CUSTOMER TRANSFORMATION PERSPECTIVE

Customer value deepens when culture shapes how organizations treat people, honor commitments, and respond to needs. Over time, institutions learned that trust, empathy, and fairness are cultural choices, not operational byproducts. Mature cultures embed customer respect into daily behavior—how teams communicate, resolve issues, and design experiences. Culture influences tone, transparency, and ethical boundaries. When employees believe in the organization's purpose, customers feel it. Maturity is achieved when customer-centricity is not a department but a shared mindset, guiding decisions across functions. Culture becomes the invisible infrastructure that sustains loyalty, credibility, and long-term relationships.

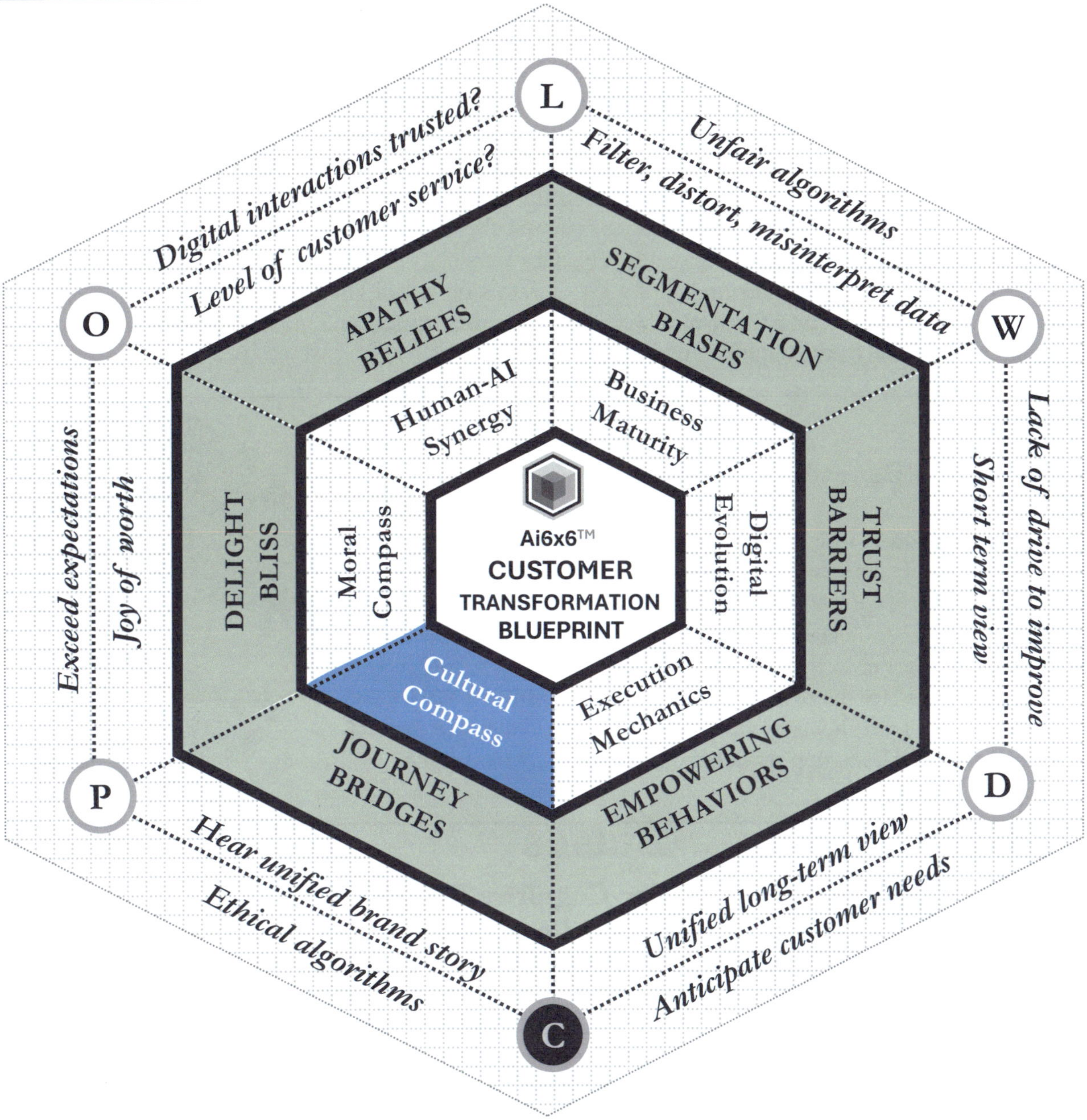

The **Ai6x6 Research Lab** chose this lens because the ultimate vitality of a brand is found in the collective "Bliss" of its customers. We reframed this dimension as the stewardship of the brand's soul, recognizing that culture is the invisible operating system that dictates whether a customer feels valued or merely managed. It is vital because it addresses the invisible shadows—the **Customer Apathy** and **Segmentation Biases**—that live in the minds of the organization. This lens is the bridge between the leader's high-tech strategy and the customer's visceral reality, moving the needle from indifference to **Customer Delight**.

We selected it to ensure leaders build a customer-centric community rather than just a profitable database. By dismantling **Barriers** to trust and building **Journey Bridges**, the leader creates a "greenhouse" for human potential where **Empowering Behaviors** become the standard. The Lab believes that in an age of automation, the human connection is the only self-sustaining source of market resilience. This lens forces a look at how leadership judgment impacts the collective behavior of the service team. When culture is treated as a discipline, the organization achieves a state of "bliss" where the customer feels fundamentally connected to the mission.

Culture has always been the invisible architecture of an organization, shaping how people think, behave, and make decisions long before any formal strategy takes hold. In the age of AI, this architecture becomes even more consequential. Technology may accelerate capability, but culture determines whether that capability becomes meaningful progress or wasted potential. The **Cultural Compass** lens forces leaders to confront the truth that transformation is not sustained by systems alone; it is sustained by the beliefs people hold, the biases they carry, the barriers they navigate, the behaviors they exhibit, the bridges they build, and the moments of profound satisfaction that anchor their commitment. Without cultural alignment, even the most sophisticated AI strategy collapses under the weight of human resistance. This lens reframes culture as a dynamic system rather than a static identity. It recognizes that culture is not a set of values on a wall but a living pattern of interpretation, interaction, and intention. As AI reshapes workflows, decision cycles, and power dynamics, leaders must understand how cultural forces amplify or undermine transformation.

Beliefs

The Narratives Beneath the Surface

Beliefs and assumptions form the first dimension of the Cultural Compass. They are the quiet forces that shape how people interpret change, how they respond to uncertainty, and how they judge the intentions behind leadership decisions. When beliefs are outdated or misaligned, they distort perception and create friction. Leaders who ignore this dimension discover that people do not resist technology—they resist the meaning they attach to it. Addressing beliefs requires leaders to surface the narratives that operate beneath the surface and reshape them with clarity, honesty, and purpose. In this dimension, cultural truth-telling becomes a leadership act rather than a communication tactic.

Biases

The Cognitive Patterns That Shape Judgment

Biases and mental shortcuts represent the second dimension. These cognitive patterns influence how people evaluate information, assign trust, and make decisions. In an AI-enabled environment, biases can distort both human judgment and the interpretation of machine-generated insights. Addressing them requires intentional design, transparent dialogue, and environments where people feel safe challenging their own assumptions. Leaders must create conditions where cognitive unlearning becomes possible. In this dimension, cultural evolution is driven by awareness, humility, and the willingness to question long-held mental models.

Barriers

The Structural Forces That Restrict Change

Barriers and systemic impediments form the third dimension. These are the structural forces—policies, hierarchies, incentives, and legacy processes—that constrain behavior even when people are willing to change. Leaders often misinterpret resistance as emotional when it is structural. Removing barriers is not about demanding more effort; it is about redesigning the environment so that the desired behaviors become the natural ones. In this dimension, organizational de-friction becomes a strategic priority rather than an operational afterthought, enabling people to move with clarity rather than struggle against constraints.

Behaviors

The Visible Expression of Culture

Behaviors and observable actions represent the fourth dimension. Culture is not what people say; it is what they do consistently. In the AI era, behaviors must evolve to reflect curiosity, adaptability, collaboration, and responsible experimentation. Leaders must model these behaviors visibly and repeatedly. They must demonstrate that learning is valued, that questioning is encouraged, and that accountability is shared. In this dimension, behavioral signaling becomes a powerful tool for shaping norms, reinforcing the idea that culture is enacted through daily choices rather than declared through slogans.

Bridges

The Pathways That Connect People and Purpose

Bridges and cultural pathways form the fifth dimension. These are the connections that allow ideas, trust, and collaboration to flow across the organization. Bridges are built through shared language, shared rituals, and shared purpose. They transform isolated teams into aligned communities. Leaders who invest in cultural pathways create environments where people feel connected to something larger than their role. In this dimension, collective identity-building becomes a catalyst for cohesion, enabling the organization to move as a unified whole rather than a collection of disconnected parts.

Bliss

The Emotional Anchor of Commitment

Bliss and profound satisfaction complete the compass. Bliss is not about comfort; it is about meaning. It is the sense of fulfillment that emerges when people feel their work matters, their contributions are recognized, and their growth is supported. Bliss is the emotional anchor that sustains commitment through uncertainty. Leaders who cultivate it create cultures where people do not merely comply—they care. In this dimension, purposeful fulfillment becomes a strategic asset, strengthening resilience and deepening engagement.

The Cultural Compass lens reminds leaders that transformation is not sustained by technology or process alone. It is sustained by the human experience of change. Culture determines whether people interpret transformation as threat or opportunity, whether they respond with resistance or curiosity, and whether they move in alignment or fragmentation. In a world where AI accelerates both progress and pressure, leaders must treat culture as the primary operating system of the enterprise. When leaders align beliefs, challenge biases, remove barriers, shape behaviors, build bridges, and cultivate meaning, they create cultures capable of evolving continuously. They build environments where people feel psychologically safe, intellectually stimulated, and emotionally invested. Culture becomes the force that amplifies strategy rather than undermining it. It becomes the stabilizing anchor in moments of disruption and the accelerating engine in moments of opportunity.

CULTURAL COMPASS ANALOGY – CULTIVATING JAPANESE GARDEN

The False Climate — BELIEFS

Dismantling "Weather Myths" ensures that unrealistic, universal expectations don't stifle individual growth. Exposing the false belief that all "plants" require the same sun, water, or timing—allowing for specialized care.

The Invasive Weeds — BIASES

Identifying and removing hidden mental shortcuts prevents old habits from choking out innovation. Surgically exposing the leadership judgment and systemic biases that create "shadow" cultures and block progress.

The Hard Clay — BARRIERS

Breaking through structural resistance allows the organizational roots to expand and stabilize. Removing the deeply embedded obstacles like data, automation and AI anxiety that act as physical walls against digital growth.

Calibrated Care — BEHAVIORS

Executing specialized nurturing routines ensures each distinct "species" receives what it needs to thrive. Moving away from "one-size-fits-all" mode to model specific watering, feeding, and light-balancing behaviors required for a diverse ecosystem.

The Trellis System — BRIDGES

Constructing pathways for upward growth provides the support needed for teams to reach their full potential. Building the specific leadership and development paths that allow the workforce to climb through the maturity stages.

The Full Harvest — BLISS

Securing a state of permanent organizational harmony results in a self-sustaining ecosystem of excellence. Reaching the ultimate destination where the "Cultural Soul" operates in a high-fidelity hum, delivering constant value.

Ai6x6™ Transformation Blueprint

Customer Apathy Beliefs

"It is not the employer who pays the wages. Employers only handle the money. It's the customer who pays the wages." - Henry Ford

Henry Ford (1863–1947) was an American industrialist and founder of the Ford Motor Company. He is best known for pioneering the moving assembly line, which revolutionized manufacturing and made cars affordable for the public.

Customer apathy beliefs are negative assumptions about customers that lead to a transactional, inward-focused mindset. These beliefs can be deeply ingrained, often stemming from a period of past success or a lack of direct customer interaction. A common belief is that "customers are primarily driven by price," which leads to an over-focus on cost-cutting and a neglect of the customer experience. Another limiting belief is that customers are a monolith, leading to a "one-size-fits-all" approach that fails to resonate with diverse customer segments. The journey to customer transformation begins by challenging these beliefs and cultivating a new mindset that sees the customer not as a transaction but as a relationship to be nurtured. This requires a commitment to deep customer understanding, empathy, and a willingness to prioritize the customer's perspective above all else.

Completing the Relational Core, this book examines Customer Apathy Beliefs, the psychological convictions that govern market reaction to new offerings, influence Engagement, and lead directly to Human Fulfillment—the ultimate external measure of successful transformation. It explores the non-product factors that drive purchasing behavior and long-term loyalty. Chapters detail how to cultivate Brand Loyalty Beliefs, ensuring the company consistently delivers on its promises across all touchpoints, and establishing fundamental Trust Beliefs through reliability, ethical conduct, and transparency. Crucially, the book analyzes how the customer perceives the true usefulness and benefit through Value Proposition Beliefs and how external validation (via Social Proof Beliefs) accelerates mass market adoption and reduces purchase friction. Finally, it addresses the psychological barriers to adoption, focusing on convictions around Beliefs in Ease-of-Use (the product will be simple to integrate) and the future commitment conveyed through Beliefs in Future Promise—the conviction that the company will continue to innovate and support the customer long-term.

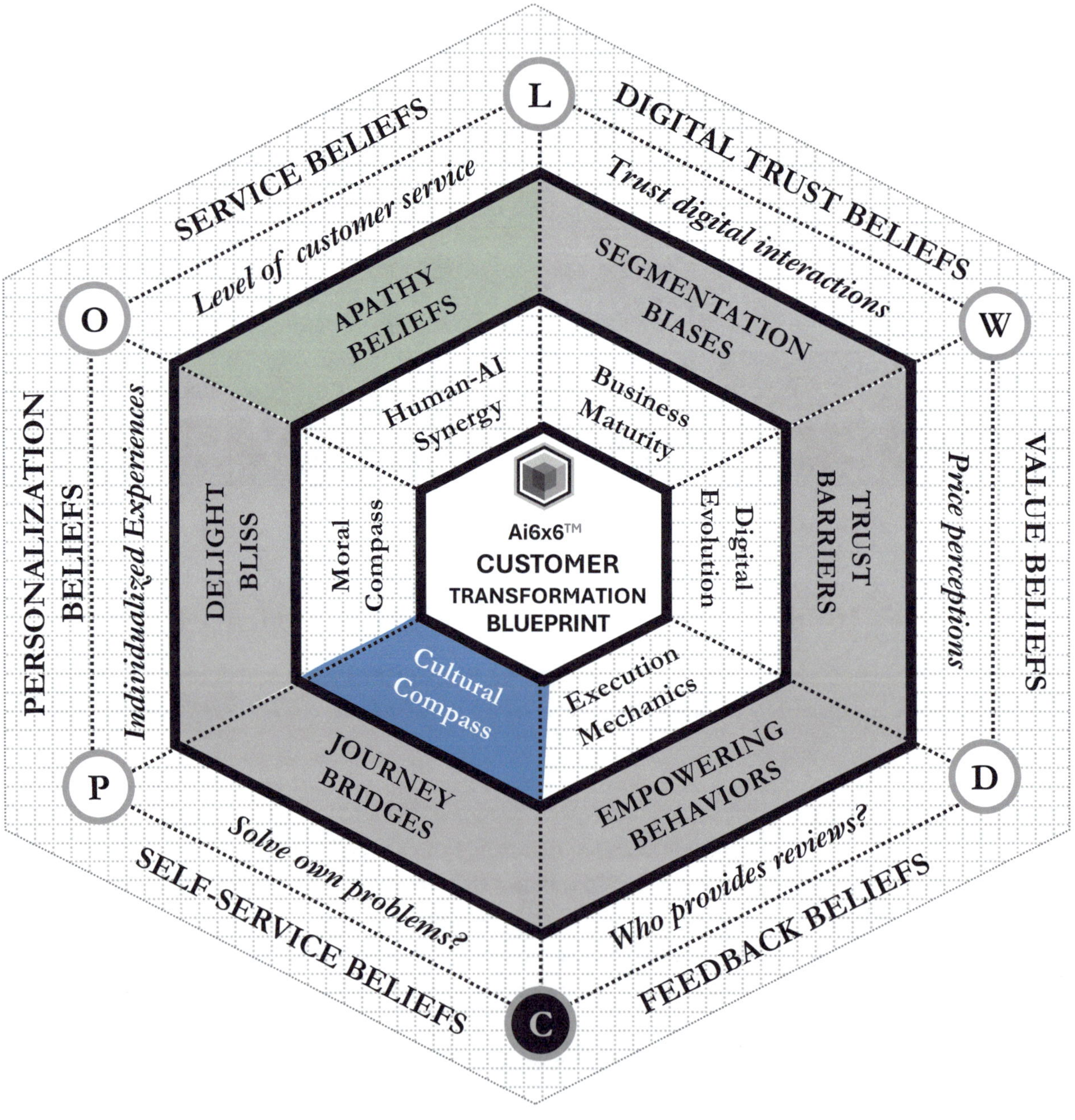

Ai6x6 CULTURAL COMPASS FRAMEWORK FOR **CUSTOMER LIFECYCLE | BELIEFS**

1. SERVICE BELIEFS | The Unseen Bar

If you hold the belief that customers only expect "functional speed," you are missing the emotional heart of the AI era. Many organizations believe that as long as the bot answers quickly, the service is successful. However, in an era of automated interactions, customers actually crave high-touch empathy more than ever before. Daily, this belief leads to "efficiently cold" service that resolves the ticket but loses the customer's heart and future loyalty. Transformation requires believing that service is a platform for deep connection, not just a utility for basic resolution. When you believe service must be both intelligent and empathetic, you move beyond mere transactions to creating memorable experiences that foster long-term advocacy in a crowded and noisy digital marketplace.

DO'S FOR THE LEADER	DON'TS FOR THE LEADER
Believe in a personalized experience.	Assume one-size-fits-all is enough.
Exceed expectations consistently.	Treat service as a necessary evil.
Empower employees to solve issues.	Rely on rigid scripts for everything.
Use customer feedback to improve.	Ignore feedback as just complaining.
Invest in high-quality service.	Cut service costs to the bare minimum.
Treat all customer issues as important.	Dismiss customer issues as trivial.

SITUATION	SUGGESTED ACTION
Customers abandon chat sessions because slow responses signal low commitment to resolving their issues	Set clear response targets, staff peak periods appropriately, use queue transparency, and reinforce reliability expectations across all digital and human service channels
Customers repeatedly ask for supervisors because frontline agents seem powerless to make meaningful decisions	Expand frontline decision rights, provide scenario playbooks, and publicly celebrate quick resolutions achieved without escalation to supervisors in customer view to build confidence
Customers call instead of using apps because they expect humans to fix issues faster	Redesign key flows for speed, surface live wait comparisons, and show successful self-service outcomes directly inside the app to prove digital efficiency
Customers downgrade plans because post-sale support feels slower and less attentive than sales interactions	Mirror sales responsiveness standards in support, track handoff quality, and review post-onboarding experience metrics in executive forums on a recurring basis to drive accountability
Customers avoid new channels because they assume service quality declines when companies introduce automation	Provide hybrid options, highlight human oversight, and reinforce that automation enhances speed without reducing care or attention
Customers escalate billing questions immediately because they believe only senior staff understand complex account details	Train frontline teams on billing scenarios, simplify explanations, and reinforce confidence by resolving common issues without escalation

If customers "automatically trust" digital systems because they use them is a dangerous belief that can lead to strategic ruin. In the AI era, trust is the scarcest and most valuable resource you possess. If you believe digital trust is a given, you will ignore the radical transparency required to explain how customer data is being used to fuel AI insights. Daily, this manifests as opaque "black box" interactions that leave customers feeling monitored and exploited rather than truly served. To transform, you must believe that digital trust is something you earn every day through honesty. By shifting this belief, you prioritize privacy by design, ensuring that every automated interaction feels like a secure, respectful conversation rather than a data-mining exercise.

DO'S FOR THE LEADER	DON'TS FOR THE LEADER
Prioritize data security and privacy.	Treat data security as an afterthought.
Be transparent about data usage.	Hide how you use customer data.
Use simple, clear digital interfaces.	Create confusing or complex systems.
Make policies easy to understand.	Use vague, long legal agreements.
Earn trust with consistent service.	Take customer trust for granted.
Proactively communicate security.	Wait for a breach to communicate.

SITUATION	SUGGESTED ACTION
Customers abandon registration when permission screens appear, assuming data will be sold or misused	Use plain-language consent, show specific uses with examples, and provide easy toggles so customers control optional data sharing at any time across channels
Customers refuse chatbot help because previous bots looped endlessly without understanding simple account questions	Route ambiguous intents quickly to humans, display escalation options upfront, and share examples where the assistant successfully resolved similar issues for comparable customers recently
Customers hesitate to store cards because they doubt encryption and fear unauthorized recurring charges	Highlight independent security certifications, explain storage design visually, and allow granular controls for disabling saved payment methods instantly from web, app, and support channels
Customers ignore security alerts because previous notifications felt generic, frequent, and rarely relevant to them	Personalize risk messages with concrete context, reduce noise, and show past incidents where timely alerts prevented real customer harm in similar account situations
Customers avoid app updates because they assume new versions introduce unnecessary complexity or hidden changes	Provide guided tours, highlight improvements, and reinforce stability by showing what remains familiar after each update
Customers distrust recommendation engines because they believe algorithms prioritize company revenue over customer benefit	Explain recommendation logic, show relevance examples, and reinforce fairness by highlighting how suggestions align with customer goals

The belief that customers perceive value only through price or basic features is significantly outdated in a hyper-connected world. In the AI era, value is increasingly defined by "relevance" and the "time saved" for the user. If you believe that your product's value is static, you will fail to use AI to dynamically adapt to a customer's changing life stages and needs. Daily, this leads to static marketing that feels "tone-deaf" to the customer's current reality. Transformation requires believing that value is a moving target that must be co-created with the customer. When you believe that value is about helping the customer achieve their specific goals at the right moment, you unlock the power of AI to provide "just-in-time" solutions.

DO'S FOR THE LEADER	DON'TS FOR THE LEADER
Understand what customers value.	Assume value is just about price.
Clearly communicate your value.	Let customers guess your product's worth.
Offer a fair, transparent price.	Use complex, hidden pricing.
Constantly add value for customers.	Let your product get stale and old.
Ask for feedback on perceived value.	Believe your product is obviously the best.
Focus on the customer's needs.	Only focus on what you're selling.

SITUATION	SUGGESTED ACTION
Customers compare prices obsessively because they believe your premium branding hides only cosmetic differences	Quantify outcome improvements, share independent benchmarks, and translate technical advantages into concrete savings or risk reduction for customers over realistic usage periods
Customers downgrade tiers because they rarely use advanced features bundled into higher-priced plans	Provide usage heatmaps, recommend right-sized configurations, and proactively suggest plan changes before renewal conversations begin to demonstrate partnership and fairness in commercial decisions
Customers question implementation fees because they assume setup is mostly automated and low effort	Decompose project activities, show human expertise involved, and connect each component to reduced risk or accelerated time-to-value for their specific environment and constraints
Customers treat pilots as free trials and resist paying for proven, ongoing optimization work	Define pilot scope, document measurable impact, and position post-pilot services as structured programs delivering continuous performance gains over agreed business horizons together
Customers hesitate to renew because they believe long-term value declines after initial deployment benefits	Share improvement roadmaps, highlight new capabilities, and reinforce ongoing value through measurable performance gains
Customers resist add-ons because they assume core offerings should already include those capabilities	Clarify product boundaries, show optional value, and reinforce modular design benefits through real customer examples

If you believe that customers only give feedback when they are angry, you are operating with a massive strategic blind spot. This belief leads to a "firefighting" culture where you only listen to the loudest, most negative voices. In reality, modern customers want to be co-creators of the brand they support. Daily, this manifests as ignoring the subtle "quiet data" of customer behavior that signals untapped opportunities and early-stage needs. Transformation requires believing that all customer signals—both loud and silent—are a gift of actionable insight. By shifting this belief, you move from simple surveys to "listening at scale," using AI to capture the sentiment of the "silent majority." This ensures your product evolution is guided by the true, holistic voice.

DO'S FOR THE LEADER	DON'TS FOR THE LEADER
Actively seek all types of feedback.	Ignore negative or critical feedback.
Believe feedback is a gift.	See feedback as a burden or complaint.
Make it easy to leave feedback.	Create a complicated feedback process.
Respond to customer feedback.	Let feedback go into a black hole.
Use feedback to drive improvements.	Dismiss feedback as a waste of time.
Share feedback with the whole team.	Keep feedback locked in one department.

SITUATION	SUGGESTED ACTION
Customers skip satisfaction surveys after support because previous submissions never produced visible changes	Publish recent improvements tied to feedback, thank contributors personally when possible, and close the loop within defined timeframes for each major channel
Customers only respond when angry because they believe positive experiences are already expected baseline	Ask targeted questions about standout moments, invite quick rating taps, and showcase positive stories customers chose to share with their permission publicly
Customers provide vague comments because forms feel generic and disconnected from their actual journey	Trigger contextual micro-surveys at key moments, prefill scenario details, and ask one precise question per interaction to capture actionable insight quickly and respectfully
Customers post long public reviews because they believe social platforms get faster attention than support	Monitor external channels actively, respond with concrete next steps, and invite structured follow-up through owned feedback pathways that feel genuinely responsive and respectful
Customers decline interviews because they assume companies only want feedback that confirms existing decisions	Explain interview purpose, show examples of real changes, and reinforce openness to challenging insights
Customers stop giving feedback because they believe companies reward promotional comments more than constructive suggestions	Provide transparent criteria, recognize meaningful contributions, and reinforce fairness in how all feedback is valued

The assumption that self-service is just a "cost-cutting tool" is a belief that sabotages the customer experience at its core. If you believe customers "have" to use self-service to save you money, you will build frustrating and rigid systems. In fact, modern customers often prefer the autonomy of self-service if it is fast, effective, and intuitive. Daily, this leads to poorly designed bots that feel like "digital hurdles" rather than helps. Transformation requires believing that self-service is a powerful empowerment tool. When you believe that self-service should be the fastest path to a "win" for the customer, you design intuitive, AI-powered portals that provide blissful autonomy, allowing customers to solve their problems with ease and leaving humans for complex work.

DO'S FOR THE LEADER	DON'TS FOR THE LEADER
Provide a comprehensive knowledge base.	Offer no self-help resources.
Make self-service intuitive and easy.	Use confusing, hard-to-find resources.
Empower customers to solve problems.	Force customers to call for help.
Keep help content up-to-date.	Have outdated and useless information.
Make human support an easy option.	Hide your phone number or email.
Use data to improve self-service.	Assume your self-service is perfect.

SITUATION	SUGGESTED ACTION
Customers avoid help centers because they believe self-service articles are outdated and rarely helpful	Update content continuously, show revision timestamps, and reinforce trust by highlighting recently solved issues similar to their current situation
Customers abandon troubleshooting flows because they assume automated steps will eventually force contacting support anyway	Provide complete resolution paths, show success rates, and reinforce confidence by demonstrating real cases resolved entirely through self-service journeys
Customers skip FAQs because they believe answers are generic and not tailored to their specific context	Personalize FAQ ordering, surface relevant scenarios, and reinforce usefulness by matching content to customer history and product configuration
Customers call immediately because they assume self-service tools cannot handle complex or unusual problems	Add branching diagnostics, show advanced options, and reinforce capability by demonstrating successful resolutions for comparable complex cases
Customers ignore in-app guidance because they believe instructions oversimplify issues and hide important details	Provide layered explanations, reveal deeper steps on demand, and reinforce transparency by showing full troubleshooting logic
Customers distrust automated returns because they assume refunds require human approval to be processed correctly	Display refund timelines, show automated verification steps, and reinforce reliability by highlighting consistent refund accuracy across similar transactions

Believing that personalization is just "putting a name in an email" is a belief that leads to "shallow relevance" and ignored messages. In the AI era, if you believe personalization is a marketing gimmick, you will fail to build the deep context required for true individualization. Daily, this manifests as "creepy" or "irrelevant" recommendations that miss the customer's true intent. Transformation requires believing that personalization is an act of genuine service. When you believe that every customer interaction should be as unique as their fingerprint, you commit to using AI to understand their history and preferences in real-time. This belief shifts your focus from "mass marketing" to "one-to-one service," ensuring that every customer feels genuinely seen and understood by your brand.

DO'S FOR THE LEADER	DON'TS FOR THE LEADER
Use data to personalize experiences.	Offer a one-size-fits-all solution.
Ask customers about their preferences.	Make assumptions about what they want.
Offer personalized product suggestions.	Spam customers with generic promotions.
Be respectful of customer privacy.	Use data in a way that feels intrusive.
Personalize to add real value.	Personalize for the sake of it.
Create different customer segments.	Treat every single customer the same.

SITUATION	SUGGESTED ACTION
Customers ignore recommendations because they believe personalization repeats the same suggestions regardless of changing needs	Refresh recommendation logic, incorporate recent behavior, and reinforce relevance by showing why each suggestion appears
Customers disable notifications because they assume personalized alerts are thinly disguised promotional messages	Separate promotional and functional alerts, clarify purpose, and reinforce trust by showing meaningful, context-specific triggers
Customers reject onboarding tips because they believe personalization requires sharing excessive personal information	Offer privacy-first onboarding, minimize required fields, and reinforce control by letting customers adjust personalization depth anytime
Customers skip preference settings because they assume personalization will not noticeably change their experience	Demonstrate immediate impact, preview personalized layouts, and reinforce value through visible before-and-after comparisons
Customers distrust tailored pricing because they believe personalization hides unfair or inconsistent discounts	Explain pricing logic, show eligibility criteria, and reinforce fairness by applying consistent rules across similar customer profiles
Customers avoid profile completion because they assume companies use data primarily for targeted upselling	Clarify data purpose, highlight service improvements, and reinforce customer benefit through transparent examples

Ai6x6™ Transformation Blueprint

Customer Segmentation Biases

"To understand the customer, walk a day in their shoes." - Japanese Proverb

This proverb is a traditional saying from Japan that encapsulates the profound importance of empathy and perspective-taking. It is a guiding principle in many aspects of Japanese culture and business.

Customer segmentation biases are cognitive shortcuts that can distort a company's understanding of its customers. These biases can lead to flawed strategies and a poor customer experience. The Framing Effect Bias, for example, highlights how a customer's decision can be heavily influenced by how information is presented, rather than its substance. Another critical issue is Selective Attention Bias, where an organization focuses on a specific type of feedback (e.g., positive reviews) while ignoring other valuable input (e.g., nuanced suggestions or complaints). This creates a distorted, overly positive view of the customer experience. Overcoming these biases requires a commitment to a holistic, evidence-based approach to understanding every customer interaction. As the Japanese Proverb suggests, "To understand the customer, walk a day in their shoes." This is about building a culture of empathy and curiosity, where every member of the organization is committed to seeing the world from the customer's perspective.

Completing the Relational domain, this chapter examines Customer Segmentation Biases, the psychological shortcuts that filter, distort, and ultimately govern market reaction to new offerings, influencing Engagement and the perception of Human Fulfillment. It provides marketers and designers with the psychological literacy needed to craft trustworthy messaging. Sections detail the manipulation of context through the Framing Effect, where the presentation of information—not the content—sways reaction, and the challenge of getting noticed due to Selective Attention Bias. The chapter explores how customer memory can be fundamentally altered by the Misinformation Effect and how the Recency/Primacy Effect weights initial and most recent experiences most heavily in overall judgment. Crucially, it explores the challenge of navigating market bubbles caused by Filter Bubbles/Echo Chambers (where customers are exposed only to reinforcing opinions) and the impact of Algorithmic Ranking Bias, where search results and recommendations fundamentally influence visibility, trust, and preference for new products, shaping customer reality before they even encounter the offering.

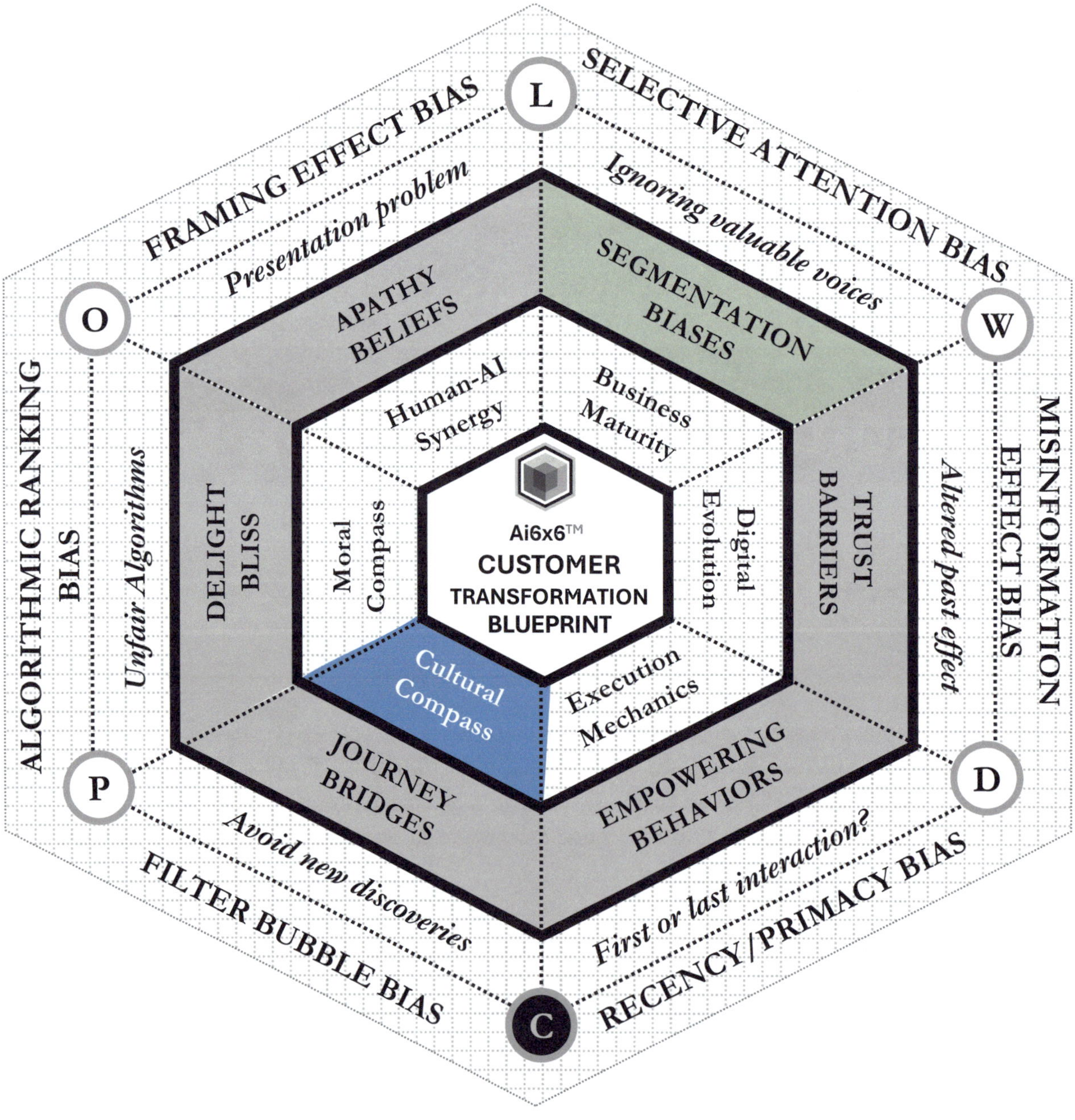

Ai6x6 CULTURAL COMPASS FOR BIASES | CUSTOMER TRANSFORMATION

1. FRAMING EFFECT BIAS | The Presentation Problem

This bias occurs when you influence a customer's decision based purely on how information is presented, rather than the objective facts. In an AI context, this means your "recommendation engines" might be nudging customers toward high-margin products by framing them as "popular choices" or "limited offers."4 While this might drive short-term sales, it erodes long-term trust. Daily, this leads to customers feeling "manipulated" by your digital interface. Transformation requires a commitment to "objective framing." By recognizing this bias, you ensure your AI provides a balanced view of options. This builds "decision integrity," ensuring that when a customer chooses your product, they do so because it truly fits their needs, not because they were nudged by a biased digital frame.

DO'S FOR THE LEADER	DON'TS FOR THE LEADER
Present info clearly and neutrally.	Use tricky wording to influence choice.
Focus on the core value of an offer.	Frame a choice to create a false urgency.
Be transparent about all options.	Hide information to make one option look good.
Use plain language in all offers.	Use complex jargon to mislead people.
Get feedback on how you present info.	Assume your presentation is fair.
Avoid emotional or loaded language.	Use emotional appeals to influence.

SITUATION	SUGGESTED ACTION
Customers choose slower shipping because the fastest option appears expensive when framed without context	Present comparative savings, show delivery impact, and reinforce clarity by reframing options using consistent metrics
Customers decline upgrades because benefits appear minor when listed after long descriptions of limitations	Lead with outcomes, simplify comparisons, and reinforce value by highlighting measurable improvements first
Customers accept unnecessary add-ons because promotional framing emphasizes discounts instead of actual usefulness	Display usage relevance, show long-term cost, and reinforce informed decisions by balancing promotional framing with practical context
Customers misinterpret risk warnings because negative framing overshadows probability details	Use neutral language, clarify likelihood, and reinforce understanding by pairing risks with clear mitigation steps
Customers choose default settings because alternative options appear complex when framed with technical jargon	Simplify descriptions, provide examples, and reinforce confidence by showing recommended scenarios for each option
Customers overvalue trial extensions because framing emphasizes “free time” instead of required future commitment	Clarify renewal terms, show total cost, and reinforce transparency by presenting commitment details upfront

2. SELECTIVE ATTENTION BIAS | The Unheard Voices:

Selective attention happens when you focus on the feedback that confirms your existing strategy while ignoring data that suggests a pivot is necessary. For an executive, this often looks like celebrating "high NPS scores" while ignoring rising "churn rates" in a specific niche segment. Daily, this bias creates a "bubble" where the leadership team believes everything is fine while a "silent exodus" of customers is occurring. Transformation requires an "omni-channel" listening strategy. By overcoming selective attention, you force yourself to look at the "uncomfortable data." This ensures your AI models are trained on the full spectrum of customer sentiment, preventing the strategic blind spots that lead to market irrelevance and ensuring your transformation is grounded in the total customer reality.

DO'S FOR THE LEADER	DON'TS FOR THE LEADER
Listen to all customer feedback.	Only focus on feedback that confirms beliefs.
Seek out a diverse range of feedback.	Ignore criticism or negative comments.
Use objective data to make choices.	Rely on anecdotal evidence and stories.
Conduct comprehensive market research.	Only use data that supports your claim.
Be mindful of your own biases.	Assume you listen to everyone.
Look for a signal in the noise.	Let a few loud voices dictate the feedback.

SITUATION	SUGGESTED ACTION
Teams overreact to one viral complaint because they ignore broader satisfaction trends across segments	Balance sentiment sources, review distribution, and reinforce decisions using complete feedback patterns
Product managers prioritize feature requests from vocal users because quieter segments receive less attention	Weight feedback by representation, analyze silent segments, and reinforce inclusive prioritization
Support teams focus on repeated issues from a small group because they overlook infrequent but severe problems	Categorize by impact, review severity, and reinforce balanced evaluation across all feedback
Marketing teams adjust messaging based on enthusiastic fans while ignoring disengaged customers' concerns	Compare engagement cohorts, analyze drop-off reasons, and reinforce messaging grounded in full-spectrum insight
Executives prioritize enterprise complaints because they overlook consistent friction experienced by smaller customers	Segment by value and volume, review patterns, and reinforce equitable attention across customer types
Teams celebrate positive comments because they overlook early warning signals hidden in neutral feedback	Analyze neutral sentiment, surface subtle patterns, and reinforce proactive interpretation of emerging issues

This bias occurs when a customer's memory of an initial interaction is altered by later, often misleading, information. For example, if a customer had a great experience, but later reads a viral "negative review" or experiences a "bad bot" interaction, their original positive memory can be overwritten. Daily, this leads to "unstable loyalty" that is easily disrupted by a single poor touchpoint. Transformation requires building a "memory-safe" customer lifecycle. By recognizing the misinformation effect, you ensure that every subsequent interaction reinforces the original brand promise. You use AI to proactively detect and correct "experience drift," ensuring that the customer's cumulative memory of your brand remains positive, consistent, and resilient to outside noise or internal service slips.

DO'S FOR THE LEADER	DON'TS FOR THE LEADER
Maintain a consistent brand narrative.	Let a customer's memory be altered.
Be honest about your product's limits.	Make false or misleading claims.
Correct any misinformation quickly.	Let misinformation spread.
Use clear and factual communication.	Use vague or ambiguous language.
Listen to how customers remember you.	Assume your story is the only one.
Be proactive about your brand's image.	Let others define your brand.

SITUATION	SUGGESTED ACTION
Customers recall long wait times because later comments from friends exaggerated delays they never experienced	Provide timestamped journey summaries, clarify actual durations, and reinforce accuracy through personalized post-interaction recaps
Customers believe support was rude because a follow-up email tone felt colder than the original conversation	Align tone across channels, personalize follow-ups, and reinforce continuity by referencing earlier interaction details
Customers misremember product performance because online reviews influenced their perception after using it successfully	Share usage analytics, highlight their actual outcomes, and reinforce confidence with personalized performance comparisons
Customers think onboarding was confusing because later instructions contradicted earlier guidance they received	Standardize instructions, remove outdated content, and reinforce clarity with a single authoritative onboarding path
Customers recall billing errors because a friend mentioned similar issues, even though their invoices were correct	Provide transparent breakdowns, show historical accuracy, and reinforce trust with clear verification steps
Customers believe they reported issues earlier because later frustrations altered their memory of initial timelines	Display interaction history, clarify timestamps, and reinforce transparency with documented event sequences

Customers tend to remember only the "first" and "last" moments of their journey—the "hello" and the "goodbye." If your onboarding is great but your cancellation process is a nightmare, the customer will leave with a bitter memory. Conversely, if the first impression is poor, they may never stick around to see the value. Daily, this leads to companies over-investing in marketing but under-investing in "offboarding" or "support." Transformation requires a "serial position audit" of the journey. By recognizing these biases, you ensure the "middle" of the journey—the long-term engagement phase—is as strong as the beginning and end, creating a "smooth arc" of excellence that sustains loyalty across the entire lifecycle, regardless of the interaction point.

DO'S FOR THE LEADER	DON'TS FOR THE LEADER
Focus on the entire customer journey.	Only focus on the first or last touchpoint.
Create a consistent, quality experience.	Let some touchpoints be better than others.
Get feedback on the whole journey.	Only ask for feedback at the end.
Make every interaction a positive one.	Let the last interaction be negative.
Use data to track the whole journey.	Focus only on the first or last data point.
Plan for the whole experience.	Only plan for the first and last moments.

SITUATION	SUGGESTED ACTION
Customers rate service poorly because their final interaction ended abruptly despite earlier positive experiences	Train agents on strong closings, confirm resolution, and reinforce consistency across the entire journey
Customers praise onboarding highly because the first call felt exceptional, despite later unresolved issues	Balance early enthusiasm with structured follow-ups, and reinforce reliability through consistent mid-journey checkpoints
Customers cancel subscriptions because the last outage overshadowed months of stable performance	Provide uptime history, contextualize incidents, and reinforce reliability with transparent recovery updates
Customers distrust recommendations because their first experience with personalization felt irrelevant and poorly targeted	Improve initial calibration, gather early preferences, and reinforce relevance through quick, visible personalization wins
Customers escalate complaints because their final support message felt rushed, even though earlier steps were thorough	Coach agents on pacing, confirm understanding, and reinforce thoughtful communication at closing moments
Customers choose competitors because their first website visit felt confusing, despite later improvements	Redesign entry flows, simplify navigation, and reinforce clarity through guided first-visit experiences

Just as in social media, AI-driven personalization can create a "filter bubble" for your customers, only showing them what they have bought before. This bias limits their discovery of your full range of products and narrows their relationship with you. Daily, this manifests as "predictive staleness," where a customer stops finding your brand exciting because they only see "more of the same." Transformation requires "serendipity by design." You must build AI systems that occasionally "nudge" the customer toward something new and complementary. By breaking the filter bubble, you expand the customer's world and their lifetime value, moving from "narrow relevance" to "expansive partnership," where the customer constantly discovers new value in your evolving and diverse brand ecosystem.

DO'S FOR THE LEADER	DON'TS FOR THE LEADER
Present diverse, relevant options.	Use an algorithm that shows only one view.
Empower customers with more control.	Create a system that reinforces beliefs.
Use a transparent recommendation engine.	Use a black box system to suggest products.
Ask customers for their preferences.	Assume what customers want to see.
Challenge customer beliefs respectfully.	Only show customers what they already like.
Acknowledge the limits of personalization.	Claim your system knows the customer best.

SITUATION	SUGGESTED ACTION
Customers only see familiar products because the system over-weights past browsing behavior excessively	Introduce exploration slots, diversify recommendations, and reinforce discovery by surfacing new categories intentionally
Customers assume your catalog is limited because they repeatedly see the same narrow product set	Expand rotation logic, highlight broader options, and reinforce variety through periodic curated collections
Customers misunderstand pricing because they only see items matching previous budget ranges	Show full price spectrum, explain filtering logic, and reinforce transparency with adjustable preference controls
Customers believe your brand lacks innovation because recommendations never include new or experimental offerings	Prioritize new releases, tag innovations clearly, and reinforce novelty through spotlight placements
Customers assume support topics are limited because help suggestions mirror their past searches too closely	Broaden suggestion models, surface uncommon topics, and reinforce completeness with category overviews
Customers think your service lacks flexibility because automated advice always mirrors their previous choices	Introduce alternative paths, highlight contrasting options, and reinforce adaptability through scenario-based suggestions

This occurs when your internal algorithms unfairly promote certain products or content based on flawed variables, like "highest margin," rather than "best fit for the customer." Daily, this leads to a customer experience that feels "self-serving" for the company rather than "customer-centric." This bias is a silent trust-killer; customers eventually realize they aren't being shown the best solution, but the best "deal" for the company. Transformation requires "ranking transparency." You must ensure your algorithms prioritize "Customer Success" as the primary variable. By building this bias out of your systems, you ensure your AI acts as a "trusted advisor," recommending products that truly solve the customer's problem, which is the only sustainable way to build long-term, high-value and honest relationships.

DO'S FOR THE LEADER	DON'TS FOR THE LEADER
Use a transparent, fair algorithm.	Use an algorithm that unfairly promotes products.
Audit your algorithm for fairness.	Assume your algorithm is unbiased.
Use diverse data to train algorithms.	Use biased data to train your algorithms.
Be transparent about your ranking logic.	Use a black box algorithm for ranking.
Test your algorithm on diverse groups.	Assume your algorithm works for everyone.
Make sure the ranking serves the user.	Let the ranking serve your business interests.

SITUATION	SUGGESTED ACTION
Customers assume top-ranked items are best because algorithmic ordering hides equally relevant alternatives	Add ranking explanations, diversify top slots, and reinforce fairness with transparent criteria
Customers overlook niche products because ranking heavily favors high-volume items with broad appeal	Introduce balanced weighting, surface niche options, and reinforce inclusivity through curated category blends
Customers distrust recommendations because sponsored items consistently appear above genuinely relevant choices	Label sponsorship clearly, separate organic rankings, and reinforce trust with transparent placement rules
Customers believe content variety is shrinking because ranking repeatedly promotes similar themes	Increase diversity thresholds, rotate themes, and reinforce breadth through periodic content refreshes
Customers misinterpret popularity signals because ranking amplifies early engagement disproportionately	Normalize engagement metrics, adjust recency weighting, and reinforce accuracy with balanced scoring
Customers feel manipulated because ranking hides lower-priced alternatives behind premium options	Display price filters prominently, show comparable choices, and reinforce fairness with visible sorting controls

Ai6x6™ Transformation Blueprint

Customer Trust Barriers

"It's worth noting that building trust takes time, but losing trust takes one poor decision." - Inc.

Inc. is a prominent American business media company and magazine. It is widely known for its journalistic content on entrepreneurship and its annual "Inc. 5000" list of fastest-growing private companies.

Customer trust barriers are structural and cultural impediments that prevent an organization from building genuine customer loyalty. These barriers are often the root cause of a poor customer experience and a lack of repeat business. The Inside-out Barrier is a common dysfunction where a company's processes are designed around its own internal needs and structure rather than the customer's journey. A Siloed Barrier is where disconnected departments, such as sales, support, and marketing, treat the customer as a separate event, leading to a fragmented and frustrating experience. A Complacency Barrier reflects a lack of urgency or drive to improve the customer experience. Overcoming these barriers requires a focused effort to redesign systems, break down departmental silos, and instill a culture of constant improvement, putting the customer journey at the very center of all decision-making.

Completing the Relational domain, this component examines Customer Trust Barriers, the psychological and practical obstacles that prevent market reception of new offerings, directly influencing Engagement and the perception of Human Fulfillment. The most powerful resistance comes from Habitual Resistance, the customer's deep-seated unwillingness to change their routine, even for a superior product. This is intensified by Perceived Risk, the fear that purchasing or using the new product is a poor financial or social investment. This section highlights the burden of High Switching Costs, detailing the effort, time, and monetary investment required for a customer to abandon their current solution for a new one. Crucially, transformation can fail due to Lack of Perceived Value, where the customer doesn't see why they need the product or how it solves their pain better than the existing alternatives. The importance of brand integrity is covered through Trust Deficit, where a lack of belief in the brand or product quality prevents adoption. Finally, the practical friction of Inconvenience is addressed, showing how even a superior product will be rejected if it is difficult to acquire, install, learn, or use daily.

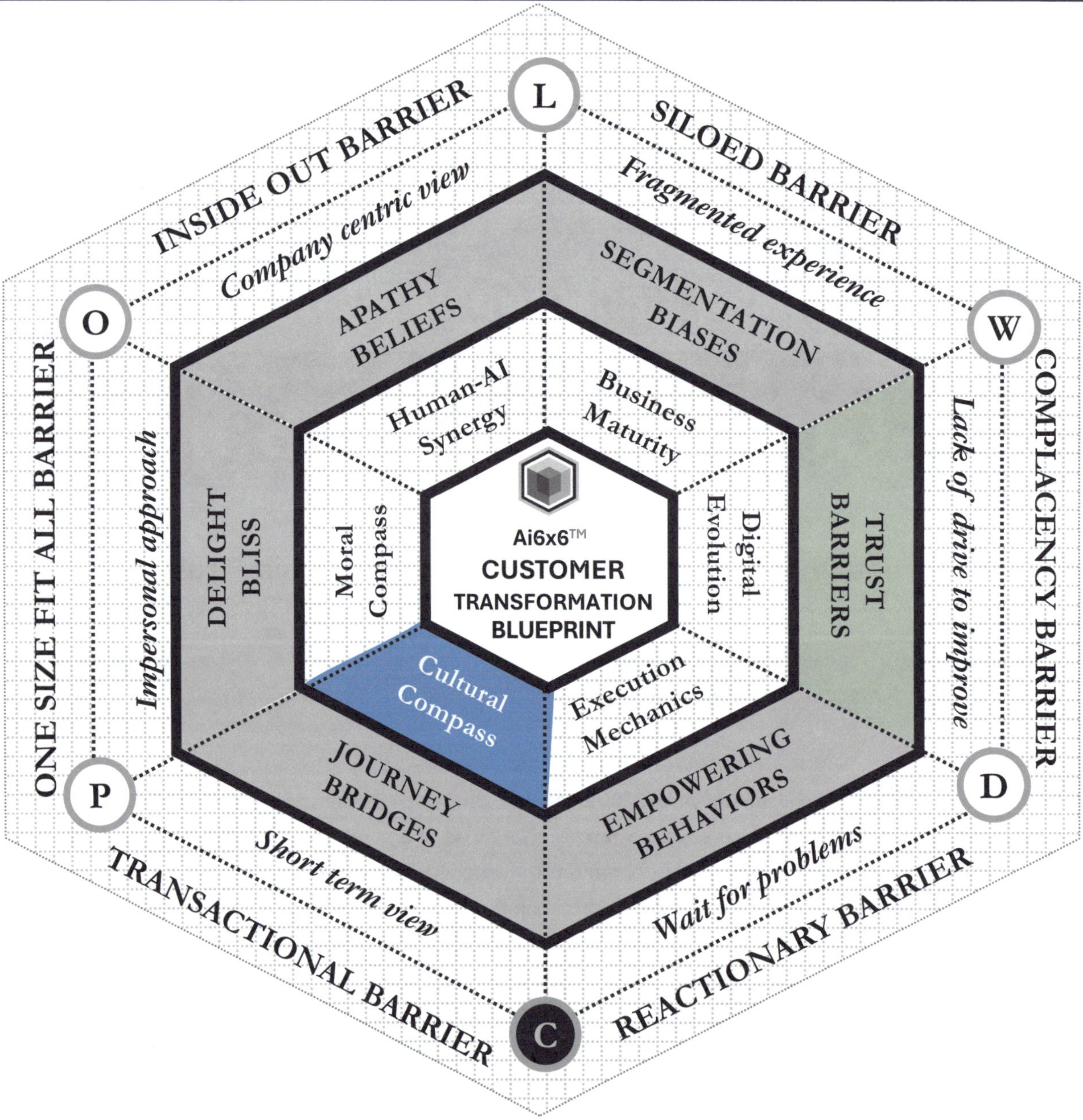

Ai6x6 CULTURAL COMPASS FOR BARRIERS | CUSTOMER TRANSFORMATION

1. INSIDE-OUT BARRIER | The Company-Centric View

The "Inside-out" barrier is the tendency to design customer experiences based on internal company processes and department structures rather than the customer's actual needs. If your website is organized by "business units" rather than "customer goals," you are hitting this barrier. Daily, this leads to a "clunky" experience where the customer has to learn your "org chart" just to get help. Transformation requires an "Outside-in" revolution. You must tear down the internal walls that dictate the journey. By breaking this barrier, you rebuild the experience from the customer's perspective. This ensures that every touchpoint—from sales to support—is designed to solve the customer's "job to be done" with the least amount of friction and technical jargon possible.

DO'S FOR THE LEADER	DON'TS FOR THE LEADER
Prioritize the customer's perspective.	Focus on your internal processes.
Use an "outside-in" approach.	Use an "inside-out" approach.
Listen to customer problems.	Focus on your internal problems.
Let customer needs drive decisions.	Let internal politics drive decisions.
Get feedback from external sources.	Only get feedback from your own staff.
Connect teams to the customer.	Let teams get disconnected from the client.

SITUATION	SUGGESTED ACTION
Customers repeat details because internal forms force them to reenter information for every department	Map end-to-end journeys, remove duplicate fields, and redesign workflows so information follows customers automatically across teams and systems without manual reentry at each step
Customers wait weeks because internal approval chains matter more than resolving simple account requests	Shorten approval paths, define clear thresholds, and automate low-risk decisions so straightforward customer requests complete within predictable timelines for every standard scenario type
Customers feel dismissed when agents defend policies instead of understanding underlying customer situations	Train agents to explore context first, document exceptions, and adjust policies where patterns show rules consistently harming reasonable customer outcomes over time
Customers experience confusing handoffs because internal ownership charts matter more than clear accountability	Assign single journey owners, publish responsibilities, and review unresolved cases where customers bounced between teams without anyone accepting final accountability for resolution
Customers receive irrelevant updates because internal reporting focuses on internal milestones, not customer impact	Redesign status communications around customer outcomes, highlight what changes for them, and retire internal milestone language from external messages customers actually receive
Customers sense resistance whenever suggestions challenge established internal processes or legacy systems	Create structured feedback reviews, require written responses to themes, and link process changes directly to recurring customer pain points surfaced through multiple channels

The Siloed barrier occurs when disconnected departments treat the customer journey as a series of separate, unrelated events. If a customer has to "repeat their story" every time they are transferred between departments, you have a silo problem. Daily, this leads to a "fragmented" experience that feels like dealing with five different companies. This barrier kills the "contextual continuity" required for AI to be effective. Transformation requires a "unified customer data platform" that serves as a single source of truth across all departments.6 By breaking the siloed barrier, you ensure that the "left hand" always knows what the "right hand" is doing, creating a "seamless handoff" that makes the customer feel known and valued at every stage of their journey.

DO'S FOR THE LEADER	DON'TS FOR THE LEADER
Create a single, unified experience.	Let departments create disjointed experiences.
Promote collaboration across teams.	Let departments operate in silos.
Create a 360-degree customer view.	Have separate customer data in each team.
Use tools that connect all teams.	Use different systems for each team.
Foster a "one company" mentality.	Pit departments against one another.
Share customer data across the company.	Hoard data in individual departments.

SITUATION	SUGGESTED ACTION
Customers repeat their story because each department keeps separate notes and disconnected interaction histories	Implement shared customer timelines, integrate core systems, and require teams to review history before engaging on any new request with that customer
Customers receive conflicting answers because product, billing, and support teams interpret policies differently	Create a single policy source, synchronize updates across functions, and audit conversations regularly for alignment on key customer topics and decisions companywide
Customers experience delays when cross-team issues wait in shared inboxes without clear ownership	Define explicit case owners, set handoff agreements, and track unresolved items until one accountable team closes the loop for customers every time
Customers see different promotions because marketing segments are not aligned with service entitlements	Align segmentation models, share eligibility rules, and ensure frontline teams can explain why specific offers appear for particular customers in detail when asked
Customers feel abandoned when handoffs between digital and physical channels break mid-journey unexpectedly	Design omnichannel flows, share context between locations, and measure drop-offs where customers switch channels during critical tasks in their journey stages
Customers receive multiple surveys from different teams about the same single interaction experience	Coordinate feedback programs, centralize survey triggers, and design one coherent questionnaire representing the entire journey instead of fragmented departmental views for customers overall

3. COMPLACENCY BARRIER | The Enemy of Improvement

The "Complacency Barrier" is a lack of internal drive to improve the customer experience because "things are good enough." This is the "incumbent's curse." If your team is satisfied with "average" CSAT scores, they will never innovate for the future. Daily, this manifests as a "status quo" culture that ignores small points of friction that eventually lead to major churn. Transformation requires a "relentless dissatisfaction" with the current experience. You must foster a culture that views every customer complaint as a "gift of insight." By breaking this barrier, you instill a "Growth Mindset" across the organization. This ensures that the customer experience is never "finished" but is a constantly evolving masterpiece that is always seeking new ways to surprise the market.

DO'S FOR THE LEADER	DON'TS FOR THE LEADER
Have a constant drive to improve service.	Become complacent with the status quo.
Be a leader in customer experience.	Ignore your competition.
Get feedback to find areas for growth.	Assume customers are already happy.
Set ambitious goals for customer joy.	Have a low bar for customer happiness.
Celebrate new innovations.	Stick to the old way of doing things.
Listen to new customer needs.	Assume old needs are the only needs.

SITUATION	SUGGESTED ACTION
Customers tolerate recurring glitches because teams treat long-standing defects as harmless annoyances operationally	Create defect backlogs with owners, set elimination targets, and review unresolved issues in leadership forums until patterns disappear for good and permanently
Customers stop complaining because nothing changes after repeated reports about the same friction	Tag recurring themes, publish remediation plans, and communicate progress visibly so customers see concrete movement on previously ignored problems over time periods
Customers accept mediocre experiences because competitors in the category feel similarly outdated too	Benchmark outside industry, set higher standards, and redesign journeys using exemplars from categories customers already admire and reference frequently in conversations elsewhere
Customers quietly churn because renewal outreach assumes satisfaction instead of testing sentiment proactively	Introduce pre-renewal health checks, analyze engagement signals, and intervene early when indicators show declining interest or unresolved frustrations for customers at risk
Customers see outdated interfaces because redesigns keep getting postponed behind internal priorities repeatedly	Ring-fence modernization capacity, commit visible timelines, and tie interface upgrades directly to measurable customer satisfaction and conversion improvements over time horizons
Customers feel unimportant because internal celebrations rarely mention customer outcomes or stories explicitly	Redesign recognition programs, require customer impact evidence, and feature real narratives whenever teams celebrate milestones or internal achievements publicly with customers included

4. REACTIONARY BARRIER | The Waiting Game

The "Reactionary Barrier" is the habit of waiting for customer problems to arise before addressing them. This "firefighting" mode is the opposite of a proactive AI strategy. If your service team is measured only on "how fast they close tickets," they will never have the time to look for the "root cause" of those tickets. Daily, this leads to a "repetitive cycle" of problems that could have been prevented. Transformation requires shifting from "reactive" to "predictive" service. You must use AI to identify patterns and intervene before the customer even knows they have a problem. By breaking the reactionary barrier, you move from "fixing things" to "ensuring success," which is the hallmark of a world-class, customer-centric organization in the modern era.

DO'S FOR THE LEADER	DON'TS FOR THE LEADER
Anticipate customer issues before they arise.	Wait for problems before addressing them.
Be proactive with communication.	Only communicate after a problem occurs.
Use data to predict future needs.	React to every new customer request.
Plan for potential customer problems.	Have no plan for customer issues.
Create a proactive support system.	Have a reactive, fire-fighting support.
Get feedback on potential issues.	Ignore all potential problems.

SITUATION	SUGGESTED ACTION
Customers report outages before monitoring systems trigger any internal alerts or responses themselves	Tighten monitoring thresholds, simulate failure scenarios, and require post-incident reviews whenever customers detect issues before internal systems do in prod. environments
Customers only hear from you after they complain publicly on social channels first	Establish proactive outreach triggers, monitor sentiment continuously, and contact customers early when signals show confusion, frustration, or emerging dissatisfaction patterns developing online
Customers repeatedly encounter the same bug because fixes only address individual tickets submitted	Cluster incidents by root cause, prioritize systemic corrections, and verify resolution across all affected journeys before closing related categories for good everywhere
Customers struggle with new features because enablement materials appear only after complaints accumulate	Design launch playbooks, prepare tutorials early, and release guidance alongside features so customers succeed from first interaction without needing support immediately
Customers face recurring billing surprises because reviews happen only after disputes escalate externally	Introduce pre-billing checks, analyze anomaly patterns, and adjust rules before customers experience repeated unexpected charges or confusing invoice structures appearing again
Customers receive apologies but no prevention plan after major service failures disrupt operations	Require documented prevention actions, share timelines publicly, and track whether similar incidents recur after commitments made to affected customers in future periods

The "Transactional Barrier" is viewing customers as a series of "one-off" sales rather than a long-term relationship. If your incentives are focused purely on "new customer acquisition" and ignore "existing customer retention," you are trapped behind this barrier. Daily, this manifests as a "churn-and-burn" culture that treats customers like statistics. Transformation requires shifting to a "Relationship Economy" mindset. You must value the "Lifetime Value" of a customer over the "Initial Transaction" value. By breaking this barrier, you build loyalty programs and engagement strategies that reward long-term partnership. This ensures that your brand becomes a "habit" in the customer's life, rather than just a "choice" they make once based on a discount or a single, urgent need.

DO'S FOR THE LEADER	DON'TS FOR THE LEADER
Build long-term relationships with customers.	View customers as a single transaction.
Focus on lifetime customer value.	Only focus on the initial purchase.
Use loyalty programs to reward customers.	Have no incentives for returning clients.
Understand your customers' needs over time.	See customers as just a credit card number.
Make it easy to return and buy again.	Make it hard for people to come back.
Support customers even after the sale.	Abandon customers after the sale.

SITUATION	SUGGESTED ACTION
Customers feel forgotten after purchase because teams focus only on closing deals, not ongoing value	Build structured post-purchase journeys, schedule proactive check-ins, and reinforce long-term relationship ownership across sales, support, and success teams
Customers switch providers because renewal conversations feel purely financial without discussing evolving goals	Train teams to review outcomes, explore future needs, and reinforce partnership by connecting renewal decisions to measurable customer progress
Customers disengage because communications only appear when payments are due or issues arise unexpectedly	Create value-led touchpoints, share helpful insights, and reinforce presence through consistent, non-transactional interactions throughout the year
Customers feel pressured because sales teams push upgrades without understanding current usage or constraints	Require usage reviews, tailor recommendations, and reinforce credibility by aligning offers with demonstrated customer needs and timing
Customers distrust incentives because discounts appear only when they threaten to cancel services	Offer lifecycle-based benefits, communicate eligibility early, and reinforce fairness by avoiding reactive, last-minute retention tactics
Customers feel unappreciated because loyalty is measured only by spend, not engagement or advocacy	Redesign loyalty metrics, recognize meaningful behaviors, and reinforce appreciation through personalized acknowledgments beyond financial contribution

This barrier is the failure to adapt products and communication for individual customers. It's the "mass production" mindset applied to a "personalized world." If you send the same generic email to a first-time buyer and a ten-year loyalist, you are hitting this barrier. Daily, this leads to a "diluted" brand message that fails to resonate with anyone deeply. Transformation requires "Hyper-personalization at Scale." You must use AI to segment and target at a granular level, treating a "segment of one" as the new standard. By breaking this barrier, you ensure that every customer receives a "tailored experience" that feels like it was made just for them. This creates a powerful emotional bond that generic, "one-size-fits-all" brands can never hope to replicate.

DO'S FOR THE LEADER	DON'TS FOR THE LEADER
Personalize experiences for customers.	Offer a one-size-fits-all solution.
Tailor products to individual needs.	Force every customer to fit the product.
Ask customers about their preferences.	Assume what your customers want.
Use data to inform personalization.	Personalize for the sake of it.
Be respectful of customer privacy.	Use data in a way that feels intrusive.
Provide flexible product options.	Offer only one rigid product.

SITUATION	SUGGESTED ACTION
Customers abandon onboarding because instructions assume identical needs instead of acknowledging different experience levels	Offer adaptive flows, ask early preference questions, and reinforce relevance by tailoring guidance to customer familiarity and goals
Customers misuse features because training materials assume uniform workflows across industries and contexts	Provide industry-specific examples, segment tutorials, and reinforce clarity by mapping features to real customer scenarios
Customers feel unseen because marketing messages ignore their unique priorities and constraints	Build persona-based messaging, test variations, and reinforce resonance by reflecting customer language and motivations
Customers reject product bundles because included components do not match their actual usage patterns	Analyze behavioral data, create modular bundles, and reinforce alignment by recommending configurations based on demonstrated needs
Customers complain about irrelevant alerts because notification rules treat all users identically	Introduce preference controls, personalize triggers, and reinforce usefulness by showing why each alert appears
Customers struggle with pricing tiers because plans assume uniform value perception across diverse customer groups	Offer flexible tiers, clarify differences, and reinforce fairness by mapping pricing to measurable outcomes

Ai6x6™ Transformation Blueprint

Customer Empowering Behaviors

"You've got to start with the customer experience and work back toward the technology, not the other way around." - Steve Jobs

Steve Jobs (1955–2011) was an American business magnate and co-founder of Apple Inc. He is considered one of the most influential figures in technology and is known for his user-centric design philosophy and innovative products.

Customer empowering behaviors are positive and proactive actions that reorient the entire organization around the customer. These are the practices that build genuine loyalty, trust, and long-lasting relationships. Outside-in Behavior is the practice of prioritizing the customer's perspective over internal processes, ensuring every decision is made with the end user in mind. Holistic Behavior involves viewing the customer journey as a single, unified experience, a direct antidote to the siloed barrier. Proactive Behavior involves anticipating customer needs and issues before they arise, which builds trust and loyalty in a way that reactive measures never can. As Steve Jobs famously said, "You've got to start with the customer experience and work back toward the technology, not the other way around." These behaviors, when adopted, create a virtuous cycle where a great customer experience leads to more business and more valuable feedback.

This framework examines Customer Empowering Behaviors, detailing the actions customers exhibit when they accept and internalize a new offering, directly fueling Engagement and validating Human Fulfillment. Initial adoption often starts with Trial Behavior, the customer's active willingness to test and experiment with a new product or service, often based on low-risk entry points. Successful integration leads to Feedback-Giving Behavior, where customers actively provide constructive comments and suggestions for improvement, indicating deep investment in the product's success and future. The ultimate validation is Advocate Behavior, recommending the product to others within their professional or social networks and generating powerful organic growth. This must be followed by Repeat Purchase Behavior, continuing to use the product over time and integrating it into their daily personal or professional routines. The framework also values Community-Building Behavior, participating in online forums and user groups centered around the product to share knowledge and build a shared experience. Finally, mature users exhibit Self-Service Behavior, independently solving their own problems using available digital resources, reducing reliance on expensive customer support channels.

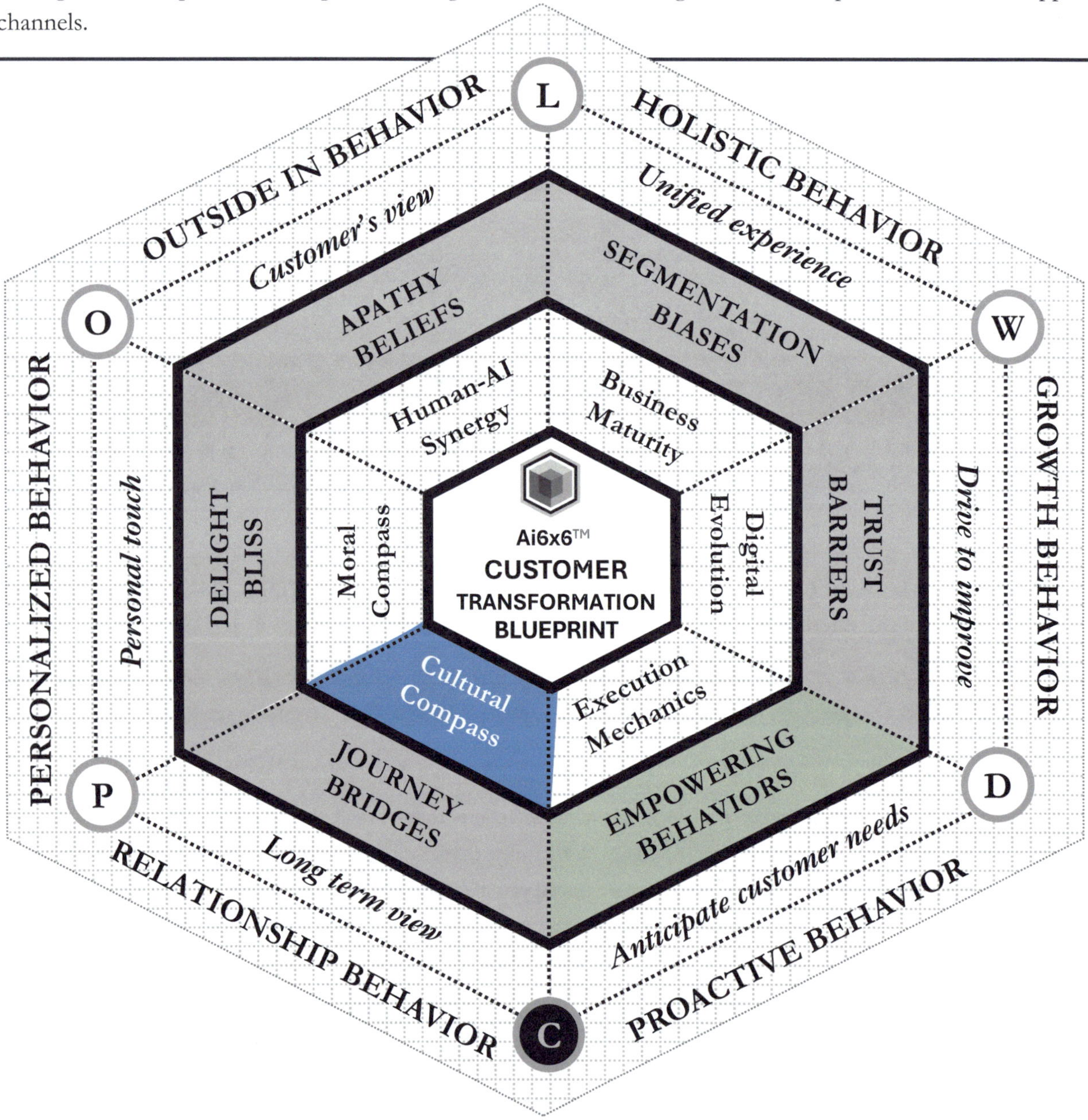

Ai6x6 CULTURAL COMPASS FOR BEHAVIORS | CUSTOMER TRANSFORMATION

1. OUTSIDE-IN BEHAVIOR | The Customer's View

In a customer-transformed organization, "Outside-in" isn't a slogan; it's a daily behavior. It means starting every meeting by asking: "How does this decision benefit the customer?" This behavior involves executives spending time "on the front lines" to hear the unfiltered customer voice. Daily, it looks like prioritizing product features based on "user pain points" rather than "internal tech feasibility." By modeling Outside-in behavior, you ensure that the company's "gravity" is centered on the customer. This behavior turns your organization into a highly empathetic machine that is naturally aligned with the market. It ensures that your AI transformation is focused on "human outcomes" rather than just "technical outputs," making your innovations far more likely to succeed and scale in the real world.

DO'S FOR THE LEADER	DON'TS FOR THE LEADER
Prioritize the customer's perspective.	Focus on your internal processes.
Use an "outside-in" approach.	Use an "inside-out" approach.
Listen to customer problems.	Focus on your internal problems.
Let customer needs drive decisions.	Let internal politics drive decisions.
Get feedback from external sources.	Only get feedback from your own staff.
Connect teams to the customer.	Let teams get disconnected from the client.

SITUATION	SUGGESTED ACTION
Teams redesign workflows after observing customers repeatedly struggle with steps considered simple internally	Conduct field studies, document friction, and reinforce redesign decisions using real customer behavior rather than internal assumptions
Leaders adjust priorities when customer interviews reveal needs not captured in existing roadmaps	Integrate qualitative insights, rebalance initiatives, and reinforce alignment by anchoring decisions in customer-expressed goals
Product teams change feature sequencing after learning customers value foundational stability over new capabilities	Reorder backlogs, communicate rationale, and reinforce trust by delivering improvements matching customer expectations
Support teams rewrite scripts after noticing customers interpret standard phrasing as dismissive or robotic	Test alternative language, gather feedback, and reinforce empathy by adopting customer-preferred communication styles
Marketing teams refine messaging after discovering customers misunderstand key value propositions	Validate comprehension, simplify claims, and reinforce clarity through iterative testing with representative audiences
Executives shift KPIs after realizing internal metrics fail to reflect actual customer outcomes	Replace vanity metrics, track impact indicators, and reinforce accountability through customer-centric performance measures

Viewing the customer journey as a single, unified experience is a behavior that eliminates the "handoff friction" between departments. Daily, this looks like cross-functional teams meeting to review the "end-to-end" journey, from the first ad to the third-year renewal. This behavior ensures that the customer "narrative" remains consistent. For you, it means breaking down the departmental egos that say "that's not my problem." Holistic behavior builds a sense of "collective responsibility" for the customer. When everyone from the coder to the closer feels responsible for the entire journey, the experience becomes seamless. This behavior is the prerequisite for "unified AI," where data flows smoothly across the entire lifecycle to provide a truly integrated and personalized customer experience for every user.

DO'S FOR THE LEADER	DON'TS FOR THE LEADER
View the customer journey as a whole.	Let departments create disjointed experiences.
Promote collaboration across teams.	Let departments operate in silos.
Create a 360-degree customer view.	Have separate customer data in each team.
Use tools that connect all teams.	Use different systems for each team.
Foster a "one company" mentality.	Pit departments against one another.
Share customer data across the company.	Hoard data in individual departments.

SITUATION	SUGGESTED ACTION
Teams coordinate handoffs because customers expect seamless transitions across digital, physical, and human channels	Build shared journey maps, define ownership boundaries, and reinforce continuity through integrated systems and cross-team rituals
Leaders review journeys end-to-end after discovering isolated improvements fail to fix overall experience	Conduct holistic audits, prioritize systemic fixes, and reinforce alignment by addressing root causes across multiple touchpoints
Product teams collaborate with support after realizing feature gaps create recurring service burdens	Share insights, co-design solutions, and reinforce unity by solving problems spanning product and service boundaries
Marketing adjusts campaigns after learning acquisition promises misalign with actual onboarding experiences	Align messaging, validate expectations, and reinforce consistency across early-stage and post-purchase interactions
Success teams coordinate with billing when customers express confusion about pricing during renewal cycles	Synchronize explanations, simplify structures, and reinforce clarity through unified communication across departments
Operations teams redesign processes after identifying journey stages where customers repeatedly encounter avoidable friction	Analyze cross-functional data, remove bottlenecks, and reinforce smooth flow through shared accountability mechanisms

A constant, restless drive to improve the customer experience is the hallmark of "Growth Behavior." This looks like rewarding employees for finding and fixing "friction points," even if they are small. Daily, it involves a "kaizen" approach to service—constant, incremental improvements that add up to a massive competitive advantage. As a leader, you foster this behavior by celebrating "service breakthroughs" and funding "experience experiments." Growth behavior ensures that your customer journey never becomes "stale." By always looking for the "next level" of delight, you stay ahead of the competition. This behavior turns your workforce into a team of "experience architects" who are constantly reimagining what is possible for the customer, ensuring long-term market leadership and deep consumer satisfaction.

DO'S FOR THE LEADER	DON'TS FOR THE LEADER
Drive to improve the customer experience.	Become complacent with the status quo.
Be a leader in customer experience.	Ignore your competition.
Get feedback to find areas for growth.	Assume customers are already happy.
Set ambitious goals for customer joy.	Have a low bar for customer happiness.
Celebrate new innovations.	Stick to the old way of doing things.
Listen to new customer needs.	Assume old needs are the only needs.

SITUATION	SUGGESTED ACTION
Teams revisit processes after noticing customers repeatedly struggle with steps considered acceptable internally	Conduct recurring journey reviews, analyze friction patterns, and reinforce continuous improvement by prioritizing fixes that deliver measurable customer impact quickly
Leaders adjust roadmaps when customer insights reveal unmet needs not captured in existing planning cycles	Integrate qualitative research, rebalance priorities, and reinforce adaptability by aligning investments with evolving customer expectations
Product teams refine features after observing customers using tools differently than originally intended	Study real usage, redesign interactions, and reinforce learning by incorporating unexpected behaviors into future iterations
Support teams improve scripts after identifying phrases customers consistently interpret as dismissive or unclear	Test alternative language, gather feedback, and reinforce empathy by adopting communication styles customers prefer
Marketing teams update messaging when customers misunderstand value propositions despite repeated explanations	Validate comprehension, simplify claims, and reinforce clarity through iterative testing with representative audiences
Executives evolve KPIs when internal metrics fail to reflect actual customer outcomes across journeys	Replace vanity indicators, track impact measures, and reinforce accountability through customer-centric performance standards

Anticipating customer needs and issues before they arise is the most transformative behavior in the AI era. Daily, this looks like using predictive analytics to "reach out" to a customer when their data signals a potential problem. It's about sending a "proactive update" or a "timely suggestion" before the customer has to ask for it. This behavior builds incredible trust; it shows the customer you are "looking out for them." Proactive behavior shifts the relationship from "reactive service" to "strategic partnership." For you, it means your service team becomes a "value engine" rather than a "cost center." By being proactive, you eliminate customer frustration at the source, creating a "smooth journey" that feels effortless and magical to the end user.

DO'S FOR THE LEADER	DON'TS FOR THE LEADER
Anticipate customer issues before they arise.	Wait for problems before addressing them.
Be proactive with communication.	Only communicate after a problem occurs.
Use data to predict future needs.	React to every new customer request.
Plan for potential customer problems.	Have no plan for customer issues.
Create a proactive support system.	Have a reactive, fire-fighting support.
Get feedback on potential issues.	Ignore all potential problems.

SITUATION	SUGGESTED ACTION
Teams notify customers early when usage patterns indicate potential service disruptions ahead of time	Build predictive alerts, monitor thresholds, and reinforce trust by communicating risks before customers experience visible issues
Leaders intervene when engagement signals show declining interest long before renewal conversations begin	Track behavioral indicators, trigger outreach, and reinforce retention by addressing concerns proactively
Product teams release guidance before launching features customers may find confusing initially	Prepare tutorials, publish walkthroughs, and reinforce readiness by supporting customers from first interaction
Support teams contact customers when repeated actions suggest they are struggling silently with a task	Detect friction signals, offer help, and reinforce care by reaching out before frustration escalates
Marketing teams adjust campaigns when early sentiment shows confusion about new offerings	Analyze feedback quickly, refine messaging, and reinforce clarity through rapid iteration
Operations teams fix emerging bottlenecks when data shows rising delays across specific journey stages	Monitor flow metrics, address constraints, and reinforce smooth experiences by resolving issues before customers notice

Building long-term relationships instead of focusing on short-term transactions is a behavior that prioritizes "trust" over "immediate revenue." Daily, this looks like a sales rep telling a customer "you don't need this extra feature yet," or a service agent spending "extra time" to truly understand a customer's life goals. This behavior builds "Brand Affinity" that is immune to competitor discounts. For you, it means measuring success by "Customer Lifetime Value" and "Net Promoter Score" rather than just "monthly sales volume." Relationship behavior turns customers into "advocates" who do your marketing for you. By valuing the "person" over the "pixel," you create a loyal "community" that sustains the company through market shifts and technological changes, providing a secure foundation.

DO'S FOR THE LEADER	DON'TS FOR THE LEADER
Build long-term relationships.	View customers as a single transaction.
Focus on lifetime customer value.	Only focus on the initial purchase.
Use loyalty programs to reward customers.	Have no incentives for returning clients.
Understand your customers' needs over time.	See customers as just a credit card number.
Make it easy to return and buy again.	Make it hard for people to come back.
Support customers even after the sale.	Abandon customers after the sale.

SITUATION	SUGGESTED ACTION
Teams schedule follow-ups because customers appreciate ongoing support beyond initial purchase moments	Create structured engagement plans, personalize outreach, and reinforce continuity through consistent relationship touchpoints
Leaders review customer goals regularly to ensure evolving needs remain aligned with delivered value	Conduct periodic check-ins, document objectives, and reinforce partnership by adjusting plans collaboratively
Product teams co-design features with customers who rely heavily on specific workflows	Invite early participation, gather detailed input, and reinforce trust by incorporating customer feedback visibly
Support teams track long-term cases to ensure recurring issues do not resurface unexpectedly	Maintain case histories, monitor patterns, and reinforce reliability through sustained oversight
Marketing teams highlight customer stories to strengthen emotional connection beyond product capabilities	Feature authentic narratives, celebrate milestones, and reinforce loyalty through shared success moments
Executives meet strategic customers periodically to understand broader challenges influencing future decisions	Facilitate executive reviews, explore long-term goals, and reinforce commitment through senior-level engagement

Tailoring every product, service, and communication to the individual is the "Personalized Behavior" that wins in the AI era. Daily, this looks like using AI to ensure that every email, notification, and dashboard is uniquely relevant to that specific user. It's about moving from "mass communication" to "individualized conversation." This behavior requires a deep respect for "customer context." For you, it means investing in the data infrastructure that allows for "segmentation of one." Personalized behavior makes the customer feel "special" and "understood." When your brand acts like a "personal concierge" rather than a "faceless corporation," you create a deep emotional connection that is incredibly difficult for competitors to break, ensuring a dominant position in the customer's heart and their long-term wallet.

DO'S FOR THE LEADER	DON'TS FOR THE LEADER
Use data to personalize experiences.	Offer a one-size-fits-all solution.
Ask customers about their preferences.	Make assumptions about what they want.
Offer personalized product suggestions.	Spam customers with generic promotions.
Be respectful of customer privacy.	Use data in a way that feels intrusive.
Use personalization to add value.	Personalize for the sake of it.
Create different customer segments.	Treat every single customer the same.

SITUATION	SUGGESTED ACTION
Teams adjust recommendations after learning customers prefer guidance based on recent behavior rather than historical patterns	Refresh models frequently, incorporate real-time signals, and reinforce relevance by explaining why suggestions appear
Leaders tailor communications when customers express preference for specific formats or levels of detail	Capture communication preferences, segment messaging, and reinforce respect by honoring individual styles
Product teams customize onboarding flows when customers indicate varying familiarity with core concepts	Offer adaptive paths, ask clarifying questions, and reinforce comfort by matching guidance to experience levels
Support teams personalize troubleshooting steps when customers describe unique configurations or constraints	Build branching diagnostics, gather context, and reinforce accuracy by tailoring instructions to specific environments
Marketing teams deliver targeted content when customers show interest in particular themes or product categories	Track engagement patterns, curate materials, and reinforce value by aligning content with demonstrated interests
Success teams design individualized growth plans when customers pursue different outcomes using the same product	Document goals, map capabilities, and reinforce partnership by aligning recommendations with each customer's objectives

Ai6x6™ Transformation Blueprint

Customer Journey Bridges

"We see our customers as invited guests to a party, and we are the hosts. It's our job every day to make every important aspect of the customer experience a little bit better." - Jeff Bezos

Jeff Bezos (b. 1964) is an American entrepreneur and the founder of Amazon. He is known for his "customer obsession" philosophy, which prioritizes a seamless and satisfying customer experience, and is a key figure in the e-commerce revolution.

Customer journey bridges are the practical frameworks and processes that connect an organization's current state to a future of customer delight. They are the strategic tools that turn good intentions into a seamless and satisfying customer experience. An Empathetic Listening Bridge is a system for actively listening to all customer feedback, from surveys to social media comments, ensuring no input is ignored due to selective attention bias. The Consistent Narrative Bridge is a strategy for maintaining a consistent and honest brand story across every channel, building trust and counteracting the framing effect bias. A Fair Digital Pathway Bridge is an algorithm designed to treat all customers fairly and without bias. As Jeff Bezos suggests, we should "see our customers as invited guests to a party." These bridges are the vital links that enable an organization to do just that, ensuring that every touchpoint is a positive one and that the customer journey is a source of delight.

This framework examines Customer Journey Bridges, detailing the mechanisms that connect the organization to its market, directly fueling Engagement and validating Human Fulfillment. Feedback Bridges are continuous, multi-channel loops for actively soliciting customer thoughts, suggestions, and frustrations, moving beyond passive surveys to real-time conversational channels. Support Bridges ensure reliable, fast, and empathetic mechanisms are in place for customer assistance, turning moments of crisis into opportunities to build trust and loyalty. Community Bridges involve establishing and nurturing online forums, user groups, and events where customers can connect with each other and the brand, fostering deep relational ties. Communication Bridges cover all marketing, service, and transactional channels, ensuring targeted, relevant, and timely messaging that builds expectation and trust. Onboarding Bridges focus on tools, tutorials, and guided experiences that minimize the user's initial learning curve and confusion, ensuring rapid product integration. Finally, Value Bridges ensure the clear and continuous link between the product's function and the customer's perceived needs and desired outcomes is always articulated, proving the product's worth repeatedly.

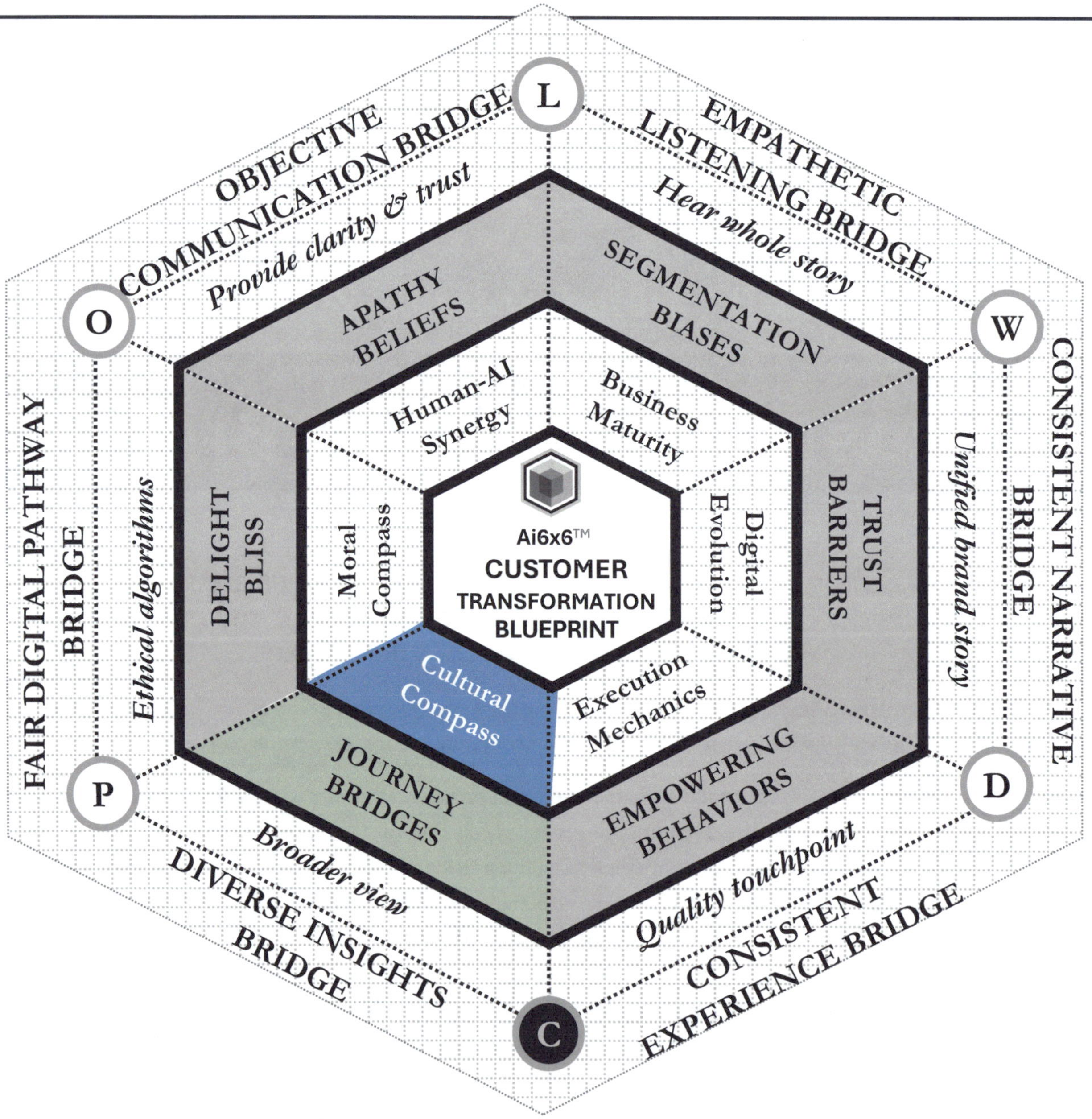

Ai6x6 CULTURAL COMPASS FOR BRIDGES | CUSTOMER TRANSFORMATION

This bridge ensures that all customer communication is clear, honest, and unbiased. It's a process for stripping away "marketing fluff" and "legal jargon" to provide customers with the facts they need to make the best decision for themselves. Daily, this bridge looks like "simplified contracts" and "transparent pricing" models. By building this bridge, you cross from the island of "persuasion" to the mainland of "partnership." It ensures that your customer communication is a "source of truth." For an executive, this bridge is a massive trust-builder; it prevents the "misunderstanding churn" that occurs when customers feel they were misled. This bridge provides the "clarity of intent" that allows for a truly honest and long-term customer-brand relationship to flourish.

DO'S FOR THE LEADER	DON'TS FOR THE LEADER
Use clear, unbiased language.	Use ambiguous or loaded language.
Be honest about your product's limits.	Make false or misleading claims.
Correct any misinformation quickly.	Let misinformation spread.
Use a consistent, factual narrative.	Use inconsistent or emotional stories.
Get feedback on your communication.	Assume your message is clear.
Be transparent about your processes.	Hide how things work.

SITUATION	SUGGESTED ACTION
Customers misinterpret updates because messages emphasize internal priorities instead of explaining customer impact clearly	Rewrite communications using neutral language, highlight customer outcomes, and reinforce clarity by testing messages with representative audiences before release
Customers feel blamed when issue notifications use phrasing suggesting user error rather than system responsibility	Remove accusatory wording, clarify root causes, and reinforce trust by acknowledging organizational accountability transparently
Customers misunderstand pricing changes because explanations focus on technical cost drivers instead of practical value	Translate drivers into benefits, simplify reasoning, and reinforce fairness by showing how changes support improved service quality
Customers escalate concerns because support emails use vague statements that avoid addressing the actual question	Train teams on direct responses, require explicit answers, and reinforce precision through quality reviews of outbound communication
Customers distrust recommendations because descriptions highlight company advantages instead of customer relevance	Reframe suggestions around customer goals, provide context, and reinforce objectivity by showing why each recommendation fits their situation
Customers feel confused when different teams describe the same issue using inconsistent terminology	Standardize vocabulary, publish shared definitions, and reinforce alignment through cross-team communication guidelines

This bridge is a systematic process for actively listening to all customer feedback—from formal surveys to casual social media mentions and "quiet data" behaviors. It involves using AI-driven "sentiment analysis" to hear the "emotion" behind the words. Daily, this bridge ensures that no customer voice is ignored. By building this bridge, you turn "noise" into "insight." It allows you to move beyond "average scores" to understand the "nuance of need." For the executive, this bridge provides the "empathy at scale" required to lead a global organization that feels like a local one. This bridge ensures that your strategic pivots are guided by the "real feelings" of your market, ensuring your transformation remains human-centric and deeply relevant to the customer's specific life.

DO'S FOR THE LEADER	DON'TS FOR THE LEADER
Actively listen to all feedback.	Ignore criticism or negative comments.
Empower staff to show empathy.	Make staff follow a rigid script.
Use feedback to drive improvements.	Dismiss feedback as just complaining.
Create channels for anonymous feedback.	Punish people who speak their mind.
Listen to understand, not to respond.	Wait for your turn to talk.
Show that you care about their issues.	Respond with a canned, unhelpful response.

SITUATION	SUGGESTED ACTION
Customers feel unheard because agents focus on scripts instead of acknowledging emotional context	Train reflective listening, validate feelings, and reinforce empathy by summarizing concerns before offering solutions
Customers repeat themselves because teams fail to capture details accurately during initial conversations	Improve note-taking standards, confirm understanding, and reinforce accuracy through playback summaries
Customers escalate quickly because early responses minimize the seriousness of their concerns	Teach de-escalation techniques, acknowledge impact, and reinforce respect by addressing urgency directly
Customers provide limited feedback because surveys never ask about issues they actually experienced	Redesign questions using journey data, personalize prompts, and reinforce relevance by matching surveys to real interactions
Customers feel dismissed when support teams respond with generic troubleshooting steps unrelated to described problems	Require contextual probing, tailor instructions, and reinforce attentiveness by referencing customer-specific details
Customers stop sharing insights because previous feedback disappeared without acknowledgment or follow-up	Implement acknowledgment workflows, share progress updates, and reinforce appreciation by closing feedback loops consistently

This bridge is a strategy for maintaining a consistent and honest brand story across every touchpoint, from an Instagram ad to a support call. It ensures that the "promise" made by Marketing is "kept" by Operations. Daily, this bridge looks like a "unified brand voice" and shared "mission values" across all departments. By building this bridge, you eliminate the "cognitive dissonance" that occurs when a brand feels "different" in different places. For the executive, this bridge is the "authenticity guardrail." It ensures that your company "walks the talk." This bridge provides the "narrative continuity" that allows a customer to build a coherent and stable relationship with your brand, turning a series of separate interactions into a single, trustworthy story.

DO'S FOR THE LEADER	DON'TS FOR THE LEADER
Maintain a consistent brand narrative.	Let a customer's memory be altered.
Be honest about your product's limits.	Make false or misleading claims.
Correct any misinformation quickly.	Let misinformation spread.
Use clear and factual communication.	Use vague or ambiguous language.
Listen to how customers remember you.	Assume your story is the only one.
Be proactive about your brand's image.	Let others define your brand.

SITUATION	SUGGESTED ACTION
Customers lose trust because marketing promises differ from actual product capabilities during onboarding	Align messaging with reality, validate claims, and reinforce credibility through cross-team content reviews
Customers question reliability because different teams describe the same feature using conflicting value propositions	Create unified messaging guidelines, synchronize updates, and reinforce consistency through shared narrative ownership
Customers hesitate to upgrade because promotional materials highlight benefits not reflected in real user experiences	Incorporate customer stories, validate claims, and reinforce authenticity by grounding messaging in measurable outcomes
Customers feel misled when support explanations contradict earlier sales commitments	Document commitments, train teams, and reinforce alignment through shared accountability mechanisms
Customers disengage when brand tone shifts dramatically between channels and departments	Standardize tone frameworks, audit communications, and reinforce cohesion through cross-channel governance
Customers doubt transparency when product roadmaps change without clear explanation of reasoning	Communicate rationale, share context, and reinforce trust by explaining how decisions support long-term customer value

This bridge is a framework for ensuring "high quality" at every customer touchpoint, regardless of the channel or department. It's about building "operational excellence" into the entire journey. Daily, this bridge looks like "Standard Operating Procedures" that are fueled by real-time AI feedback to ensure no interaction falls below the "brand standard." By building this bridge, you eliminate the "experience lottery" where a customer's satisfaction depends on which agent they get. For the executive, this bridge is the "reliability engine." It ensures that your brand is a "safe bet" for the customer. This bridge provides the "predictable excellence" that is the hallmark of a world-class organization, building the "unshakeable loyalty" that comes from consistently meeting and exceeding customer expectations.

DO'S FOR THE LEADER	DON'TS FOR THE LEADER
Ensure high quality at every touchpoint.	Let some touchpoints be better than others.
Use a consistent design language.	Use a jumble of different styles.
Train all teams on customer experience.	Let departments operate in silos.
Get feedback on the entire journey.	Only ask for feedback at one point.
Create a seamless, end-to-end journey.	Let the journey feel disjointed.
Maintain a single, unified customer view.	Have separate customer data in each team.

SITUATION	SUGGESTED ACTION
Customers abandon journeys because digital and human channels provide different answers to identical questions	Align knowledge bases, synchronize updates, and reinforce consistency through unified training across all teams
Customers feel frustrated when service quality varies significantly between locations or representatives	Standardize procedures, monitor performance, and reinforce reliability through shared operational benchmarks
Customers distrust support because resolution quality depends heavily on which agent handles their case	Implement quality reviews, provide coaching, and reinforce fairness by ensuring consistent resolution standards
Customers hesitate to adopt new features because onboarding quality differs across departments	Create shared playbooks, coordinate training, and reinforce alignment through cross-functional launch processes
Customers complain about unpredictable wait times because staffing models differ across channels	Harmonize capacity planning, publish expectations, and reinforce transparency through real-time availability indicators
Customers feel disoriented when visual design and terminology differ across product surfaces	Unify design systems, standardize labels, and reinforce familiarity through consistent interface patterns

5. DIVERSE INSIGHTS BRIDGE | A Broader View:

This bridge is a system for gathering feedback from a wide range of customers, including the "extreme users," the "silent majority," and the "non-customers." It involves proactively seeking out "dissenting voices" to challenge the internal status quo. Daily, this bridge looks like "Customer Advisory Boards" that are intentionally diverse. By building this bridge, you avoid the "echo chamber" of satisfied users. It allows you to see the "future of the market" before it arrives. For the executive, this bridge is the "innovation radar." It ensures that your product roadmap is guided by a comprehensive view of human needs. This bridge provides the "perspective diversity" that prevents strategic obsolescence and ensures your transformation is resilient to changing social and global market dynamics.

DO'S FOR THE LEADER	DON'TS FOR THE LEADER
Gather feedback from a wide range.	Only listen to a few trusted people.
Use diverse methods to get feedback.	Only use one feedback method.
Talk to customers from different segments.	Only talk to customers who love you.
Use objective data, not personal stories.	Base decisions on a single, vivid story.
Challenge your own assumptions.	Assume you know what everyone wants.
Empower a diverse team to listen.	Let a homogeneous team get insights.

SITUATION	SUGGESTED ACTION
Customers from smaller segments feel ignored because feedback programs overrepresent high-volume groups	Expand sampling frames, weight responses appropriately, and reinforce inclusion by ensuring every segment influences decisions meaningfully
Customers in rural areas feel unseen because surveys assume stable connectivity and long online sessions	Offer offline options, shorten forms, and reinforce accessibility by designing feedback tools for varied connectivity conditions
Customers with disabilities struggle to contribute because feedback channels lack proper accessibility features	Add assistive support, test with diverse users, and reinforce equity by meeting accessibility standards consistently
Customers from emerging markets feel misrepresented because questions assume mature product usage patterns	Localize prompts, adapt scenarios, and reinforce relevance by reflecting regional realities accurately
Customers with niche workflows feel unheard because feedback forms focus on mainstream use cases	Add scenario branches, invite targeted interviews, and reinforce depth by capturing specialized needs
Customers distrust surveys because they rarely see representation of their demographic or industry context	Publish participation summaries, highlight diverse voices, and reinforce trust by showing broad contributor inclusion

This bridge is an algorithm designed to treat all customers fairly, regardless of their "lifetime value" or "demographic." It ensures that even "low-margin" customers receive a high-quality, respectful experience. Daily, this bridge looks like "fairness audits" of your routing and pricing algorithms to ensure they aren't systematically penalizing certain groups. By building this bridge, you ensure that your "digital heart" is in the right place. For the executive, this bridge is the "ethical foundation" of your AI strategy. It ensures that your digital transformation is "inclusive" rather than "exclusive." This bridge provides the "universal respect" that builds a powerful and inclusive brand reputation, ensuring that your company is seen as a force for good in the AI-driven world.

DO'S FOR THE LEADER	DON'TS FOR THE LEADER
Use an algorithm that is transparent.	Use an algorithm that unfairly promotes.
Audit your algorithm for fairness.	Assume your algorithm is unbiased.
Use diverse data to train algorithms.	Use biased data to train your algorithms.
Be transparent about your ranking logic.	Use a black box algorithm for ranking.
Test your algorithm on diverse groups.	Assume your algorithm works for everyone.
Make sure the ranking serves the user.	Let the ranking serve your business interests.

SITUATION	SUGGESTED ACTION
Customers receive fewer recommendations because algorithms over-prioritize high-spending profiles unfairly	Rebalance weighting models, audit bias sources, and reinforce fairness by validating outcomes across diverse customer groups
Customers see limited content because personalization narrows options excessively over time	Introduce exploration slots, diversify signals, and reinforce openness by intentionally surfacing unfamiliar categories periodically
Customers feel disadvantaged because pricing suggestions differ subtly across similar profiles	Standardize logic, publish criteria, and reinforce transparency by explaining how recommendations are generated
Customers distrust search results because sponsored items consistently outrank more relevant alternatives	Separate sponsored placements, label clearly, and reinforce integrity by prioritizing relevance in organic rankings
Customers experience slower support routing because algorithms misclassify urgency for certain issue types	Improve training data, refine detection, and reinforce reliability by testing models against real escalation patterns
Customers feel excluded when automated decisions overlook contextual factors important to their situation	Add human review paths, capture nuance, and reinforce fairness by blending automation with judgment

Ai6x6™ Transformation Blueprint

Customer Delight Bliss

"The goal as a company is to have customer service that is not just the best but legendary." - Sam Walton

Sam Walton (1918–1992) was an American businessman and entrepreneur. He founded the retail giants Walmart and Sam's Club and is known for his focus on low prices and friendly customer service.

Customer delight bliss is the ultimate state of harmony where customer satisfaction is a source of profound joy and success. This is the culmination of a successful customer transformation journey. Blissful Service is the satisfaction of knowing you have exceeded a customer's expectations. This isn't just about meeting needs but about creating moments of genuine surprise and delight. Blissful Security is the confidence that comes from providing a secure and trusted experience, knowing that customers feel safe and valued. Blissful Insights is the pleasure of using customer feedback to constantly improve and innovate. This state of bliss is not about a single happy customer but about a culture where excellence is the norm. It is the reward for the hard work of introspection and systemic change, creating a company that is not just successful but also deeply valued by its customers.

Completing the **Relational domain**, this framework explores Customer Delight Bliss, detailing the psychological and behavioral outcomes that cement customer Engagement and validate Human Fulfillment. The immediate outcome is High Satisfaction, where customers are consistently pleased with their interactions and the product's performance across all touchpoints. The highest form of loyalty is Brand Loyalty, where customers repeatedly choose the brand over competitors, even when faced with alternatives, becoming immune to competitive pricing. Bliss generates Positive Word-of-Mouth, with customers enthusiastically recommending the product to their networks, serving as the most authentic and cost-effective marketing channel possible. The goal is a Seamless Experience, ensuring a frictionless, intuitive user journey across all touchpoints, from discovery to support, drastically reducing customer churn. The ultimate relational outcome is Customer Advocacy, turning satisfied users into outspoken champions of the brand's mission and value, creating a powerful external voice. Finally, success requires achieving Perceived Value, where customers genuinely believe the product is worth significantly more than the price they paid, securing long-term price tolerance, and maximizing transaction margins.

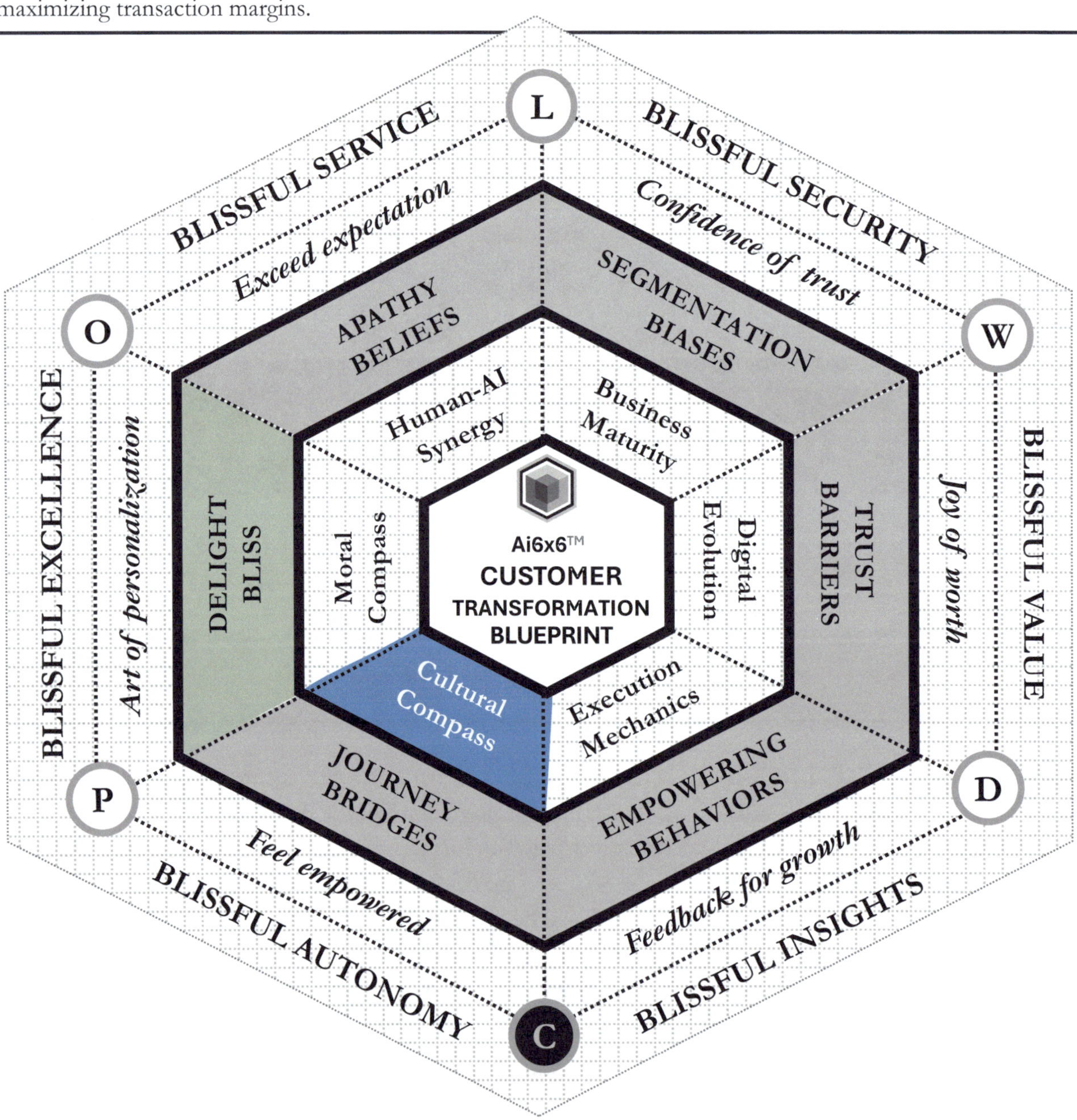

Ai6x6 CULTURAL COMPASS FOR BLISS | CUSTOMER TRANSFORMATION

There is a profound sense of fulfillment that comes from knowing you have not just "met" but "exceeded" a customer's expectations. This bliss is the "joy of the wow." In your daily life as an executive, it's the feeling of reading a customer story where your team went the "extra mile" to solve an impossible problem. This bliss turns work into a "mission of care." It provides a deep sense of purpose and pride across the entire organization. This feeling of "making someone's day better" is the ultimate motivator for excellence. This bliss is the reward for a truly customer-centric culture; it transforms the company from a "money-making machine" into a "value-giving community" that is genuinely loved by its customers.

DO'S FOR THE LEADER	DON'TS FOR THE LEADER
Exceed customer expectations.	Meet only the minimum standard.
Create a feeling of surprise and delight.	Be predictable and uninspired.
Empower staff to solve problems.	Make staff follow a rigid script.
Respond to feedback and act on it.	Ignore feedback as just complaining.
Treat all customer issues as important.	Dismiss customer issues as trivial.
Build trust with consistent service.	Be inconsistent and unreliable.

SITUATION	SUGGESTED ACTION
Customers express delight when teams resolve complex issues faster than expected without requiring escalation steps	Document successful behaviors, replicate patterns, and reinforce excellence by training teams to deliver consistently fast, high-quality resolutions across similar scenarios
Customers feel valued when agents anticipate needs before customers fully articulate underlying concerns or frustrations	Train situational awareness, study behavioral cues, and reinforce proactive care by coaching teams to identify emerging needs early during conversations
Customers share praise when support teams follow up unprompted to ensure issues remain resolved long after closure	Build structured follow-up routines, track outcomes, and reinforce reliability by confirming stability after initial resolution across relevant touchpoints
Customers recommend services when onboarding feels effortless, personalized, and clearly aligned with their specific goals	Design adaptive onboarding flows, simplify steps, and reinforce confidence by guiding customers through tailored pathways matching their context
Customers express loyalty when teams handle exceptions gracefully without unnecessary friction or rigid policy enforcement	Empower frontline judgment, define flexible guardrails, and reinforce trust by enabling thoughtful decisions that prioritize fairness
Customers feel surprised when service teams provide helpful insights unrelated to the original request they submitted	Encourage value-added guidance, train cross-domain awareness, and reinforce holistic support by sharing relevant suggestions beyond immediate issues

The confidence that comes from providing a secure, transparent, and trusted experience is the bliss of "peace of mind." It's the feeling of knowing that your customers' data is as safe as your own. This bliss is the "death of anxiety" for both the company and the customer. Daily, it means you can innovate with AI knowing your "ethical shield" is strong. It provides a sense of "integrity and honor." This bliss is the realization that trust is the ultimate competitive advantage. When your customers don't have to "worry" about doing business with you, they become more open, more experimental, and more loyal. This bliss is the foundation of a "fearless brand" that can lead the market through disruption.

DO'S FOR THE LEADER	DON'TS FOR THE LEADER
Prioritize data security and privacy.	Be careless with customer information.
Be transparent about data usage.	Hide how you use customer data.
Proactively communicate security.	Wait for a breach to communicate.
Use simple, clear digital interfaces.	Create confusing or complex systems.
Make policies easy to understand.	Use vague, long legal agreements.
Build trust with consistent service.	Take customer trust for granted.

SITUATION	SUGGESTED ACTION
Customers feel reassured when security alerts explain risks clearly without causing unnecessary alarm or confusion	Use calm language, provide context, and reinforce trust by outlining protective steps taken automatically on their behalf
Customers adopt features confidently when privacy settings are transparent, intuitive, and easy to adjust anytime	Simplify permission controls, highlight impacts, and reinforce autonomy by showing real-time effects of preference changes
Customers trust transactions when payment flows display visible safeguards, independent certifications, and clear verification steps	Surface security indicators, explain protections, and reinforce safety by demonstrating how systems prevent unauthorized activity
Customers feel protected when suspicious activity triggers immediate, personalized guidance rather than generic automated warnings	Tailor alerts to context, offer actionable steps, and reinforce confidence by showing specific actions already taken
Customers remain loyal when data breaches are handled transparently with clear remediation timelines and honest communication	Share verified details, outline recovery plans, and reinforce accountability through consistent updates until full resolution
Customers relax when identity verification processes feel secure without being unnecessarily burdensome or repetitive	Streamline verification steps, use adaptive checks, and reinforce balance by matching verification rigor to actual risk levels

The joy of knowing that a customer perceives your product as the "best possible value" for their life is the bliss of "perfect fit." It's the feeling of "relevance achieved." This bliss happens when your AI-driven personalization perfectly aligns your solution with the customer's specific "job to be done." Daily, it manifests as high "retention rates" and low "price sensitivity." It provides a sense of "market mastery." This bliss is the reward for listening deeply and adapting constantly. When you know you are providing the "maximum utility" for the customer's time and money, you lead with a sense of "unshakeable value." This bliss turns the competitive struggle into a "partnership for success," where both the company and customer thrive.

DO'S FOR THE LEADER	DON'TS FOR THE LEADER
Understand what customers truly value.	Assume value is just about price.
Constantly add value to products.	Let your product get stale and old.
Offer a fair, transparent price.	Use complex, hidden pricing.
Clearly communicate your value.	Fail to show how your product helps.
Ask for feedback on perceived value.	Assume your value is obvious.
Focus on the customer's needs.	Only focus on what you're selling.

SITUATION	SUGGESTED ACTION
Customers express confidence when pricing feels fair, transparent, and clearly connected to measurable outcomes	Share value breakdowns, highlight savings, and reinforce trust by linking costs directly to demonstrated customer benefits
Customers renew enthusiastically when they see continuous improvements without unexpected increases or hidden charges	Communicate enhancements, show impact metrics, and reinforce fairness through predictable pricing practices
Customers recommend services when they experience meaningful results quickly without requiring extensive customization or additional fees	Showcase early wins, document outcomes, and reinforce value by emphasizing efficiency
Customers feel satisfied when product capabilities exceed expectations during real-world usage scenarios	Capture success stories, share comparisons, and reinforce differentiation through evidence-based performance
Customers upgrade willingly when new tiers clearly align with evolving needs and deliver tangible improvements	Provide tailored recommendations, explain benefits, and reinforce alignment with customer goals
Customers stay loyal when long-term value compounds through consistent reliability and ongoing enhancements	Maintain improvement roadmaps, communicate progress, and reinforce commitment through transparent updates

The satisfaction of using customer feedback to "constantly improve" is the bliss of "infinite evolution." It's the feeling of a brand that is "getting better every day." This bliss turns "complaints" into "fuel" and "data" into "direction." Daily, it means your team is excited by what they learn from the market. It provides a sense of "intellectual momentum." This bliss is the realization that you are in a "continuous conversation" with your customers. When you use AI to close the "feedback loop" in real-time, you create a "responsive brand" that is always at the cutting edge.10 This bliss is the reward for being "humbly brilliant"—always learning, always growing, and always guided by the true needs of the people.

DO'S FOR THE LEADER	DON'TS FOR THE LEADER
Use feedback to constantly improve.	Ignore feedback as a waste of time.
Actively seek all types of feedback.	Only listen to positive feedback.
Share feedback with the whole team.	Keep feedback locked in one department.
Use data to find areas for growth.	Assume customers are already happy.
Empower teams to act on feedback.	Require approval for every small change.
Thank customers for their input.	Make customers feel their feedback is wasted.

SITUATION	SUGGESTED ACTION
Customers feel appreciated when their feedback directly shapes product improvements visible in subsequent releases	Publish change logs, highlight contributor impact, and reinforce partnership by acknowledging customer influence transparently
Customers share deeper insights when teams demonstrate genuine curiosity about underlying motivations and challenges	Ask exploratory questions, document themes, and reinforce learning through structured analysis that informs decisions
Customers trust the process when feedback loops close quickly with clear explanations of implemented actions	Provide timely updates, outline decisions, and reinforce transparency by showing progress and rationale
Customers contribute regularly when feedback channels feel simple, accessible, and respectful of their time	Streamline forms, personalize prompts, and reinforce ease through thoughtful design that minimizes effort
Customers offer strategic suggestions when teams invite them into early concept reviews and prototype sessions	Facilitate co-creation workshops, gather reactions, and reinforce collaboration by integrating insights visibly
Customers feel invested when they see long-term improvements shaped by their earlier recommendations	Track multi-release themes, share evolution stories, and reinforce continuity through visible progress across versions

The feeling of empowerment that comes from a customer being able to "easily solve their own problems" is the bliss of "effortless partnership." It's the joy of a "frictionless world." This bliss happens when your self-service AI is so intuitive that it feels like "magic." Daily, it manifests as "speed and ease" for the customer and "efficiency and focus" for the company. It provides a sense of "mutual respect." This bliss is the realization that the best service is the one the customer doesn't even have to "ask" for. When you provide blissful autonomy, you give the customer the gift of "time," which is precious. This bliss builds a "deep and quiet loyalty" that is incredibly resilient and high-value.

DO'S FOR THE LEADER	DON'TS FOR THE LEADER
Provide a comprehensive knowledge base.	Offer no self-help resources.
Make self-service intuitive and easy.	Use confusing, hard-to-find resources.
Empower customers to solve problems.	Force customers to call for help.
Keep help content up-to-date.	Have outdated and useless information.
Use data to improve self-service.	Assume your self-service is perfect.
Make human support an easy option.	Hide your phone number or email.

SITUATION	SUGGESTED ACTION
Customers feel empowered when self-service tools resolve issues quickly without requiring human intervention	Build intuitive flows, simplify steps, and reinforce confidence through clear guidance and predictable outcomes
Customers adopt features readily when instructions appear contextually at the exact moment they need help	Trigger contextual tips, personalize timing, and reinforce usability through adaptive assistance aligned with behavior
Customers trust automation when outcomes consistently match or exceed human-assisted resolutions	Monitor accuracy, refine logic, and reinforce reliability by validating automated decisions across scenarios
Customers complete tasks independently when interfaces reduce cognitive load and highlight essential actions clearly	Use progressive disclosure, streamline layouts, and reinforce clarity through thoughtful design patterns
Customers explore advanced capabilities when learning resources feel approachable, structured, and easy to navigate	Provide modular tutorials, surface examples, and reinforce mastery through guided practice opportunities
Customers rely on self-service confidently when recovery paths exist for mistakes without punitive consequences	Add undo options, clarify risks, and reinforce safety through reversible actions and clear explanations

The satisfaction of creating a "truly personalized experience" for every single customer is the bliss of "individualized mastery." It's the feeling of a brand that "scales intimacy." This bliss happens when your AI transformation allows you to treat a million customers like a million individuals. Daily, it manifests as "profound connection" and "magical relevance." It provides a sense of "vision realized." This bliss is the goal of the "Customer Transformation"—the moment when technology and humanity perfectly merge to create a world of "personalized excellence" that delights everyone it touches. This is the pinnacle of organizational achievement, where every interaction is a masterpiece of intent and execution, solidifying your brand as a leader in the human-AI era.

DO'S FOR THE LEADER	DON'TS FOR THE LEADER
Tailor products to individual needs.	Offer a one-size-fits-all solution.
Use data to personalize experiences.	Make assumptions about what they want.
Be respectful of customer privacy.	Use data in a way that feels intrusive.
Offer personalized product suggestions.	Spam customers with generic promotions.
Use personalization to add value.	Personalize for the sake of it.
Create different customer segments.	Treat every single customer the same.

SITUATION	SUGGESTED ACTION
Customers feel recognized when experiences adapt seamlessly to their preferences across every channel consistently	Sync preference data, unify profiles, and reinforce continuity through coordinated personalization across touchpoints
Customers appreciate relevance when recommendations reflect real-time behavior rather than outdated assumptions	Refresh models frequently, incorporate recent signals, and reinforce accuracy through transparent explanations of logic
Customers feel understood when communications match their tone, pace, and preferred level of detail	Capture style preferences, segment messaging, and reinforce respect by honoring individual communication norms
Customers engage deeply when product surfaces highlight features aligned with their specific goals and workflows	Personalize layouts, prioritize relevant tools, and reinforce alignment through contextual suggestions tailored to needs
Customers feel valued when service teams remember past interactions and anticipate future needs naturally	Maintain unified histories, analyze patterns, and reinforce foresight through proactive preparation and thoughtful follow-ups
Customers stay loyal when personalization evolves continuously as their circumstances and priorities change	Monitor shifts, adjust experiences, and reinforce adaptability through ongoing refinement based on updated signals

Ai6x6™ Transformation Blueprint

Customer Diversity

"We need diversity of thought in the world to face the new challenges."

Tim Berners-Lee, Inventor of the World Wide Web. The complexity of modern challenges—especially in the digital realm—cannot be solved by homogenous thinking, making diverse thought crucial for the future.

Customer Transformation requires deep diversity to achieve genuine market relevance and develop digital experiences that resonate globally. A customer strategy designed by a team with a narrow viewpoint is almost guaranteed to suffer from **cultural myopia**, leading to digital products, interfaces, and communications that are generic, poorly localized, or alienating to specific customer segments. A diverse internal team—one that includes individuals who personally understand the varied cultural, linguistic, and socioeconomic nuances of the target market—is the only way to establish true customer empathy. This insider knowledge allows the business to move beyond simple translation to genuine **cultural adaptation**, ensuring that digital interactions feel personal, authentic, and inclusive in every region. By gathering, valuing, and integrating these varied perspectives, the company can design frictionless customer journeys, personalize offerings effectively, and successfully address accessibility needs. Diversity in this context prevents costly market missteps and transforms the transformation into a powerful growth engine, effectively unlocking new and underserved international and domestic market segments.

In today's interconnected world, diversity is a fundamental pillar of strength and innovation. Often simplified to visible differences, true diversity is a rich tapestry of attributes shaping how individuals perceive and contribute. Understanding this multifaceted nature is crucial for fostering environments where every voice thrives. This essay explores diversity through six critical dimensions: **demographic, cognitive, experiential, geographic, functional, and perspective diversity**. By examining these distinct facets, we can better appreciate how a holistic approach to diversity enriches teams and communities, driving robust solutions.

First, **Demographic Diversity** encompasses varied inherent and acquired characteristics. This includes age, bringing diverse generational insights, and gender, race, ethnicity, sexual orientation, ensuring a wide range of lived experiences. Socioeconomic status, religion, disability, and nationality further contribute to a richer collective understanding. Valuing demographic diversity ensures equitable representation and leverages unique insights from varied personal journeys.

Secondly, **Cognitive Diversity** highlights how people think, process information, and approach problems. This includes varied thinking styles like analytical vs. intuitive approaches, and different problem-solving or learning methods. When a group embraces cognitive diversity, it tackles problems from multiple angles, leading to more innovative solutions and a thorough understanding of complex issues, free from a single dominant thought process.

Next, **Experiential Diversity** focuses on the breadth of life and work experiences individuals bring. This covers varied work histories across industries and roles, providing a rich pool of professional knowledge. Diverse educational backgrounds, from different disciplines or institutions, also offer distinct frameworks. Beyond formal credentials, personal life experiences like travel and cultural immersion contribute significantly, equipping teams with a wide array of skills and adaptable approaches.

Geographic Diversity emphasizes the impact of place on an individual's perspective. This includes differences from regional or local backgrounds, shaping values and priorities. National backgrounds introduce distinct cultural norms, while urban versus rural upbringings offer varying insights into community dynamics and societal needs. These varied geographical lenses provide a broader understanding of global and local markets and behaviors.

Furthermore, **Functional Diversity** pertains to the variety of professional roles, specialized areas, and departmental expertise within a team. This dimension recognizes contributions from different functions like engineering, marketing, or finance, each bringing specific domain knowledge. Variations in hierarchical roles and industry experience also inform cross-pollination of best practices. Leveraging functional diversity ensures all necessary technical skills are present, fostering comprehensive problem-solving.

Finally, **Perspective Diversity** encompasses the fundamental differences in viewpoints and mental models individuals hold, often influenced by other diversity forms. This includes varied ideological stances, cultural interpretations, and distinct value systems. Understanding diverse stakeholder perspectives—from customers to employees—is crucial for holistic decision-making. Ultimately, perspective diversity allows for problems to be framed in multiple ways, challenging assumptions and encouraging empathetic, effective outcomes.

By intentionally cultivating and integrating these six dimensions—demographic, cognitive, experiential, geographic, functional, and perspective—organizations and societies can harness the profound power of true diversity. This comprehensive approach not only fosters equity but also enhances creativity, resilience, and innovation, demonstrating that a multitude of viewpoints is the clearest path to collective success.

Embracing **demographic diversity** within customer-facing teams, from sales and marketing to support and product development, means intentionally cultivating a talent pool that brings varied life experiences to understanding, serving, and engaging diverse customer segments. This goes beyond mere representation; it's about leveraging the inherent value that comes from professionals with varied ages, genders, racial and ethnic backgrounds, physical abilities, socio-economic upbringings, and sexual orientations. These visible and non-visible differences are crucial for identifying nuanced customer needs, preventing unintended biases in service delivery, ensuring inclusive communication, and developing experiences that resonate with a global and varied customer base. A demographically diverse customer team brings distinct insights into market segments, cultural sensitivities in communication, and potential blind spots in customer journey mapping, which are crucial for effective customer transformation. When customer teams lack this breadth, they risk developing strategies that are narrow in appeal, fail to address the needs of significant customer populations, or perpetuate existing inequalities in service. True demographic diversity fosters more empathetic, relevant, and ultimately, more successful customer relationships, better equipped to address complex customer challenges and connect authentically with diverse individuals and communities.

Topic	Tangible steps	Challenges you may face
AGE DIVERSITY IN CUSTOMER SERVICE		
Varied age groups among customer service staff bring different communication styles and empathy for diverse customer needs. Enhances rapport and broadens service effectiveness.	Form cross-generational customer service teams. Implement intergenerational mentorship for customer support roles. Provide continuous training on diverse communication styles and channels.	Resistance to new service approaches from established staff. Overcoming biases favoring specific age groups in customer interactions. Managing different customer expectations based on age.
GENDER DIVERSITY IN PERSONALIZED MARKETING		
Balanced gender representation in marketing influences messaging, product positioning, and customer segmentation. Diverse gender perspectives lead to more inclusive and appealing campaigns.	Set targets for gender balance in marketing/CX teams. Implement inclusive language and imagery guidelines for customer communications. Conduct gender-focused customer testing and feedback loops.	Unconscious biases in customer persona creation. Lack of female customer leadership. Resistance to changing established marketing norms.
RACIAL & ETHNIC DIVERSITY IN CUSTOMER ENGAGEMENT		
Diverse racial & ethnic backgrounds foster critical insights into cultural relevance in customer engagement, brand trust, and ethical marketing. Improves global customer relationships.	Develop cultural sensitivity training for customer engagement teams. Establish ERGs focused on customer inclusion. Prioritize diverse representation in customer research panels.	Systemic bias in target customer assumptions. Difficulty fostering psychological safety to discuss cultural nuances. Insufficient understanding of cultural implications for service delivery.

Cognitive diversity within customer teams is paramount for navigating the inherent complexities of customer transformation. It acknowledges that true innovation and robust problem-solving in customer experience stem not just from varied experiences or demographics, but fundamentally from differences in how people think, process information, and approach customer challenges. A customer team composed solely of analytical thinkers might excel at optimizing existing customer service metrics but could struggle with identifying novel customer needs or designing truly disruptive customer experiences. Conversely, a team too reliant on intuitive thinking might miss critical data points or overlook practical constraints in customer relationship management. By fostering a mix of analytical, creative, systemic, and neurodiverse cognitive styles across all customer-facing roles, organizations are better equipped to challenge assumptions in customer strategy, identify unseen customer segments, predict unforeseen shifts in customer behavior, and design more comprehensive and empathetic customer engagement models. It's about building a collective customer intelligence where different cognitive strengths complement each other, leading to more thorough customer evaluations, more innovative customer solutions, and a greater capacity for adaptive learning in the face of evolving customer demands. This richness in thought ensures that customer decisions are thoroughly vetted from multiple angles, preventing blind spots and fostering intellectual agility crucial for sustained customer transformation success.

Topic	Tangible steps	Challenges you may face
ANALYTICAL VS. INTUITIVE THINKING IN CX DESIGN		
Balancing data-driven & gut-feeling approaches in CX design shapes customer journeys. Leads to nuanced decision-making, combining quantitative insights with empathetic foresight.	Design CX workflows requiring diverse cognitive input. Encourage CX strategists to articulate both data-backed and intuitive reasoning. Provide training on leveraging different thinking styles in projects.	Friction between data rigor & rapid intuition in CX. Difficulty quantifying intuitive customer insights. Culture valuing one CX design style over another hinders collaboration.
SYSTEMIC THINKERS IN CUSTOMER ECOSYSTEMS		
Customer professionals seeing interconnectedness of customer touchpoints & long-term implications drive comprehensive customer experience improvement.	Provide systems thinking training for CX architects/managers. Assign systemic thinkers to cross-functional customer journey initiatives. Use "customer journey mapping" to identify impacts.	Difficulty identifying/developing systemic thinking in CX roles. Resistance to long-term CX views. Overwhelm from too much interconnected customer information.
CREATIVE PROBLEM-SOLVERS IN CUSTOMER ENGAGEMENT		
Customer professionals with imaginative thinking inspire novel solutions to complex customer problems. Crucial for breakthrough customer engagement strategies.	Establish customer experience "hackathons" or ideation labs. Implement brainstorming sessions for unconventional engagement ideas. Create dedicated time for "blue-sky" customer thinking.	Risk aversion stifling new engagement ideas. Difficulty translating creative customer solutions into actionable plans. Perceived lack of "seriousness" for certain customer functions.

Experiential diversity among customer professionals is a powerful catalyst for successful customer transformation, providing a rich tapestry of practical insights drawn from varied career paths, industries, educational backgrounds, and personal life stages. Customer professionals who have navigated different organizational cultures, worked across diverse sectors (e.g., B2B to B2C, non-profit to corporate), or gained international exposure bring a broader repertoire of customer solutions and a more nuanced understanding of complex customer challenges. This breadth of experience prevents insular thinking in customer strategy, encourages the adoption of "best practices" from seemingly unrelated customer fields, and fosters resilience by drawing on lessons learned from past customer successes and failures in varied contexts. For instance, a customer professional with experience in high-volume call centers can bring invaluable composure and clearheadedness during periods of intense customer surges. By intentionally cultivating a customer team with diverse professional and personal journeys, an organization gains a formidable collective wisdom, enhancing its ability to foresee customer opportunities, mitigate customer risks, and execute customer transformation initiatives with greater adaptability and effectiveness. It promotes a learning culture where varied practical customer wisdom is valued and applied to new strategic customer challenges.

Topic	Tangible steps	Challenges you may face
CROSS-INDUSTRY INSIGHTS IN CUSTOMER EXPERIENCE		
Customer professionals from different industries bring fresh perspectives & transferable solutions to CX challenges. Broadens organizational customer problem-solving.	Recruit CX talent with diverse industry backgrounds. Organize peer-learning forums for cross-industry customer insights. Encourage participation in external CX communities.	Difficulty translating customer insights to new industry contexts. Resistance to "outsider" CX approaches. Lack of understanding of new industry customer nuances for integration.
B2B VS. B2C CUSTOMER APPROACHES		
Impact of customer professionals from B2B vs. B2C environments on relationship building & scale. Blending these fosters both personalized engagement & efficient service.	Integrate B2B customer talent into B2C initiatives (and vice versa). Create joint ventures pairing B2B/B2C customer teams. Foster internal "customer-centricity" programs across segments.	Culture clash between personalized & volume-driven customer approaches. Difficulty integrating different customer sales cycles. High expectations for immediate customer ROI.
GLOBAL VS. LOCAL CUSTOMER RELATIONSHIP MANAGEMENT		
Customer professionals with diverse international & local market exposure balance global customer standardization & regional adaptation. Optimizes service efficiency & market relevance.	Implement global customer talent rotation programs. Establish regional customer knowledge hubs with local market insights. Encourage dual reporting lines for international customer projects.	Complex balance of global vs. local customer needs. Resistance from local customer teams to global mandates. Misinterpretation of cultural nuances in customer communication.

Geographic diversity within customer teams is indispensable for organizations operating in a globalized world, especially during periods of significant customer transformation. Customer professionals deeply rooted in varied regions, countries, or even urban versus rural settings bring critical local market understanding, cultural nuances, and insights into unique regulatory environments and logistical challenges of customer service delivery. Without this localized customer perspective, global customer strategies risk being homogenous, culturally insensitive, or simply impractical in execution. For example, a customer professional experienced in emerging markets might bring crucial insights into scalable, low-cost customer engagement solutions, while someone from a highly regulated economy understands the complexities of customer data privacy. This diversity ensures that customer transformation initiatives are not just conceptualized centrally but are designed with built-in adaptability for regional customer variations. It empowers the organization to navigate geopolitical risks affecting customer sentiment, understand localized customer behaviors for service delivery, and foster stronger relationships with diverse regional customer stakeholders. Customer professionals with varied geographic exposure contribute to customer strategies that are globally coherent yet locally relevant, maximizing impact and minimizing missteps in a fragmented customer world.

Topic	Tangible steps	Challenges you may face
REGIONAL CUSTOMER MARKET UNDERSTANDING		
Customer professionals from different regions offer unique insights into local market dynamics, consumer behaviors, & competition affecting customer engagement. Vital for global CX strategies.	Recruit CX talent with local market knowledge. Mandate global CX strategies be reviewed/adapted by local customer teams. Support travel/immersion in diverse regions for customer insights.	Overlooking customer nuances for standardization. Difficulty adapting to rapid changes in local markets. Resistance from central CX teams to delegate.
NAVIGATING GLOBAL CUSTOMER REGULATORY LANDSCAPES		
Customer professionals familiar with diverse international legal/compliance frameworks are crucial for multi-national customer operations. Ensures adherence & mitigates customer risks.	Create global customer compliance guidelines for CX teams. Invest in continuous education on international customer regulations (e.g., data privacy). Consult local legal experts for every new CX initiative.	Keeping up with rapidly changing customer laws. Reconciling conflicting requirements. Risk of non-compliance due to inadequate local customer understanding.
LOCAL INFRASTRUCTURE AND SERVICE DELIVERY KNOWLEDGE		
Customer professionals understanding local infrastructure (digital connectivity, physical access) & resources impact service delivery. Optimizes local CX implementation.	Empower local customer teams for infrastructure investments impacting service. Develop centralized knowledge of regional customer capabilities. Foster local service provider partnerships for CX.	Underestimating varied local infrastructure impact on CX. Inaccurate resource assumptions for customer service delivery. Difficulty scaling global CX solutions.

Functional diversity within customer teams is critical for executing comprehensive and interconnected customer transformations. It ensures that customer strategic and operational decisions are informed by the deep expertise and unique perspectives of various professional disciplines, rather than being confined to a single departmental silo. A customer team comprising experts from sales, marketing, support, product development, operations, and legal functions can collectively identify customer opportunities, anticipate customer challenges, and design customer solutions that are holistic, practical, and sustainable. For instance, a sales leader identifies growth opportunities, while a marketing specialist champions personalized outreach, and a legal expert ensures data privacy compliance. Without this blend, customer transformation efforts might optimize one area at the expense of another, leading to fragmented implementation or unintended negative consequences in customer experience. By fostering cross-functional collaboration and valuing diverse functional expertise, an organization can break down traditional barriers, enhance teamwork, and create truly integrated customer strategies that leverage the strengths of every part of the business, ensuring that the customer transformation is not just conceptually sound but also executable across all organizational facets.

Topic	Tangible steps	Challenges you may face
SALES & ACCOUNT MANAGEMENT IN CUSTOMER RETENTION		
Customer professionals with strong sales backgrounds drive strategies for customer acquisition & retention. Ensures customer transformation fuels revenue growth.	Appoint sales leaders to CX strategy teams to define customer growth metrics. Establish clear retention targets & metrics for CX initiatives. Implement CRM systems aligned with customer journey stages.	Over-emphasis on short-term sales cycles. Resistance if sales goals overshadow customer experience. Difficulty integrating sales incentives with broader CX objectives.
MARKETING-DRIVEN PERSONALIZATION AT SCALE		
Customer professionals with deep marketing expertise champion personalized customer journeys. Essential for driving targeted engagement throughout transformation.	Elevate marketers to strategic roles in CX projects. Foster "customer data literacy" for all teams. Create cross-functional teams combining marketing/tech for personalized CX initiatives.	Resistance to new marketing automation technologies. Rapid evolution challenges marketing skills. Difficulty integrating personalized marketing with legacy systems and traditional channels.
CUSTOMER SUPPORT'S ROLE IN PROACTIVE SERVICE		
Customer support professionals focus on understanding customer pain points & proactive resolution. Ensures human touch & problem-solving are central to CX success.	Position customer support teams as strategic partners in CX design. Develop robust feedback loops from support to product/operations. Implement self-service tools informed by support insights.	Business units resist prioritizing support over other areas. Difficulty measuring ROI of proactive support. Challenges addressing systemic customer issues identified by support.

Perspective diversity within customer teams is arguably one of the most subtle yet profound drivers of successful customer transformation. It goes beyond demographics, experience, or function, delving into the varied worldviews, core values, ethical frameworks, and risk tolerances that shape how customer professionals interpret information and prioritize customer objectives. A customer team with diverse perspectives challenges inherent biases in customer segmentation, avoids groupthink in experience design, and considers a wider spectrum of customer possibilities and potential consequences. For example, balancing customer professionals who prioritize short-term sales gains with those focused on long-term customer loyalty, or those who are risk-seeking with personalization experimentation with those who are risk-averse with customer data privacy, leads to more robust and balanced customer strategic decisions. This diversity ensures that customer transformation is not just about efficiency or growth in customer numbers, but also about ethical responsibility, data security, personalized value, and genuine stakeholder relationships with customers. It encourages customer professionals to approach customer problems from fundamentally different angles, leading to more resilient customer strategies, better customer engagement, and a more adaptive and morally grounded organization capable of navigating complex and ambiguous future customer landscapes. It cultivates a culture of constructive challenge and holistic customer-driven decision-making.

Topic	Tangible steps	Challenges you may face
SHORT-TERM SALES VS. LONG-TERM CUSTOMER LOYALTY		
Customer professionals balance immediate sales targets with building lasting customer relationships. Ensures customer transformation fuels both revenue and sustainable growth.	Develop customer loyalty programs with clear long-term metrics. Assign customer teams with varied focuses. Use scenario planning to visualize impacts of sales strategies on customer retention.	Pressure for immediate sales results. Customer professionals may prioritize short-term gains. Difficulty articulating long-term customer value to all stakeholders.
RISK-SEEKING VS. RISK-AVERSE APPROACHES IN PERSONALIZATION		
Varied risk tolerance impacts pursuit of disruptive vs. incremental customer experience improvements. A healthy balance drives calculated customer growth and controlled experimentation.	Foster culture where customer risks in personalization are encouraged. Establish "fail fast, learn faster" practices for customer pilots. Create customer project balancing high-risk/reward with safer bets.	Decision paralysis when customer risk tolerances conflict. Dominance of extreme customer risk views. Difficulty setting clear, unified risk parameters across customer teams.
ETHICAL DATA USE VS. MARKETING AGGRESSIVENESS		
Tension when customer professionals prioritize social responsibility & ethics alongside aggressive marketing tactics. Shapes corporate reputation & long-term customer trust.	Develop robust ethical framework for all customer data use. Implement ethical impact assessments for new marketing campaigns. Integrate ESG customer metrics into performance goals.	Shareholder pressure for profit-only from marketing. Difficulty quantifying ethical long-term customer benefits. Customer professionals struggle making decisions reducing profits

BLUEPRINT SERIES

SECTION G

Moral Compass

Ai6x6™ Transformation Blueprint

Story 1 — The Nuremberg Trials and Ethical Accountability (1945–1946): After World War II, the world confronted atrocities committed under the guise of legal authority. The Nuremberg Trials established that individuals and institutions are accountable for actions that violate fundamental human rights. Legal frameworks were strengthened by moral reasoning, not replaced by it. Judges balanced evidence with ethical principles to define new global norms. The trials demonstrated that morality must guide power. They became a foundation for modern ethical governance.

Story 2 — The Tuskegee Syphilis Study Exposure (1932–1972): For decades, researchers withheld treatment from Black men to observe disease progression, violating consent and dignity. Public exposure triggered national outrage and forced reforms in medical ethics. Oversight boards, informed consent, and patient rights became mandatory. Machines and data had played roles in tracking outcomes, but human ethics had been absent. The scandal showed how systems fail when morality is excluded. It reshaped global standards for ethical research.

Story 3 — The Environmental Movement and Earth Day (1970): Industrial growth created pollution that threatened ecosystems and public health. Scientists provided data, but moral arguments mobilized action. Citizens demanded accountability, and governments introduced regulations to protect air, water, and wildlife. The movement reframed environmental protection as a shared ethical duty. It demonstrated how moral clarity can transform policy, industry, and public behavior. The cultural shift established sustainability as a long-term societal obligation.

Story 4 — The Anti-Apartheid Global Solidarity Movement (1950s–1994: Apartheid enforced racial segregation and oppression in South Africa. Activists, governments, and institutions worldwide united to challenge the regime through sanctions, protests, and moral pressure. Human stories amplified the injustice, while global networks coordinated action. The movement showed how collective moral conviction can influence political systems. It proved that ethical solidarity can reshape international norms. The end of apartheid became a landmark in moral mobilization.

Story 5 — The Rise of Corporate Social Responsibility (1990s–2000s): Globalization exposed labor abuses, environmental harm, and opaque supply chains. Consumers demanded transparency, and companies adopted codes of conduct, sustainability reporting, and ethical sourcing. Machines tracked compliance, but humans defined the standards. CSR shifted from public relations to strategic responsibility. Organizations learned that long-term value depends on ethical behavior. It marked a turning point where morality became integral to corporate identity.

Story 6 — The AI Ethics Movement (2015–2025): As AI systems influenced decisions in hiring, policing, healthcare, and finance, concerns grew about bias, transparency, and accountability. Researchers, policymakers, and civil society pushed for ethical frameworks to guide development. Machines provided capability, but humans defined boundaries. The movement emphasized fairness, explainability, and human oversight. It demonstrated that emerging technologies require moral governance as much as technical innovation. AI ethics became a global priority shaping the future of intelligent systems.

LESSONS FROM A CUSTOMER TRANSFORMATION PERSPECTIVE

Customer trust deepens when organizations treat people not as transactions but as individuals deserving fairness, respect, and transparency. Over time, institutions learned that ethical lapses—whether in privacy, safety, or representation—erode credibility and long-term value. Mature organizations embed moral principles into product design, communication, and service delivery. They prioritize honesty, informed consent, and equitable treatment. Customers feel the difference when ethics guide decisions, not just marketing. Maturity is achieved when customer well-being becomes a core organizational obligation, shaping choices even when no one is watching. Ethical commitment becomes a competitive advantage rooted in integrity.

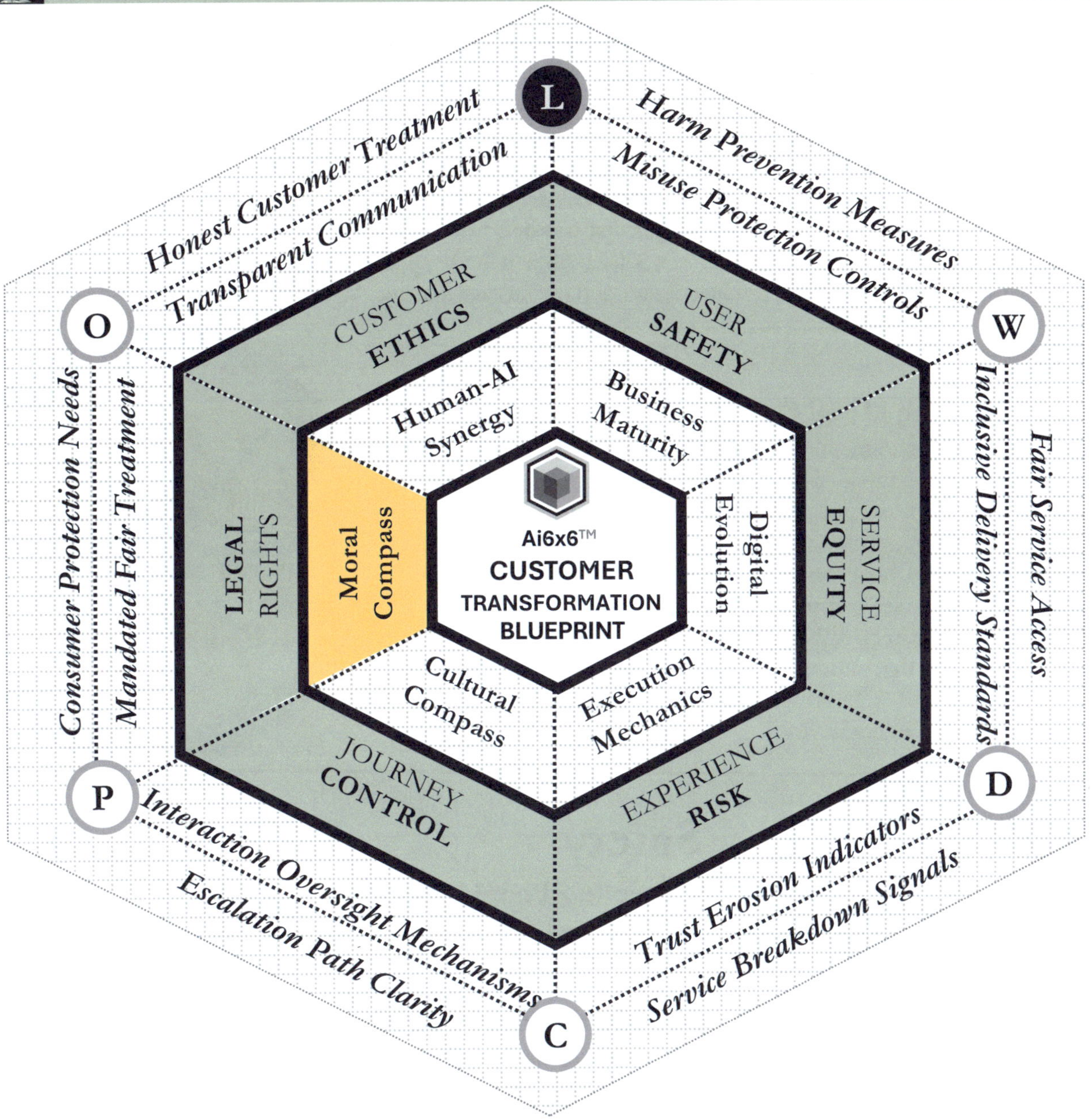

The **Ai6x6 Research Lab** identified the Moral Compass as the most vital lens because it serves as the final guardian of the brand's honor and the customer's rights. This is the ultimate filter for every decision, ensuring the drive for conversion never outpaces the bedrock of **Customer Ethics**. We chose this because a brand's legacy is built on how it protected its users during moments of vulnerability, not just the revenue it generated. It is vital because it provides the moral authority required to lead through the ethical "fog" of predictive marketing and autonomous fulfillment.

This lens is the heart's final check—a rigorous audit of **User Safety**, **Equity**, and **Accessibility**. It ensures the transformed enterprise is not only a titan of its industry but a beacon of public trust. The Lab believes that in the 2026 landscape, the greatest risk to a brand is not a product failure, but a moral breach in the customer journey. By focusing on **Experience Risk** and **Escalation Control**, we provide the tools to navigate the heavy responsibility of influence. This lens is about the courage to be honorable when the temptation for growth is highest, ensuring that the legacy of the brand is one of integrity and respect.

Morality has always shaped how organizations behave, but in the age of AI, it becomes a structural requirement rather than an aspirational ideal. The **Moral Compass** lens forces organizations to confront a truth that is often ignored: intelligence—whether human or machine—does not create ethical behavior. It amplifies whatever moral foundation already exists. When an organization is grounded in integrity, clarity, and principled action, AI strengthens it. When the foundation is inconsistent, opaque, or misaligned, AI exposes those weaknesses with unforgiving speed. Morality is no longer a philosophical preference; it is the architecture that determines whether accelerated decisions, automated systems, and scaled consequences lead to progress or harm. This lens reframes morality as an enterprise-wide discipline, not a leadership trait. It applies equally to workforce practices, data governance, customer interactions, product design, and operational systems.

Ethics

Establishing the Moral Center

Ethics forms the first dimension of the Moral Compass. Ethics is not simply about avoiding misconduct; it is about establishing a moral center that guides every decision, every interaction, and every use of technology. Across all six pillars, ethics determines how data is collected, how employees are treated, how customers are served, how products are built, and how operations are executed. Organizations that prioritize ethics create cultures where integrity is expected, not encouraged. Ethical clarity becomes the anchor that ensures AI is deployed with intention, restraint, and respect. In this dimension, principled behavior becomes the foundation of moral authority.

Safety

Protecting People, Systems, and Society

Safety represents the second dimension. As AI accelerates decision-making and expands operational reach, the potential for harm increases. Safety is not limited to physical environments; it includes psychological safety for the workforce, data safety for customers, model safety for AI systems, and operational safety for mission-critical processes. Safety ensures that no individual, team, or automated system can cause harm through unchecked influence or unintended consequences. It is not about limiting innovation; it is about ensuring that innovation does not outpace responsibility. In this dimension, responsible protection becomes essential to moral operation.

Equity

Ensuring Fairness Across Stakeholders

Equity forms the third dimension. Influence in modern organizations is no longer tied solely to hierarchy; it flows through data pipelines, algorithmic recommendations, and automated workflows. Equity ensures that these systems do not privilege certain groups, perspectives, or outcomes at the expense of others. It applies to workforce advancement, customer treatment, product accessibility, and data representation. Equity is not about equalizing results; it is about ensuring that the system itself is not biased. In this dimension, fairness becomes a structural requirement rather than a social aspiration.

Risk

Navigating Uncertainty With Discipline

Risk represents the fourth dimension. AI-enabled decisions carry greater speed, greater reach, and greater uncertainty. Risk is not about avoiding uncertainty; it is about navigating it with discipline. Organizations must anticipate consequences, stress-test assumptions, and understand the risks embedded in both human and machine judgment. This applies to operational risks, product risks, customer risks, workforce risks, and data risks. Organizations that embrace this dimension act decisively without acting blindly. In this dimension, risk-aware decision-making becomes a moral responsibility.

Control

Defining Governance and Accountability

Control forms the fifth dimension. As organizations adopt AI, authority becomes more complex: who decides, who approves, who monitors, and who intervenes when systems drift? Control ensures that governance structures are clear, decision rights are defined, and accountability is transparent. It applies to data governance, product governance, operational governance, and workforce governance. Control is not about bureaucracy; it is about ensuring that power—human or machine—is exercised with clarity and accountability. In this dimension, structured oversight becomes the backbone of moral operation.

Legal

Upholding Lawful and Responsible Conduct

Legal completes the Moral Compass. In a world where data, automation, and intelligence shape every action, legal obligations become inseparable from organizational obligations. Compliance is not a constraint; it is a safeguard. Legal integrity ensures that decisions align with regulatory expectations, societal norms, and ethical standards across all six pillars. It protects the organization from missteps and reinforces public trust. In this dimension, lawful stewardship becomes the final test of moral credibility.

The Moral Compass lens reminds organizations that moral authority is not a byproduct of good intentions; it is the result of deliberate, disciplined choices. Morality must be engineered into systems, not assumed in people. When organizations prioritize ethics, enforce safety, ensure equity, manage risk, establish control, and uphold legal accountability, they create environments capable of navigating complexity with confidence and integrity. They build systems where people feel protected, customers feel respected, and society feels considered. They create the moral conditions for transformation to take root and endure. In this maturity, morality becomes an enterprise capability—one that strengthens trust, accelerates adoption, and stabilizes transformation. It becomes the differentiator that determines whether AI amplifies value or amplifies harm. Organizations that master the Moral Compass do not merely deploy intelligent systems; they deploy them responsibly. They create futures where intelligence and integrity advance together, ensuring that progress is not only possible but principled.

MORAL COMPASS ANALOGY – NURTURING THE FAMILY HOME

Family Values — ETHICS

The foundational moral code—religious, cultural, and social—that dictates the ground rules for "Right vs. Wrong." Establishing the core integrity of the home so that all who belong understand the non-negotiable standards of conduct.

Home Security — SAFETY

Integrated protection—cameras, sensors, and smoke detectors—that provide total awareness of the premises. Deploying proactive warning systems and delegation guardrails to protect the inhabitants from "hidden fires."

Balanced Contribution — EQUITY

Ensuring everyone helps with the cooking and upkeep, with expenses shared equitably by each earning member. Maintaining participation and representation so that no member feels excluded from the collective success.

The Insurance Shield — RISK

Comprehensive fire, flood, and auto coverage to provide long-term peace of mind for the entire household. Using stress testing and "what-if" scenarios to protect the family assets from unforeseen market storms.

Access & Keys — CONTROL

Ensuring all adults have keys and codes, with clear emergency rules and a defined line of command for disasters. Establishing rigid RACI controls to ensure that a human always has the final word on the most critical locks.

The Estate Will — LEGAL

Recording Wills, Trusts, and Medical Mandates with a legal custodian to safeguard the valuables for the future. Securing the organization's legacy through regulatory accountability, ensuring the estate is legally unassailable for generations.

Ai6x6™ Transformation Blueprint

Customer Ethics

"Trustworthy AI begins with uncomfortable questions and honest answers."

Timnit Gebru, AI ethics researcher known for work on bias and accountability.

Customer Ethics requires principled engagement, and Honest Customer Treatment defines the standards for fairness, clarity, and truthfulness in every interaction. It establishes expectations that prevent manipulation, misrepresentation, or deceptive practices. This matters because customer trust is built on consistent ethical behavior. Honest Customer Treatment prevents harm by ensuring customers receive accurate information and respectful service. It also reinforces fairness by ensuring all customers are treated with the same ethical rigor. When ignored, customer relationships deteriorate quickly. Teams experience confusion about acceptable communication boundaries. Over time, Customer Ethics weakens as trust erodes and behavior becomes inconsistent. Transparent Communication Practices ensure customers receive clear, complete, and understandable information.

The relationship with the customer is a sacred trust that must be protected from short-term greed. Leaders review Honest Marketing Reviews and Ethical Targeting Screens to ensure that immediate gains do not lead to long-term "Trust Erosion." By studying "Customer Respect Audits" in a print format, the leader can focus on the long-term health of the brand. This topic is about moving beyond "Dark Patterns" to a state of radical transparency. Ethical customer engagement is the only way to secure the "Bliss Zone," where the customer's loyalty becomes an impenetrable wall against market predators. In the print blueprint, you can look at "Transparency Notices" and "Complaint Logs" together, seeing the whole narrative of customer trust. This clinical focus ensures that your "Customer Transformation" is built on integrity, preventing your organization from being a "Museum" waiting for its last visitor by ensuring your brand remains morally relevant to your customers.

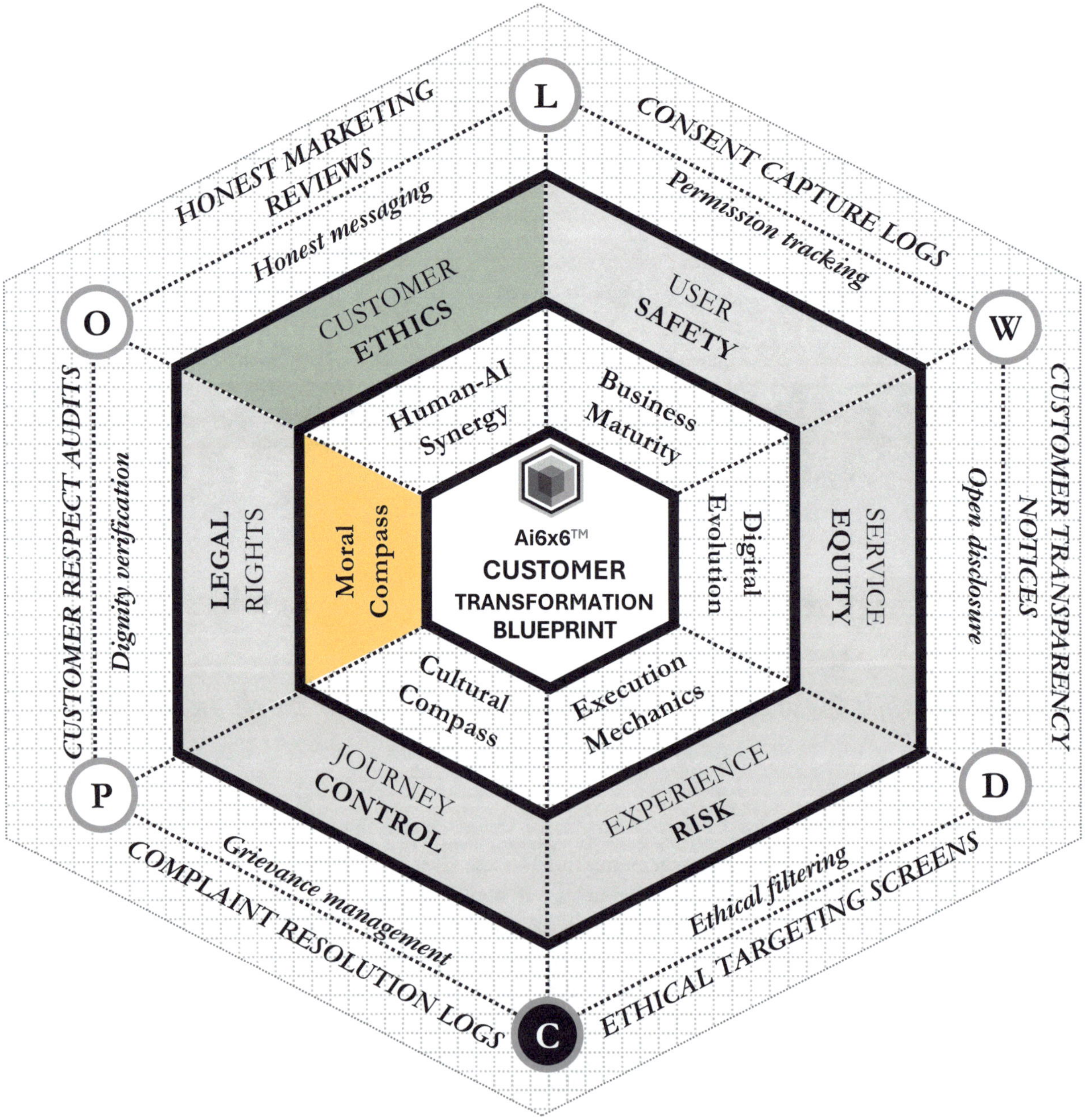

Ai6x6 CUSTOMER ETHICS FOR MORAL COMPASS | CUSTOMER TRANSFORMATION

1. HONEST MARKETING REVIEWS

Customer Ethics requires truthfulness in every representation, and Honest Marketing Reviews ensure that claims, benefits, and messages reflect verifiable reality. They evaluate whether marketing materials align with actual product performance rather than aspirational positioning. This matters because customers rely on accuracy to make informed decisions. Honest Marketing Reviews prevent distortion by forcing teams to validate every assertion before release. They also reinforce fairness by ensuring no customer is misled through exaggeration or omission. When ignored, marketing becomes a vehicle for deception and long term trust erosion. Customers begin to question the organization's integrity. Over time, Customer Ethics collapses as credibility deteriorates and reputational risk escalates.

When executed well, Honest Marketing Reviews strengthen Customer Ethics by aligning communication with truth. Customers trust that the organization respects their intelligence and autonomy. This improves loyalty, reduces disputes, and elevates brand reputation. Reviews reinforce the principle that honesty is a strategic asset, not a constraint. They remind leaders that ethical marketing requires discipline, humility, and verification. Ultimately, they build a customer environment where truth is the foundation of every interaction.

MORAL COMPASS SCENARIO	LEADERSHIP ACTION
Maribel notices a campaign exaggerates product capabilities by implying guaranteed results. She realizes the claim crosses into misleading territory.	The leader pulls the campaign immediately and rewrites the messaging to reflect verifiable outcomes. They also implement a pre-launch honesty review for all marketing assets.
Darius sees that a testimonial used in an ad was heavily edited to remove negative context. He senses the manipulation undermines trust.	The leader restores the full, accurate sentiment and updates testimonial-editing rules. They also train the marketing team on authenticity standards.
Leena discovers that a comparison chart omits a competitor's superior feature. She recognizes the omission as deceptive by omission.	The leader updates the chart to include all relevant facts. They also require competitive-comparison audits before publication.
Omar notices that influencer partners are not disclosing sponsorships. He fears regulatory violations.	The leader enforces mandatory disclosure language and pauses all non-compliant posts. They also add disclosure checks to influencer contracts.
Serena sees that a "limited time offer" has been running for six months. She realizes the scarcity claim is false.	The leader removes the misleading label and resets promotional rules. They also require expiration-date validation for all time-bound offers.
Hugo finds that marketing materials imply data privacy guarantees the product cannot deliver. He senses a legal and ethical risk.	The leader corrects the claims and aligns messaging with actual capabilities. They also require legal review for all privacy-related statements.

2. CONSENT CAPTURE LOGS

Customer Ethics requires respect for autonomy, and Consent Capture Logs document when, how, and under what terms customers authorize data use or service actions. They verify that permissions were granted knowingly, transparently, and without coercion. This matters because customers must trust that their choices are honored. Consent Capture Logs prevent unauthorized use by proving that customers understood and accepted the terms. They also reinforce fairness by ensuring consent is explicit rather than implied. When ignored, organizations risk violating customer rights and eroding trust. Customers feel exploited when their data or preferences are used without permission. Over time, Customer Ethics deteriorates as consent becomes inconsistent or unverifiable.

When executed well, Consent Capture Logs strengthen Customer Ethics by ensuring customer choices guide organizational behavior. Customers trust that their autonomy is protected and respected. This improves transparency, compliance, and ethical credibility. Logs reinforce the principle that permission must be documented, not assumed. They remind leaders that ethical relationships require clarity and accountability. Ultimately, they build a customer environment where trust is earned through respect for individual agency.

MORAL COMPASS SCENARIO	LEADERSHIP ACTION
Aditi discovers that customers who opted out of email marketing are still receiving messages. She realizes consent logs are not synchronized.	The leader fixes the sync pipeline and validates all opt-out flags. They also implement real-time consent propagation across systems.
Rowan notices that consent is bundled into a single checkbox covering multiple unrelated uses. He knows this violates informed consent.	The leader redesigns the consent flow with granular options. They also require explicit opt-in for each data use category.
Serena sees that verbal consents from customer calls are not being logged. She fears compliance gaps.	The leader adds call-recording consent capture and links it to customer profiles. They also train agents on proper disclosure scripts.
Omar finds that consent withdrawals are not being honored in downstream analytics. He senses a governance failure.	The leader enforces downstream deletion and updates the consent-revocation workflow. They also add automated revocation alerts.
Lila notices that consent logs are stored in spreadsheets with no audit trail. She recognizes the risk of tampering.	The leader migrates logs to a secure, immutable system. They also restrict editing rights to authorized personnel.
Andre discovers that minors' consents are being captured without parental verification. He fears regulatory exposure.	The leader deletes improperly collected data and implements parental-consent workflows. They also add age-verification checks.

3. CUSTOMER TRANSPARENCY NOTICES

Customer Ethics requires clarity, and Customer Transparency Notices explain how data is collected, used, shared, and protected. They provide customers with accessible, understandable information about organizational practices. This matters because customers cannot make informed decisions without visibility. Transparency Notices prevent confusion by revealing what happens behind the scenes. They also reinforce fairness by ensuring all customers receive equal insight into data practices. When ignored, organizations appear deceptive or evasive. Customers lose confidence in the integrity of the relationship. Over time, Customer Ethics collapses as opacity becomes normalized.

When executed well, Customer Transparency Notices strengthen Customer Ethics by creating informed, empowered customers. Customers trust that the organization communicates honestly and consistently. This improves loyalty, reduces disputes, and enhances ethical alignment. Notices reinforce the principle that clarity is a moral obligation. They remind leaders that trust grows through openness. Ultimately, they build a customer environment where information is shared responsibly and respectfully.

MORAL COMPASS SCENARIO	LEADERSHIP ACTION
Rhea sees that the privacy notice uses dense legal jargon. She realizes customers cannot understand how their data is used.	The leader rewrites the notice in plain language and adds visual explanations. They also require readability testing for all notices.
Omar notices that the notice omits third-party data-sharing practices. He senses a transparency gap.	The leader updates the notice to include all sharing categories. They also require quarterly notice reviews.
Lila discovers that the notice claims data is deleted after 30 days, but logs show retention for 180 days. She recognizes a compliance breach.	The leader corrects the notice and aligns retention practices with policy. They also enforce retention-policy audits.
Andre sees that the notice is not displayed before data collection begins. He fears customers are being misled.	The leader updates workflows to present notices upfront. They also add mandatory notice-acknowledgment steps.
Serena notices that regional notices are identical despite differing legal requirements. She senses regulatory risk.	The leader localizes notices by jurisdiction. They also require legal review for all regional variants.
Hugo finds that product updates affecting data use are not reflected in the notice. He worries customers are uninformed.	The leader updates the notice and adds a change-notification protocol. They also require notice updates for all product changes.

4. ETHICAL TARGETING SCREENS

Customer Ethics requires restraint in influence, and Ethical Targeting Screens ensure that marketing outreach avoids manipulation, exploitation, or discriminatory segmentation. They evaluate whether targeting criteria align with fairness, respect, and customer well being. This matters because targeting can easily cross ethical boundaries when driven solely by conversion metrics. Ethical Targeting Screens prevent harm by blocking segments that take advantage of vulnerability or asymmetry. They also reinforce fairness by ensuring all customers are treated with dignity. When ignored, targeting becomes predatory and corrosive to trust. Customers feel surveilled, profiled, or exploited. Over time, Customer Ethics deteriorates as influence becomes detached from responsibility.

When executed well, Ethical Targeting Screens strengthen Customer Ethics by aligning outreach with moral boundaries. Customers trust that the organization respects their autonomy and avoids manipulative tactics. This improves brand integrity, reduces regulatory risk, and enhances long term loyalty. Screens reinforce the principle that influence must be principled, not opportunistic. They remind leaders that ethical marketing requires intentional guardrails. Ultimately, they build a customer environment where persuasion is grounded in respect.

MORAL COMPASS SCENARIO	LEADERSHIP ACTION
Keisha discovers that targeting rules exclude customers from low-income ZIP codes. She recognizes discriminatory segmentation.	The leader removes the exclusion and audits all targeting rules for bias. They also implement fairness checks before campaign approval.
Damon sees that targeting criteria include inferred health conditions. He fears ethical and legal violations.	The leader deletes the criteria and bans health-related inference in marketing. They also retrain teams on sensitive-attribute restrictions.
Liora notices that ads for high-interest loans are disproportionately shown to young adults. She senses predatory targeting.	The leader suspends the campaign and reviews all financial-product targeting. They also add ethical-risk scoring for sensitive categories.
Mateo finds that lookalike audiences are built from biased historical data. He realizes the bias is being amplified.	The leader rebuilds the audience using balanced seed data. They also require bias audits for all lookalike models.
Serena sees that targeting rules rely on outdated demographic assumptions. She worries about stereotyping.	The leader updates the rules with behavior-based criteria. They also require periodic targeting-criteria reviews.
Omar discovers that customers who opted out of personalization are still being targeted. He senses a consent violation.	The leader enforces opt-out filtering and validates compliance. They also add personalization-opt-out checks to campaign workflows.

5. COMPLAINT RESOLUTION LOGS

Customer Ethics requires accountability, and Complaint Resolution Logs document issues raised by customers and how they were addressed. They track patterns, timelines, and outcomes to ensure concerns receive responsible follow through. This matters because unresolved complaints signal neglect and disrespect. Logs prevent dismissal by ensuring every concern is visible and actionable. They also reinforce fairness by ensuring similar issues receive consistent responses. When ignored, complaints accumulate and customer frustration intensifies. Trust erodes as customers feel unheard. Over time, Customer Ethics collapses into systemic dissatisfaction.

When executed well, Complaint Resolution Logs strengthen Customer Ethics by creating visibility and disciplined response. Customers trust that their concerns matter and lead to action. This improves loyalty, reduces churn, and enhances organizational learning. Logs reinforce the principle that accountability is essential to ethical service. They remind leaders that customer voice must be honored. Ultimately, they build a customer environment where issues are resolved with integrity and consistency.

MORAL COMPASS SCENARIO	LEADERSHIP ACTION
Nadia notices that complaints are marked "resolved" without customer confirmation. She realizes the logs reflect internal convenience, not customer reality.	The leader requires customer-verified closure and updates the resolution workflow. They also audit past cases for premature closure.
Trevor sees that complaints from non-English speakers take twice as long to resolve. He suspects language-access gaps.	The leader adds multilingual support and tracks resolution times by language. They also train agents on inclusive communication.
Aisha discovers that recurring complaints about billing errors are logged individually but never analyzed. She senses systemic failure.	The leader aggregates complaint themes and assigns a root-cause task force. They also require trend reviews in monthly operations meetings.
Marco notices that high-severity complaints are buried among low-severity ones. He fears delayed escalation.	The leader adds severity scoring and triage workflows. They also require daily review of high-severity cases.
Evelyn sees that complaint logs lack timestamps. She worries about SLA compliance.	The leader updates the logging system to enforce timestamps. They also track SLA adherence publicly.
Jonah finds that complaints from older adults are dismissed as "user error." He recognizes age bias.	The leader retrains agents on respectful handling and audits dismissals. They also add age-equity checks to complaint reviews.

Customer Ethics requires dignity in every interaction, and Customer Respect Audits evaluate whether policies, communications, and service behaviors uphold that standard. They examine tone, treatment, and the implicit messages embedded in customer experiences. This matters because disrespect—subtle or overt—erodes trust faster than operational failure. Respect Audits prevent cultural drift by identifying where customers feel minimized or dismissed. They also reinforce fairness by ensuring all customers receive dignified treatment. When ignored, disrespect becomes normalized and customers disengage. Complaints rise, loyalty declines, and reputational damage accelerates. Over time, Customer Ethics collapses as dignity is replaced by transactional indifference.

When executed well, Customer Respect Audits strengthen Customer Ethics by embedding dignity into the organization's operating system. Customers trust that they will be treated with humanity and consideration. This improves satisfaction, retention, and brand strength. Audits reinforce the principle that respect is a moral and strategic imperative. They remind leaders that dignity must be measured, not assumed. Ultimately, they build a customer environment where respect is the default posture.

MORAL COMPASS SCENARIO	LEADERSHIP ACTION
Rhea discovers that customer-service scripts interrupt callers after eight seconds. She realizes the script is disrespectful and rushed.	The leader rewrites the script to prioritize listening. They also train agents on empathy-based communication.
Omar notices that support agents refer to customers by account numbers instead of names. He senses dehumanization.	The leader updates protocols to require name-based interaction. They also reinforce respect standards in coaching sessions.
Lila finds that automated emails use condescending language when customers make mistakes. She recognizes tone bias.	The leader rewrites templates with respectful, neutral language. They also add tone reviews to content workflows.
Andre sees that customers with disabilities face longer support times due to inaccessible channels. He worries about inequity.	The leader adds accessible support options and audits channel performance. They also train agents on disability-inclusive service.
Serena notices that loyalty-program members receive preferential treatment in support queues. She senses fairness issues.	The leader equalizes queue prioritization and documents the change. They also add fairness metrics to service dashboards.
Hugo discovers that customer feedback is being mocked in internal chats. He recognizes a cultural breach.	The leader enforces conduct rules and disciplines offenders. They also reinforce a culture of respect through team training.

Ai6x6™ Transformation Blueprint

User Safety

"Trust is earned in drops and lost in floods — especially in AI."

Indra Nooyi, former CEO of PepsiCo known for ethical and sustainable leadership.

User Safety requires proactive protection, and Harm Prevention Measures define the safeguards that prevent customers from experiencing physical, digital, or emotional harm. They establish the standards that ensure products and services operate safely under expected conditions. This matters because unsafe experiences create immediate risk and long term distrust. Harm Prevention Measures prevent incidents by identifying and mitigating potential hazards. They also reinforce fairness by ensuring all users receive equal protection regardless of background or usage pattern. When ignored, safety failures become more frequent and severe. Teams experience uncertainty about risk thresholds. Over time, User Safety deteriorates as vulnerabilities accumulate. Misuse Protection Controls ensure systems prevent harmful or unintended use.

Identifying "Kill Zones" in the customer journey is a survival requirement. This topic uses Abuse Detection Logs and Interface Safety Reviews to protect the user experience from being exploited by internal fragmentation or external threats. By mapping "Misuse Prevention Checklists," a leader takes proactive responsibility for the customer's well-being. User safety is the foundation of "Brand Vision." If the customer feels unsafe, they will flee to the competition. By clinicalizing safety through the blueprint, the leader ensures that the organization's digital frontiers remain a "Safe Zone" for growth rather than a liability for litigation. Using the print framework, you can physically flip between "Safety Default Settings" and "User Protection Reports," engrossing yourself in the diagnostic that prevents brand-destroying abuse. User safety is the clinical moat that ensures your customer journey is a "Bliss Zone" of security, protecting your legacy from the "Kill Zones" of digital harm and platform misuse.

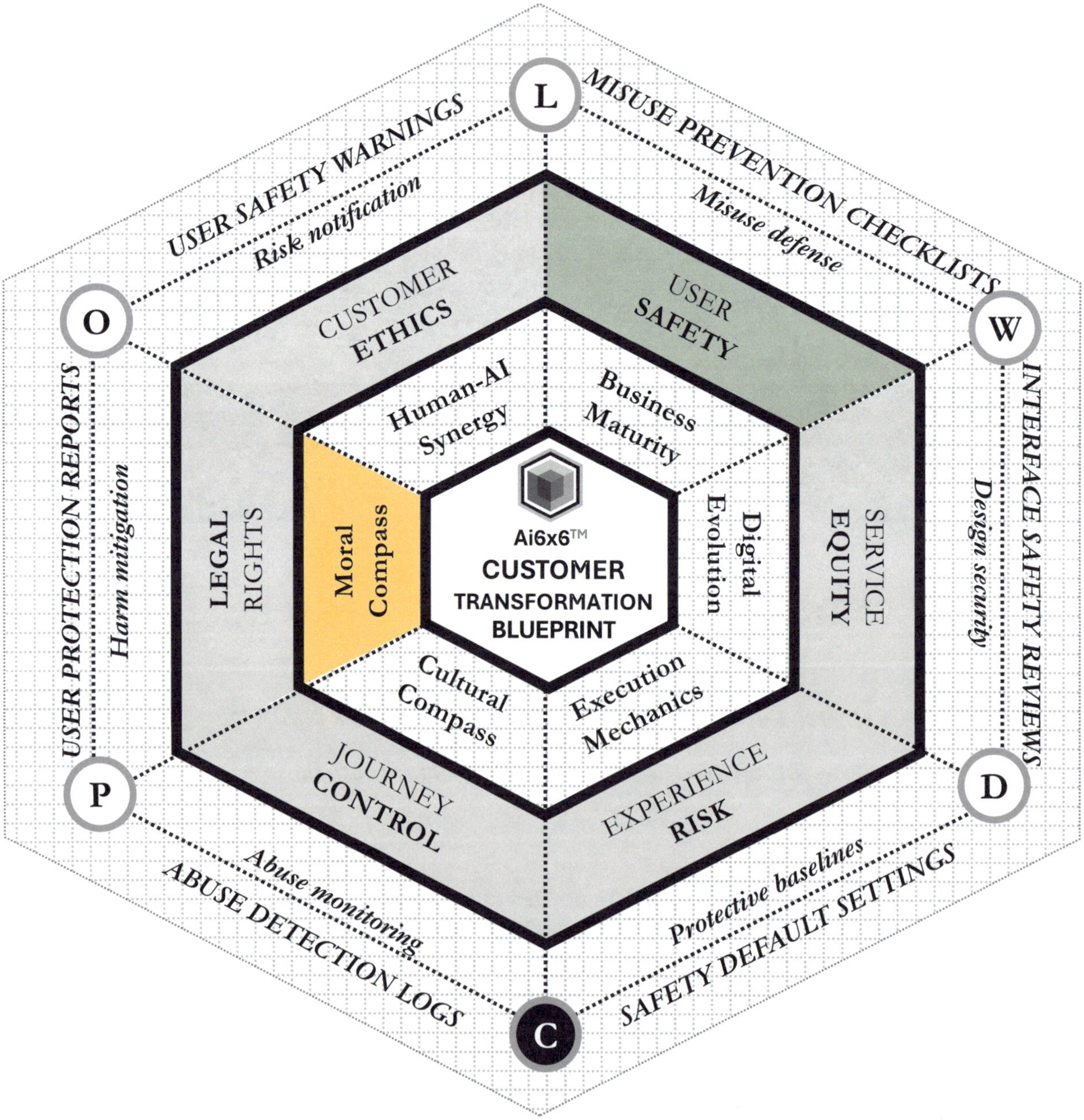

Ai6x6 USER SAFETY FOR MORAL COMPASS | CUSTOMER TRANSFORMATION

1. USER SAFETY WARNINGS

User Safety requires proactive communication, and User Safety Warnings alert customers to risks, limitations, or misuse conditions that could cause harm. They ensure that customers understand the boundaries of safe operation before engaging with a product or service. This matters because uninformed use is one of the most common sources of preventable incidents. User Safety Warnings prevent harm by making risk visible at the moment of decision. They also reinforce fairness by ensuring all users receive equal access to critical safety information. When ignored, customers operate without awareness of hazards and face unnecessary danger. Organizations appear negligent when warnings are incomplete or unclear. Over time, User Safety deteriorates as preventable incidents accumulate.

When executed well, User Safety Warnings strengthen User Safety by embedding protection into the customer experience. Customers trust that the organization prioritizes their well being. This improves satisfaction, reduces liability, and enhances brand credibility. Warnings reinforce the principle that safety requires transparency. They remind leaders that risk must be communicated, not hidden. Ultimately, they build an environment where customers can make informed, safe choices.

MORAL COMPASS SCENARIO	LEADERSHIP ACTION
Maren notices that safety warnings appear only after users trigger risky actions. She realizes the warnings come too late to prevent harm.	The leader moves warnings to pre-action screens and tests comprehension with real users. They also require proactive safety prompts for all high-risk flows.
Dorian sees that warnings use vague language like "be careful." He knows users won't understand the actual risk.	The leader rewrites warnings with explicit, actionable guidance. They also add clarity checks to the content-review workflow.
Selene discovers that warnings are missing for features involving financial transfers. She recognizes a major exposure.	The leader adds mandatory warnings for all money-movement actions. They also require legal review for financial-risk messaging.
Marcus notices that warnings are inconsistent across platforms. He fears users will miss critical cues.	The leader standardizes warning formats and synchronizes them across devices. They also enforce cross-platform consistency checks.
Fiona sees that warnings are displayed in small, low-contrast text. She worries accessibility is compromised.	The leader updates visual standards and tests warnings with accessibility tools. They also require WCAG-aligned safety messaging.
Idris finds that warnings are dismissed instantly because users are habituated. He senses warning fatigue.	The leader redesigns warnings with contextual triggers and varied formats. They also monitor dismissal rates to adjust frequency.

2. MISUSE PREVENTION CHECKLISTS

User Safety requires anticipating harmful behavior, and Misuse Prevention Checklists identify actions, configurations, or scenarios that could lead to unsafe outcomes. They guide teams in designing safeguards that prevent customers from inadvertently causing harm. This matters because misuse often arises from ambiguity, complexity, or poor interface design. Misuse Prevention Checklists prevent incidents by forcing teams to consider how real users behave under real conditions. They also reinforce fairness by ensuring protections apply to all users, not just experts. When ignored, products become vulnerable to predictable misuse. Customers face risks they cannot reasonably foresee. Over time, User Safety collapses as preventable failures multiply.

When executed well, Misuse Prevention Checklists strengthen User Safety by embedding foresight into product and service design. Customers trust that the organization has anticipated their needs and limitations. This improves reliability, reduces accidents, and enhances ethical responsibility. Checklists reinforce the principle that safety must be engineered, not assumed. They remind leaders that preventing harm requires structured thinking. Ultimately, they build an environment where misuse is minimized through intentional design.

MORAL COMPASS SCENARIO	LEADERSHIP ACTION
Naomi discovers that teams skip misuse checks during tight deadlines. She realizes harmful edge cases are slipping through.	The leader enforces mandatory checklist completion before release. They also add automated blockers for missing misuse reviews.
Dorian sees that the checklist doesn't cover new features involving user-generated content. He senses a blind spot.	The leader updates the checklist with UGC-specific misuse scenarios. They also require periodic checklist expansion.
Selene notices that teams treat the checklist as a formality. She worries they aren't thinking critically about misuse.	The leader adds scenario-based prompts to force deeper analysis. They also train teams on misuse-pattern recognition.
Marcus finds that the checklist ignores cross-platform misuse patterns. He recognizes fragmented oversight.	The leader expands the checklist to include platform-specific risks. They also require cross-platform misuse reviews.
Fiona sees that the checklist doesn't include social-engineering risks. She fears users could be manipulated.	The leader adds social-engineering scenarios and tests them with red-team simulations. They also require mitigation plans for high-risk flows.
Idris discovers that the checklist is stored in outdated documents. He worries teams are using old versions.	The leader centralizes the checklist in a version-controlled system. They also enforce update notifications for all teams.

User Safety requires clarity in interaction, and Interface Safety Reviews evaluate whether screens, controls, and workflows guide customers toward safe behavior. They examine whether the interface prevents confusion, misclicks, or unsafe sequences. This matters because poor interface design can create hidden hazards. Interface Safety Reviews prevent harm by ensuring that safety critical actions are intuitive and error resistant. They also reinforce fairness by ensuring all users—regardless of skill level—can operate safely. When ignored, interfaces become sources of accidental misuse. Customers experience frustration, risk, and unintended consequences. Over time, User Safety deteriorates as design flaws accumulate.

When executed well, Interface Safety Reviews strengthen User Safety by aligning design with human behavior. Customers trust that the interface protects them from mistakes. This improves usability, reduces incidents, and enhances brand reliability. Reviews reinforce the principle that safety must be built into every interaction. They remind leaders that design is a safety mechanism, not just an aesthetic choice. Ultimately, they build an environment where safe operation is the default experience.

MORAL COMPASS SCENARIO	LEADERSHIP ACTION
Maren notices that a button for irreversible actions looks identical to routine actions. She realizes users could make catastrophic mistakes.	The leader redesigns the interface with clear visual differentiation. They also require irreversible-action reviews in every UI audit.
Dorian sees that error messages blame users instead of guiding them. He senses psychological harm.	The leader rewrites messages with supportive, neutral language. They also add tone-of-voice checks to UI reviews.
Selene discovers that a navigation flow traps users in a loop when they try to exit. She recognizes a safety-escape failure.	The leader adds clear exit paths and tests them with real users. They also require escape-route validation for all flows.
Marcus finds that touch targets are too small on mobile, causing accidental taps. He worries about unintended actions.	The leader updates UI spacing standards and tests ergonomics across devices. They also enforce mobile-specific safety reviews.
Fiona sees that confirmation dialogs appear inconsistently. She fears users won't know when actions are final.	The leader standardizes confirmation patterns and documents them in design guidelines. They also require consistency checks in every review.
Idris notices that safety-critical information is buried below the fold. He senses visibility risk.	The leader elevates critical content to primary positions. They also require visibility scoring for all safety elements.

User Safety requires protective baselines, and Safety Default Settings ensure that products and services begin in the safest possible configuration. They minimize risk by preventing customers from starting in unsafe modes. This matters because default settings shape user behavior more than any other design choice. Safety Default Settings prevent harm by making safety the path of least resistance. They also reinforce fairness by protecting all users equally, including those who never adjust settings. When ignored, customers unknowingly operate in risky configurations. Incidents rise, and trust declines. Over time, User Safety collapses as unsafe defaults become normalized.

When executed well, Safety Default Settings strengthen User Safety by embedding protection into the starting experience. Customers trust that the organization has prioritized their well being from the first moment of use. This improves satisfaction, reduces liability, and enhances ethical alignment. Defaults reinforce the principle that safety must be proactive. They remind leaders that the safest choice should require no effort. Ultimately, they build an environment where customers begin every interaction protected.

MORAL COMPASS SCENARIO	LEADERSHIP ACTION
Naomi discovers that data-sharing defaults are set to "on." She realizes users are unknowingly exposing personal information.	The leader switches defaults to privacy-protective settings. They also require justification for any non-protective default.
Dorian sees that parental controls are disabled by default. He fears children are unprotected.	The leader enables protective defaults for child accounts. They also add onboarding prompts for parental setup.
Selene notices that location tracking is enabled automatically. She recognizes a safety and privacy risk.	The leader changes the default to opt-in and updates onboarding flows. They also require location-use explanations before activation.
Marcus finds that default notification settings overwhelm users with alerts. He senses cognitive overload.	The leader reduces default notifications and categorizes them by importance. They also test user comprehension before rollout.
Fiona sees that default visibility settings expose user profiles publicly. She worries about stalking risks.	The leader sets profiles to private by default. They also add safety education during onboarding.
Idris discovers that default password requirements are weak. He recognizes a security vulnerability.	The leader strengthens default password policies and enforces MFA. They also add password-strength indicators.

5. ABUSE DETECTION LOGS

User Safety requires vigilance, and Abuse Detection Logs track patterns of harmful, malicious, or unsafe behavior within a system. They identify misuse that threatens customers or the integrity of the service. This matters because abuse often escalates when left unmonitored. Abuse Detection Logs prevent harm by revealing early signs of dangerous activity. They also reinforce fairness by ensuring all users are protected from malicious actors. When ignored, abuse becomes invisible and unchecked. Customers experience harm that could have been prevented. Over time, User Safety deteriorates as threats multiply.

When executed well, Abuse Detection Logs strengthen User Safety by enabling rapid intervention. Customers trust that the organization actively monitors for harmful behavior. This improves security, reduces incidents, and enhances ethical responsibility. Logs reinforce the principle that safety requires continuous observation. They remind leaders that threats evolve and must be tracked. Ultimately, they build an environment where harmful behavior is detected and addressed quickly.

MORAL COMPASS SCENARIO	LEADERSHIP ACTION
Maren sees a spike in harassment reports but no corresponding entries in abuse logs. She realizes detection systems are failing.	The leader audits detection pipelines and fixes broken triggers. They also add redundancy checks to prevent silent failures.
Dorian notices that abuse logs only track explicit keywords. He fears subtle harassment is going undetected.	The leader expands detection models to include contextual signals. They also require periodic model retraining.
Selene discovers that abuse logs are not linked to user-blocking tools. She senses slow response times.	The leader integrates logs with automated blocking workflows. They also assign moderators to review high-severity cases.
Marcus finds that abuse logs are stored without timestamps. He worries investigations will be impossible.	The leader enforces timestamping and updates logging standards. They also validate historical entries.
Fiona sees that abuse logs are reviewed only monthly. She fears harmful behavior persists too long.	The leader mandates daily reviews and assigns a rotating moderation team. They also track response times as a KPI.
Idris notices that abuse logs don't distinguish between false positives and confirmed cases. He senses data pollution.	The leader adds classification labels and triage workflows. They also require accuracy audits for detection models.

6. USER PROTECTION REPORTS

User Safety requires accountability, and User Protection Reports document how the organization safeguards customers from harm. They summarize incidents, interventions, and improvements across safety critical areas. This matters because customers expect transparency about how their well being is protected. User Protection Reports prevent complacency by making safety performance visible. They also reinforce fairness by ensuring all users benefit from consistent protection standards. When ignored, safety becomes opaque and unmeasured. Customers question whether the organization takes their well being seriously. Over time, User Safety collapses as oversight weakens.

When executed well, User Protection Reports strengthen User Safety by creating transparency and continuous improvement. Customers trust that the organization monitors, measures, and enhances safety over time. This improves credibility, reduces risk, and reinforces ethical alignment. Reports reinforce the principle that safety must be documented, not assumed. They remind leaders that protection requires evidence. Ultimately, they build an environment where user well being is continuously evaluated and strengthened.

MORAL COMPASS SCENARIO	LEADERSHIP ACTION
Naomi discovers that protection reports show rising incidents of impersonation. She realizes identity safeguards are insufficient.	The leader adds stronger verification steps and updates anti-impersonation rules. They also monitor impersonation trends weekly.
Dorian sees that reports highlight repeated harm from the same user accounts. He suspects enforcement gaps.	The leader enforces stricter penalties and adds automated repeat-offender detection. They also audit moderator actions.
Selene notices that vulnerable users are disproportionately affected by scams. She recognizes a protection inequity.	The leader adds targeted safety education and strengthens scam-detection models. They also track vulnerability-specific risk metrics.
Marcus finds that protection reports lack geographic breakdowns. He senses regional blind spots.	The leader adds region-level reporting and investigates high-risk areas. They also tailor protections by region.
Fiona sees that reports show slow response times for safety escalations. She worries users are left exposed.	The leader redesigns escalation workflows and adds time-to-action KPIs. They also assign dedicated safety responders.
Idris discovers that protection reports are not shared with product teams. He realizes safety insights aren't shaping design.	The leader integrates reports into product planning. They also require safety-impact reviews for all new features.

Ai6x6™ Transformation Blueprint

Service Equity

"Ethics is the infrastructure of every technology that hopes to endure."
Bill Gates, Microsoft co-founder and global philanthropist.

Service Equity requires balanced treatment, and Fair Service Access ensures all customers can obtain services without discrimination or unnecessary barriers. It establishes the conditions that prevent exclusion based on identity, geography, or circumstance. This matters because unequal access creates structural disadvantage. Fair Service Access prevents inequity by ensuring services are available consistently and transparently. It also reinforces trust by demonstrating that fairness is embedded in service delivery. When ignored, disparities widen, and customer frustration increases. Teams experience misalignment as service decisions become inconsistent. Over time, Service Equity deteriorates as access becomes uneven. Inclusive Delivery Standards ensure services accommodate diverse customer needs.

"Bliss" is only achieved when operational dominance is inclusive across all segments. Accessibility Audits and Non-Discrimination Screens allow a leader to see where the brand is failing its promise to a diverse market. This topic involves using "Service Fairness Reports" to ensure that AI-driven personalization doesn't become a tool for exclusion. For the leader, "Service Equity" is a growth strategy. By ensuring "Inclusive Design," you open doors to new markets and prevent the "Kill Zones" that silos create. Equity ensures that every customer touchpoint is an extension of the organization's core moral compass. Within the print series, you can map "Accessibility Scorecards" against "Customer Representation Maps," seeing the whole map of service fairness. This clinical focus allows you to move capital to underserved areas, ensuring that your "Customer Transformation" is as broad as it is deep, making your brand unassailable by competitors who lack your ethical and operational inclusivity.

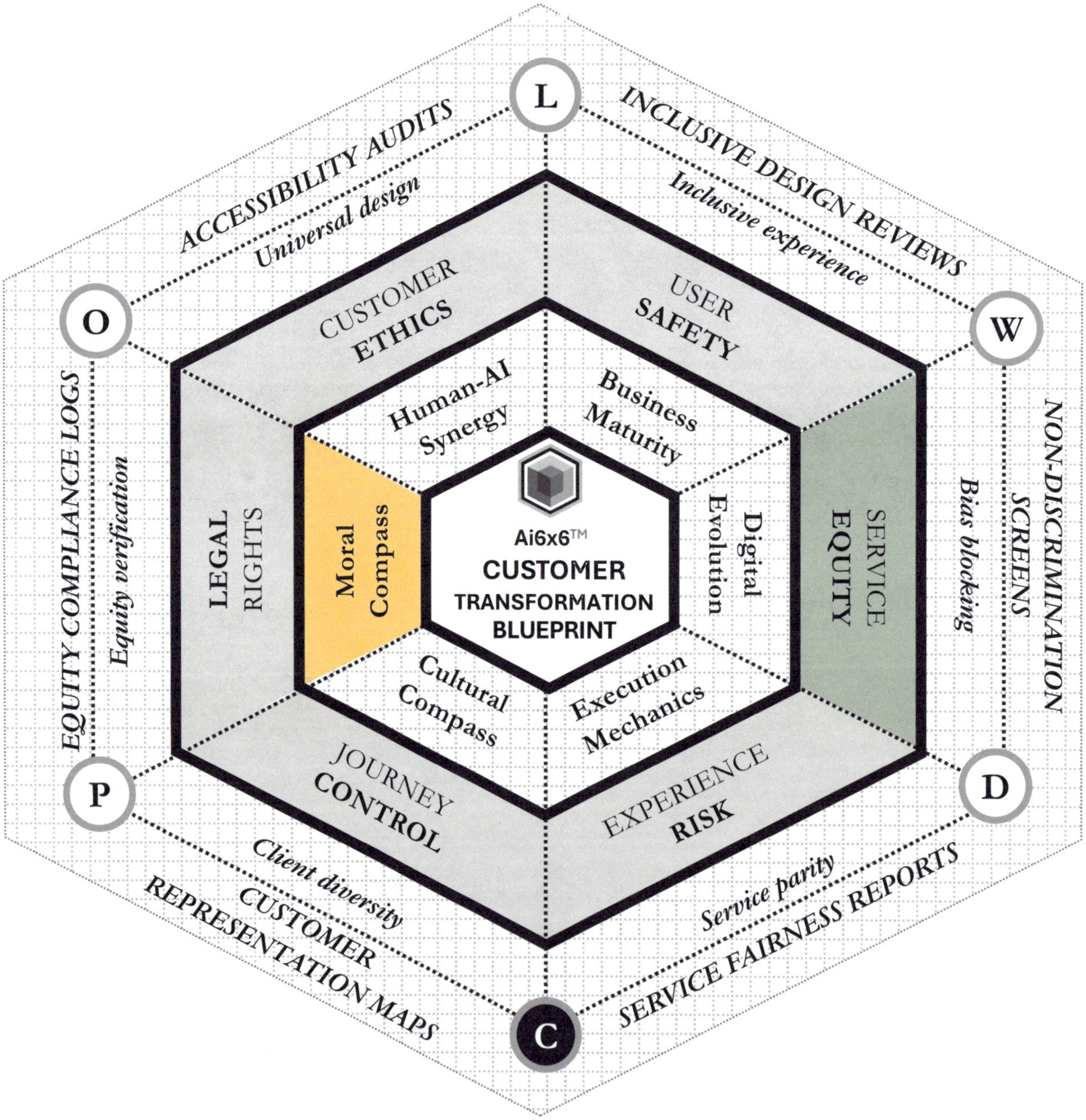

Ai6x6 SERVICE EQUITY FOR MORAL COMPASS | CUSTOMER TRANSFORMATION

1. ACCESSIBILITY AUDITS

Service Equity requires universal access, and Accessibility Audits evaluate whether products, services, and experiences can be used effectively by people with diverse abilities. They examine barriers in design, communication, and interaction that may exclude certain customer groups. This matters because inequitable access creates structural disadvantage. Accessibility Audits prevent exclusion by identifying where systems fail to accommodate real world needs. They also reinforce fairness by ensuring all customers receive equal opportunity to participate. When ignored, accessibility gaps widen and customers feel marginalized. Complaints rise as usability issues accumulate. Over time, Service Equity deteriorates as exclusion becomes normalized.

When executed well, Accessibility Audits strengthen Service Equity by embedding inclusion into every operational layer. Customers trust that the organization recognizes and respects diverse needs. This improves satisfaction, reduces friction, and enhances brand credibility. Audits reinforce the principle that access is a moral obligation, not an optional enhancement. They remind leaders that equity begins with design. Ultimately, they build an environment where every customer can engage fully and confidently.

MORAL COMPASS SCENARIO	LEADERSHIP ACTION
Mara discovers that screen-reader users cannot navigate key checkout steps. She realizes the flow violates basic accessibility standards.	The leader rebuilds the checkout path with WCAG-aligned markup. They also require quarterly accessibility audits for all critical flows.
Devin notices that color contrast fails on multiple pages. He fears visually impaired users are being excluded.	The leader updates the design system with compliant contrast ratios. They also enforce automated contrast checks in CI pipelines.
Selah finds that video content lacks captions. She recognizes a barrier for deaf and hard-of-hearing customers.	The leader mandates captioning for all videos and retrofits existing content. They also add accessibility checks to content publishing.
Mateo sees that keyboard navigation breaks on mobile web. He senses mobility-related exclusion.	The leader fixes navigation traps and tests with assistive-tech users. They also add device-specific accessibility validation.
Farah notices that error messages are not announced by screen readers. She worries users cannot recover from mistakes.	The leader updates ARIA attributes and tests with real users. They also require semantic-structure reviews for all forms.
Idris discovers that accessibility bugs remain open for months. He realizes they're deprioritized.	The leader elevates accessibility issues to P1 severity. They also track accessibility SLA compliance.

Service Equity requires intentional inclusion, and Inclusive Design Reviews evaluate whether products and services reflect the needs of diverse customer groups. They examine assumptions, biases, and design choices that may privilege one group over another. This matters because exclusion often originates in early design decisions. Inclusive Design Reviews prevent inequity by ensuring diverse perspectives shape the experience. They also reinforce fairness by ensuring no customer group is overlooked. When ignored, products become optimized for a narrow audience. Customers outside that group experience frustration and disengagement. Over time, Service Equity collapses as exclusion becomes embedded in the design process.

When executed well, Inclusive Design Reviews strengthen Service Equity by ensuring diversity informs every design decision. Customers trust that the organization considers their unique contexts and needs. This improves usability, expands reach, and enhances ethical alignment. Reviews reinforce the principle that inclusion must be engineered, not assumed. They remind leaders that equitable design requires deliberate attention. Ultimately, they build an environment where every customer feels recognized and supported.

MORAL COMPASS SCENARIO	LEADERSHIP ACTION
Nia sees that illustrations depict only one demographic group. She realizes the product imagery excludes many users.	The leader updates the asset library with diverse representations. They also require inclusive-imagery checks in design reviews.
Corin notices that onboarding assumes high digital literacy. He fears older adults will struggle.	The leader redesigns onboarding with simpler steps and optional guidance. They also test flows with low-literacy users.
Elara discovers that forms require gender selection with only two options. She recognizes exclusion of non-binary users.	The leader expands gender options and updates data schemas. They also add inclusive-language guidelines to design standards.
Mateo finds that time-zone settings default to U.S. regions only. He senses global exclusion.	The leader expands global settings and updates localization workflows. They also require inclusive-region testing.
Soren notices that error messages assume cultural norms not shared globally. He worries about cultural bias.	The leader rewrites messages with culturally neutral language. They also add cultural-sensitivity checks to reviews.
Hana sees that iconography relies on metaphors unfamiliar to international users. She recognizes usability risk.	The leader replaces ambiguous icons with universal symbols. They also test icon comprehension across regions.

3. NON-DISCRIMINATION SCREENS

Service Equity requires fairness in every interaction, and Non Discrimination Screens ensure that policies, processes, and service decisions do not disadvantage specific groups. They evaluate whether criteria, workflows, or rules create unequal outcomes. This matters because discrimination often emerges unintentionally through structural bias. Non Discrimination Screens prevent inequity by identifying where systems treat customers differently without justification. They also reinforce fairness by ensuring all customers receive equal consideration. When ignored, discriminatory patterns persist and harm affected groups. Customers lose trust in the organization's integrity. Over time, Service Equity deteriorates as bias becomes institutionalized.

When executed well, Non Discrimination Screens strengthen Service Equity by ensuring fairness is embedded in every decision. Customers trust that they will be treated consistently and respectfully. This improves loyalty, reduces complaints, and enhances ethical credibility. Screens reinforce the principle that equity requires vigilance. They remind leaders that fairness must be verified, not assumed. Ultimately, they build an environment where discrimination is actively prevented.

MORAL COMPASS SCENARIO	LEADERSHIP ACTION
Priya discovers that a support triage tool routes customers from certain ZIP codes to slower queues. She recognizes geographic discrimination.	The leader removes the biased rule and audits all triage logic. They also require fairness checks before rule deployment.
Rowan notices that eligibility rules exclude customers with non-traditional credit histories. He fears systemic bias.	The leader redesigns eligibility criteria to include alternative data. They also require non-discrimination validation for all rule sets.
Serena finds that customer-service scripts differ by perceived accent. She senses linguistic discrimination.	The leader standardizes scripts and trains agents on accent-neutral communication. They also monitor interactions for bias.
Omar sees that refund approvals vary by demographic group. He recognizes hidden bias in manual decisions.	The leader implements standardized refund criteria and audits past decisions. They also add fairness monitoring to refund workflows.
Lila discovers that product recommendations exclude older adults. She worries about age-based exclusion.	The leader updates recommendation logic and adds age-equity checks. They also test recommendations across age groups.
Andre notices that customers with disabilities face longer verification steps. He senses inequitable burden.	The leader redesigns verification flows to remove unnecessary friction. They also add disability-equity metrics to service dashboards.

Service Equity requires measurable accountability, and Service Fairness Reports evaluate whether customers receive consistent treatment across channels, teams, and contexts. They analyze disparities in wait times, resolutions, outcomes, and service quality. This matters because inequity often appears in operational patterns rather than individual actions. Service Fairness Reports prevent hidden bias by making performance differences visible. They also reinforce fairness by ensuring all customers benefit from the same standards. When ignored, inequities persist unnoticed and undermine trust. Customers perceive favoritism or inconsistency. Over time, Service Equity collapses as fairness becomes unpredictable.

When executed well, Service Fairness Reports strengthen Service Equity by creating transparency and corrective action. Customers trust that the organization monitors and improves fairness continuously. This improves satisfaction, reduces risk, and enhances brand integrity. Reports reinforce the principle that equity must be measured to be managed. They remind leaders that fairness requires data, not intuition. Ultimately, they build an environment where equitable service is consistently delivered.

MORAL COMPASS SCENARIO	LEADERSHIP ACTION
Nadia sees that wait times differ drastically by language preference. She realizes multilingual users are underserved.	The leader adds multilingual staffing and tracks fairness metrics. They also require language-equity reviews monthly.
Trevor notices that premium customers receive faster issue resolution even for safety-critical problems. He fears fairness violations.	The leader equalizes safety-related SLAs across all tiers. They also update service-tier policies.
Aisha discovers that rural customers experience more outages. She recognizes geographic inequity.	The leader investigates infrastructure gaps and prioritizes rural fixes. They also add region-level fairness indicators.
Marco finds that fairness reports show higher complaint rates among first-time users. He senses onboarding friction.	The leader improves onboarding flows and tracks fairness improvements. They also assign owners for first-time-user experience.
Evelyn notices that fairness metrics are missing for accessibility. She worries accessibility inequities are invisible.	The leader adds accessibility fairness KPIs. They also require accessibility-specific fairness reviews.
Jonah sees that fairness reports are not shared with frontline teams. He realizes insights aren't driving behavior.	The leader distributes reports and trains teams on fairness interpretation. They also integrate fairness into performance metrics.

Service Equity requires visibility into who is being served, and Customer Representation Maps illustrate how different demographic groups engage with products and services. They reveal which groups are well represented and which are underserved. This matters because inequity often emerges when certain populations are overlooked. Customer Representation Maps prevent blind spots by showing where outreach, design, or service delivery fails to reach diverse customers. They also reinforce fairness by ensuring all groups are considered in strategic decisions. When ignored, representation gaps widen and certain customers feel excluded. Opportunities for improvement remain hidden. Over time, Service Equity deteriorates as service becomes skewed toward dominant groups.

When executed well, Customer Representation Maps strengthen Service Equity by making inclusion visible and actionable. Customers trust that the organization recognizes and values their presence. This improves engagement, expands reach, and enhances ethical alignment. Maps reinforce the principle that equity requires awareness. They remind leaders that representation is a prerequisite for fairness. Ultimately, they build an environment where all customer groups are seen and supported.

MORAL COMPASS SCENARIO	LEADERSHIP ACTION
Rhea discovers that the customer base is diverse, but feedback panels are not. She realizes insights are skewed.	The leader rebuilds panels to reflect actual demographics. They also track representation coverage quarterly.
Omar notices that product testers come mostly from urban areas. He fears rural needs are ignored.	The leader expands tester recruitment to rural regions. They also add geographic-representation metrics.
Lila finds that accessibility-focused users are missing from beta programs. She recognizes a major gap.	The leader recruits accessibility testers and updates beta-program criteria. They also require accessibility representation in all tests.
Andre sees that customer personas rely on outdated stereotypes. He senses bias in product decisions.	The leader rebuilds personas using real data and inclusive research. They also require persona audits annually.
Serena notices that representation maps exclude older adults. She worries product decisions skew young.	The leader adds age-distribution mapping and updates research sampling. They also test features with older users.
Hugo discovers that representation maps don't include socioeconomic diversity. He recognizes blind spots.	The leader adds income-range mapping and updates segmentation. They also require socioeconomic diversity in research samples.

6. EQUITY COMPLIANCE LOGS

Service Equity requires disciplined oversight, and Equity Compliance Logs document whether teams follow established fairness standards. They track adherence to policies, guidelines, and equity focused procedures. This matters because equity fails when compliance becomes inconsistent. Equity Compliance Logs prevent drift by ensuring fairness practices are applied uniformly. They also reinforce accountability by making deviations visible. When ignored, inequitable behaviors become normalized. Customers experience inconsistent treatment and lose trust. Over time, Service Equity collapses as standards erode.

When executed well, Equity Compliance Logs strengthen Service Equity by ensuring fairness is enforced, not optional. Customers trust that the organization maintains consistent ethical standards. This improves reliability, reduces complaints, and enhances brand credibility. Logs reinforce the principle that equity requires structure and discipline. They remind leaders that fairness must be monitored continuously. Ultimately, they build an environment where equitable service is sustained through rigorous compliance.

MORAL COMPASS SCENARIO	LEADERSHIP ACTION
Keisha sees that equity compliance logs show repeated violations in one department. She realizes leadership is ignoring fairness standards.	The leader escalates the issue and mandates corrective training. They also track compliance improvements monthly.
Damon notices that logs are missing root-cause analysis for violations. He senses superficial remediation.	The leader adds mandatory RCA fields and assigns owners. They also require follow-up verification.
Liora discovers that logs are updated inconsistently across regions. She recognizes fragmented governance.	The leader standardizes logging procedures and trains regional leads. They also audit regional compliance quarterly.
Mateo finds that logs show recurring accessibility failures. He worries systemic issues aren't being addressed.	The leader launches a cross-functional accessibility task force. They also track remediation progress in the log.
Serena sees that logs don't include demographic impact analysis. She senses inequity is being hidden.	The leader adds demographic-impact fields and updates reporting rules. They also require impact reviews for all violations.
Omar notices that logs are stored in unsecured folders. He recognizes confidentiality risk.	The leader secures storage and restricts access. They also enforce compliance with data-handling standards.

Ai6x6™ Transformation Blueprint

Experience Risk

"AI must serve justice, not efficiency alone."

Ruth Bader Ginsburg, former U.S. Supreme Court Justice known for equality jurisprudence.

Experience Risk requires awareness of customer vulnerabilities, and Trust Erosion Indicators identify early signs that confidence in the organization is weakening. They establish visibility into emotional and behavioral signals that precede customer loss. This matters because trust declines long before customers disengage. Trust Erosion Indicators prevent deterioration by revealing where expectations are not being met. They also reinforce fairness by ensuring all customers receive consistent, reliable experiences. When ignored, dissatisfaction grows silently. Teams experience confusion as customer sentiment becomes unpredictable. Over time, Experience Risk increases as breakdowns accumulate. Service Breakdown Signals highlight where processes or interactions are failing.

Mapping friction points is the only way to prevent customer churn. By studying Trust Erosion Indicators and Service Failure Reports, leaders can see the "Silos" that competitors hunt before they reach the customer. This topic involves using "Customer Impact Assessments" to understand the psychological fallout of operational gaps. Experience risk is the "early warning system" of the blueprint. By engrossing yourself in the data of "Experience Variability," you can identify the "Kill Zones" in the customer journey. It allows the leader to "Shift" the organization toward a more stable and high-value customer relationship. The print format is essential for this diagnostic; it lets you see the whole "Experience Map" at once, identifying where trust is being lost before the customer realizes it themselves. Experience risk management is the clinical protocol for maintaining "Market Defensibility," ensuring that your "Safe Zone" doesn't become a "Danger Zone" of customer stagnation and eventual brand obsolescence.

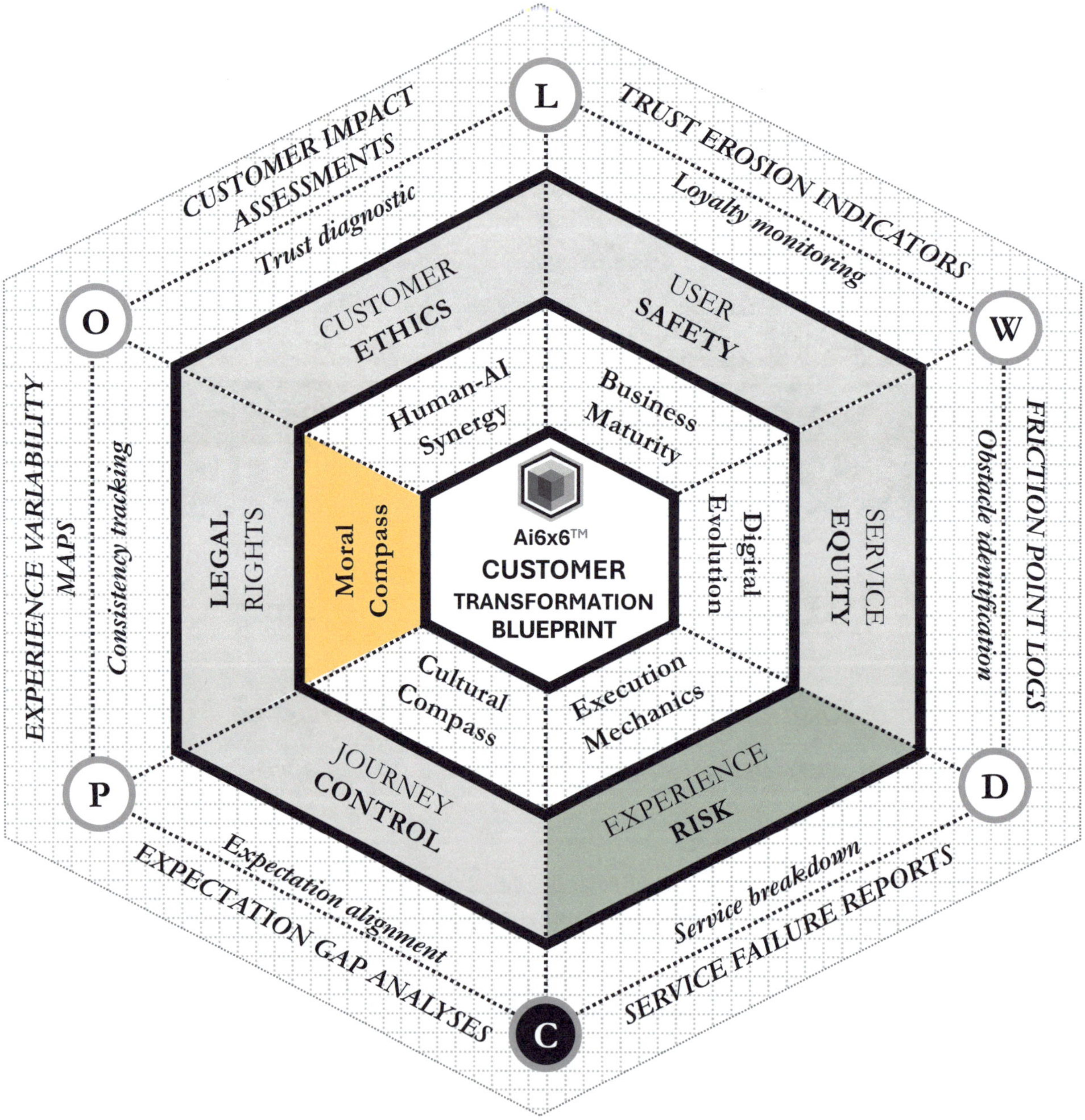

Ai6x6 EXPERIENCE RISK FOR MORAL COMPASS | CUSTOMER TRANSFORMATION

Experience Risk requires foresight, and Customer Impact Assessments evaluate how decisions, changes, or new features affect customers across emotional, functional, and operational dimensions. They analyze potential friction, confusion, or unintended harm before implementation. This matters because customer experience failures often originate from untested assumptions. Customer Impact Assessments prevent missteps by revealing how real customers will interpret and navigate change. They also reinforce fairness by ensuring all customer groups are considered, not just the most vocal or profitable. When ignored, organizations introduce changes that create frustration, distrust, or service disruption. Customers feel blindsided and unsupported. Over time, Experience Risk escalates as decisions prioritize efficiency over impact.

When executed well, Customer Impact Assessments strengthen Experience Risk management by aligning decisions with customer well being. Customers trust that the organization anticipates their needs and protects them from unintended consequences. This improves satisfaction, reduces churn, and enhances ethical alignment. Assessments reinforce the principle that experience quality must be engineered, not assumed. They remind leaders that customer perception is a strategic variable. Ultimately, they build an environment where every change is evaluated through the lens of customer impact.

MORAL COMPASS SCENARIO	LEADERSHIP ACTION
Mara discovers that a new feature increases cognitive load for older adults. She realizes the team never assessed age-specific impact.	The leader runs a full impact assessment and redesigns the feature for accessibility. They also require demographic-specific impact reviews for all launches.
Devin notices that a pricing change disproportionately affects low-income customers. He senses unintended harm.	The leader pauses the rollout and models alternative pricing structures. They also add equity-impact scoring to pricing decisions.
Selah finds that a UI update breaks assistive-tech compatibility. She recognizes a high-severity impact.	The leader reverts the update and fixes compatibility issues. They also require assistive-tech testing in all impact assessments.
Mateo sees that a new workflow increases steps for first-time users. He worries about onboarding friction.	The leader simplifies the workflow and validates improvements with new users. They also add first-time-user impact checks.
Farah notices that a policy change creates confusion for non-native English speakers. She senses linguistic inequity.	The leader updates communication materials with multilingual support. They also require language-impact reviews.
Idris discovers that a feature unintentionally exposes sensitive user data in shared environments. He recognizes privacy risk.	The leader redesigns the feature with privacy-by-default controls. They also require privacy-impact assessments for all sensitive flows.

Experience Risk requires early detection of declining confidence, and Trust Erosion Indicators track signals that customers are losing faith in the organization. They monitor sentiment shifts, behavioral changes, and patterns of disengagement. This matters because trust rarely collapses suddenly; it erodes gradually through small failures. Trust Erosion Indicators prevent surprise by revealing early warning signs. They also reinforce fairness by ensuring all customers—not just high value segments—are monitored for declining trust. When ignored, dissatisfaction grows silently until it becomes irreversible. Customers disengage, complain, or switch providers. Over time, Experience Risk becomes a structural threat to loyalty and reputation.

When executed well, Trust Erosion Indicators strengthen Experience Risk management by enabling timely intervention. Customers trust that the organization listens and responds to emerging concerns. This improves retention, reduces reputational damage, and enhances long term stability. Indicators reinforce the principle that trust must be measured continuously. They remind leaders that confidence is fragile and must be protected. Ultimately, they build an environment where trust is monitored, maintained, and restored proactively.

MORAL COMPASS SCENARIO	LEADERSHIP ACTION
Nia sees a spike in users disabling personalization features. She realizes trust in data handling is declining.	The leader investigates root causes and updates transparency messaging. They also track trust indicators weekly.
Corin notices that customers increasingly avoid linking external accounts. He senses fear of over-collection.	The leader clarifies data-use boundaries and simplifies permission prompts. They also add trust-impact scoring to integrations.
Elara discovers that users frequently abandon flows after seeing legal disclaimers. She recognizes a trust-trigger issue.	The leader rewrites disclaimers in plain language and tests comprehension. They also monitor abandonment rates post-fix.
Mateo finds that customers are contacting support to verify basic product claims. He senses credibility erosion.	The leader audits all claims and aligns messaging with reality. They also add trust-sensitivity checks to marketing workflows.
Soren notices that users increasingly opt out of surveys. He worries they no longer believe feedback matters.	The leader redesigns feedback loops with visible follow-through. They also publish "you said, we did" updates.
Hana sees that trust indicators drop after every major update. She recognizes change-management failures.	The leader improves update communication and adds pre-launch trust testing. They also track trust deltas after each release.

3. FRICTION POINT LOGS

Experience Risk requires visibility into customer struggle, and Friction Point Logs document where customers encounter delays, confusion, or unnecessary effort. They capture the moments where experience quality breaks down. This matters because friction is one of the strongest predictors of dissatisfaction. Friction Point Logs prevent escalation by identifying where processes, interfaces, or policies create avoidable obstacles. They also reinforce fairness by ensuring all customers—not just those who complain—benefit from smoother experiences. When ignored, friction accumulates and becomes normalized. Customers feel exhausted by interactions that should be simple. Over time, Experience Risk grows as frustration becomes embedded in the journey.

When executed well, Friction Point Logs strengthen Experience Risk management by enabling targeted improvements. Customers trust that the organization is committed to removing barriers. This improves satisfaction, increases efficiency, and enhances brand loyalty. Logs reinforce the principle that ease is a moral and strategic imperative. They remind leaders that friction is a design failure, not a customer problem. Ultimately, they build an environment where experiences are consistently intuitive and effortless.

MORAL COMPASS SCENARIO	LEADERSHIP ACTION
Priya discovers repeated friction logs around password resets. She realizes the flow is overly complex.	The leader redesigns the reset process with fewer steps. They also add friction-testing to authentication workflows.
Rowan notices that users struggle with payment verification. He senses unnecessary barriers.	The leader simplifies verification and tests alternatives. They also track friction severity scores.
Serena finds that mobile users frequently abandon forms due to layout issues. She recognizes device-specific friction.	The leader rebuilds mobile forms and tests them across screen sizes. They also add mobile-specific friction logging.
Omar sees that friction spikes during peak hours. He suspects performance bottlenecks.	The leader optimizes backend performance and adds load-specific friction monitoring. They also test peak-hour resilience.
Lila discovers that friction logs show confusion around privacy settings. She senses poor UX communication.	The leader redesigns settings with clearer explanations. They also add comprehension testing.
Andre notices that friction logs are not categorized by severity. He worries critical issues are buried.	The leader adds severity scoring and triage workflows. They also require weekly friction-review meetings.

4. SERVICE FAILURE REPORTS

Experience Risk requires accountability for breakdowns, and Service Failure Reports document incidents where the organization did not meet customer expectations. They analyze root causes, patterns, and the severity of impact. This matters because failures, when unexamined, repeat themselves. Service Failure Reports prevent recurrence by forcing teams to confront the underlying issues. They also reinforce fairness by ensuring all customers receive consistent remediation. When ignored, failures become systemic and predictable. Customers lose confidence in the organization's reliability. Over time, Experience Risk escalates as failures accumulate without correction.

When executed well, Service Failure Reports strengthen Experience Risk management by creating transparency and continuous improvement. Customers trust that the organization learns from mistakes. This improves resilience, reduces repeat incidents, and enhances ethical alignment. Reports reinforce the principle that accountability is essential to experience quality. They remind leaders that failure is only acceptable when it leads to learning. Ultimately, they build an environment where service reliability is continuously strengthened.

MORAL COMPASS SCENARIO	LEADERSHIP ACTION
Nadia sees repeated failures in refund processing. She realizes the workflow is brittle.	The leader rebuilds the refund pipeline and assigns a reliability owner. They also track failure recurrence.
Trevor notices that failures spike during new-feature rollouts. He suspects insufficient integration testing.	The leader adds pre-launch service-continuity tests. They also require rollback plans for all releases.
Aisha discovers that failures disproportionately affect customers in low-bandwidth regions. She recognizes infrastructure inequity.	The leader optimizes performance for low-bandwidth environments. They also add region-specific failure monitoring.
Marco finds that failures are logged without root-cause analysis. He senses shallow remediation.	The leader mandates RCA documentation for every failure. They also assign owners for corrective actions.
Evelyn sees that failures occur mostly during off-hours. She suspects gaps in monitoring.	The leader adds 24/7 monitoring and on-call escalation. They also track detection latency.
Jonah notices that failure reports are not shared with customer-facing teams. He realizes insights aren't reaching the front line.	The leader distributes reports and trains teams on failure patterns. They also integrate failure insights into support scripts.

Experience Risk requires alignment between promises and reality, and Expectation Gap Analyses evaluate where customer expectations differ from actual experience. They examine mismatches created by marketing, onboarding, or service delivery. This matters because unmet expectations are a primary driver of dissatisfaction. Expectation Gap Analyses prevent disappointment by revealing where the organization overpromises or underdelivers. They also reinforce fairness by ensuring all customers receive experiences consistent with what they were led to believe. When ignored, expectation gaps widen and trust erodes. Customers feel misled even when the product itself is functional. Over time, Experience Risk becomes a reputational liability.

When executed well, Expectation Gap Analyses strengthen Experience Risk management by aligning commitments with capabilities. Customers trust that the organization sets honest expectations. This improves satisfaction, reduces complaints, and enhances brand integrity. Analyses reinforce the principle that clarity is essential to experience quality. They remind leaders that expectations shape perception more than features. Ultimately, they build an environment where promises and delivery remain tightly synchronized.

MORAL COMPASS SCENARIO	LEADERSHIP ACTION
Rhea discovers customers expect real-time updates, but the product only refreshes hourly. She realizes the gap is driving dissatisfaction.	The leader updates communication to set accurate expectations and explores real-time capabilities. They also track expectation-alignment metrics.
Omar notices that customers assume data is private when it is actually shared with partners. He senses a transparency gap.	The leader updates disclosures and adds explicit consent prompts. They also run expectation-alignment testing.
Lila finds that customers expect refunds within days, but processing takes weeks. She recognizes a service-speed gap.	The leader accelerates refund workflows and updates timelines. They also track refund-time expectations.
Andre sees that customers expect human support but receive automated responses. He senses a channel-expectation mismatch.	The leader clarifies support channels and adds human-escalation options. They also test expectation clarity.
Serena notices that customers expect feature parity across platforms. She realizes mobile lags behind.	The leader aligns platform roadmaps and communicates differences clearly. They also track parity-gap closure.
Hugo discovers that customers expect proactive alerts for outages. He recognizes a communication gap.	The leader adds automated outage notifications. They also track alert-timeliness metrics.

6. EXPERIENCE VARIABILITY MAPS

Experience Risk requires consistency, and Experience Variability Maps illustrate how customer experiences differ across channels, locations, teams, or time periods. They reveal where quality fluctuates and why. This matters because inconsistency undermines trust even when average performance is strong. Experience Variability Maps prevent uneven service by making disparities visible. They also reinforce fairness by ensuring all customers—not just those in favorable segments—receive reliable experiences. When ignored, variability becomes entrenched and unpredictable. Customers feel uncertain about what to expect. Over time, Experience Risk escalates as inconsistency becomes the norm.

When executed well, Experience Variability Maps strengthen Experience Risk management by driving standardization and reliability. Customers trust that they will receive the same quality every time. This improves loyalty, reduces operational risk, and enhances brand stability. Maps reinforce the principle that consistency is a core component of experience excellence. They remind leaders that variability is a signal of systemic weakness. Ultimately, they build an environment where experience quality is uniform, predictable, and dependable.

MORAL COMPASS SCENARIO	LEADERSHIP ACTION
Keisha sees that experience quality varies drastically between desktop and mobile. She realizes platform inconsistency is harming trust.	The leader standardizes design patterns and tests cross-platform parity. They also track variability deltas.
Damon notices that new users experience more friction than returning users. He senses onboarding gaps.	The leader redesigns onboarding flows and monitors new-user variability. They also assign onboarding owners.
Liora discovers that experience quality differs by region due to infrastructure gaps. She recognizes geographic inequity.	The leader improves regional infrastructure and adds region-level variability metrics. They also prioritize high-variance regions.
Mateo finds that variability spikes during peak hours. He suspects load-related degradation.	The leader optimizes backend performance and adds peak-hour monitoring. They also test load-resilience regularly.
Serena sees that variability maps show inconsistent support quality across agents. She senses training gaps.	The leader standardizes support scripts and adds agent-level quality metrics. They also train agents on consistency.
Omar notices that variability increases after major updates. He recognizes change-management issues.	The leader improves rollout testing and adds post-update variability checks. They also track variability trends after each release.

BLUEPRINT SERIES

Ai6x6™ Transformation Blueprint

Journey Control

"Sustainability is the long memory of a species that often forgets."
Jane Jacobs, urban theorist known for community-centered planning.

Journey Control requires structured oversight, and Interaction Oversight Mechanisms define how customer touchpoints are monitored, guided, and evaluated. They establish the controls that prevent inconsistent or harmful interactions. This matters because unmanaged customer journeys create confusion and frustration. Interaction Oversight Mechanisms prevent drift by ensuring every step aligns with service standards. They also reinforce fairness by ensuring all customers receive consistent treatment across channels. When ignored, interactions become unpredictable and difficult to manage. Teams experience misalignment as customer journeys diverge. Over time, Journey Control deteriorates as oversight weakens. Escalation Path Clarity ensures customers know how to seek help when issues arise.

The customer journey is a moral narrative that must be orchestrated with precision. Interaction Governance Maps and Touchpoint Oversight Logs ensure the brand's integrity is maintained across every digital and physical touchpoint. By mapping the "Escalation Path" through a physical artifact, the leader eliminates the confusion that causes customer frustration. It's about maintaining "Journey Integrity." Control in this context isn't about restriction; it is about providing the "Personalization Rules" that ensure every interaction is a meaningful, ethically sound step toward the "Bliss Zone," keeping the customer relationship evolving through the scaling loop. In the print blueprint, you can flip between "Journey Mapping Artifacts" and "Interaction Governance," seeing the whole map of customer intent. This clinical focus ensures that the "How" and the "Why" of the customer experience are linked, preventing the "Kill Zone" of fragmented interactions and ensuring that your customer operations move as fast and as cleanly as your strategic thoughts.

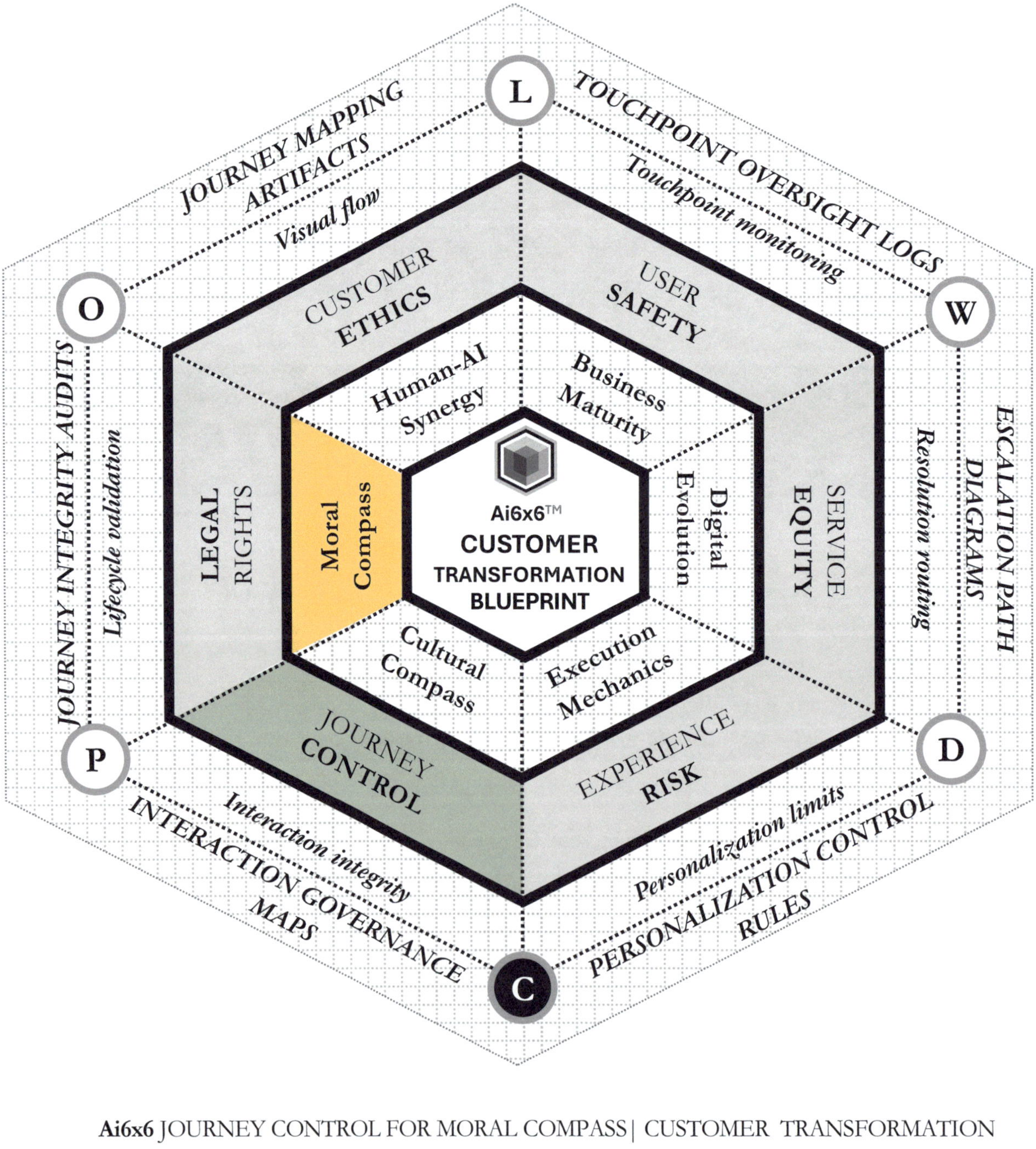

Ai6x6 JOURNEY CONTROL FOR MORAL COMPASS | CUSTOMER TRANSFORMATION

1. JOURNEY MAPPING ARTIFACTS

Journey Control requires visibility into the full customer experience, and Journey Mapping Artifacts document every step, transition, and emotional inflection point across the customer lifecycle. They reveal how customers move through channels, processes, and decision moments. This matters because organizations often misunderstand the real journey customers take. Journey Mapping Artifacts prevent blind spots by exposing friction, confusion, and unmet expectations. They also reinforce fairness by ensuring all customer segments—not just the most vocal—are represented. When ignored, journeys become fragmented and inconsistent. Customers experience unpredictable transitions and unclear pathways. Over time, Journey Control deteriorates as the organization loses sight of how customers actually navigate its ecosystem.

When executed well, Journey Mapping Artifacts strengthen Journey Control by providing a unified view of customer movement. Customers trust that the organization understands their needs and designs experiences accordingly. This improves satisfaction, reduces friction, and enhances operational alignment. Artifacts reinforce the principle that journeys must be intentional, not accidental. They remind leaders that experience quality depends on understanding the full path, not isolated touchpoints. Ultimately, they build an environment where customer journeys are coherent, predictable, and well orchestrated.

MORAL COMPASS SCENARIO	LEADERSHIP ACTION
Mara discovers that the journey map only reflects ideal paths and ignores real user detours. She realizes the artifact hides friction points.	The leader rebuilds the map using real behavioral data. They also require quarterly journey-map refresh cycles.
Devin notices that the journey map excludes accessibility-specific flows. He fears the team is designing for a narrow audience.	The leader adds accessibility journeys and tests them with assistive-tech users. They also require inclusive journey mapping for all features.
Selah finds that the map ends at purchase and ignores post-purchase support. She recognizes a lifecycle gap.	The leader extends the journey to include support, returns, and loyalty. They also assign owners for each lifecycle stage.
Mateo sees that the map is created by designers alone. He senses missing insights from support and operations.	The leader mandates cross-functional mapping workshops. They also require multi-team validation before finalizing artifacts.
Farah notices that emotional states are missing from the journey map. She worries the team is ignoring user sentiment.	The leader adds emotional-state layers and validates them with user research. They also require sentiment mapping for all critical flows.
Idris discovers that the journey map is outdated after recent product changes. He realizes teams are designing from stale information.	The leader updates the map and enforces version control. They also require updates after every major release.

Journey Control requires disciplined monitoring, and Touchpoint Oversight Logs track the quality, consistency, and performance of every customer interaction point. They document how each touchpoint behaves under real conditions. This matters because even a single weak touchpoint can undermine the entire journey. Touchpoint Oversight Logs prevent degradation by making performance issues visible early. They also reinforce fairness by ensuring all customers receive consistent treatment across channels. When ignored, touchpoints drift out of alignment and create unpredictable experiences. Customers encounter variability that erodes trust. Over time, Journey Control collapses as the journey becomes disjointed.

When executed well, Touchpoint Oversight Logs strengthen Journey Control by ensuring every interaction meets defined standards. Customers trust that the organization manages each touchpoint with care. This improves reliability, reduces complaints, and enhances brand cohesion. Logs reinforce the principle that journeys are only as strong as their weakest link. They remind leaders that oversight must be continuous, not episodic. Ultimately, they build an environment where touchpoints operate as a coordinated system.

MORAL COMPASS SCENARIO	LEADERSHIP ACTION
Nia sees that several touchpoints have no assigned owners. She realizes accountability gaps are causing inconsistent experiences.	The leader assigns owners for each touchpoint and documents responsibilities. They also track oversight performance monthly.
Corin notices that logs show repeated failures at the same touchpoint. He senses systemic design flaws.	The leader launches a root-cause investigation and redesigns the touchpoint. They also add monitoring alerts for future failures.
Elara discovers that touchpoints involving third-party vendors are missing from oversight logs. She recognizes a blind spot.	The leader integrates vendor touchpoints into the log. They also require vendor-performance reviews.
Mateo finds that logs only track failures, not successes. He worries teams lack visibility into what works well.	The leader adds positive-signal tracking and shares insights across teams. They also use success patterns to refine weak touchpoints.
Soren notices that logs are updated inconsistently across regions. He senses fragmented governance.	The leader standardizes logging procedures and trains regional leads. They also audit regional compliance quarterly.
Hana sees that logs don’t include user-impact severity. She realizes prioritization is arbitrary.	The leader adds severity scoring and triage workflows. They also require weekly review of high-severity touchpoints.

3. ESCALATION PATH DIAGRAMS

Journey Control requires clarity in problem resolution, and Escalation Path Diagrams define how customer issues move through support tiers, decision authorities, and resolution channels. They ensure that customers know what happens when problems arise. This matters because unclear escalation creates frustration and delays. Escalation Path Diagrams prevent confusion by mapping the exact flow of responsibility. They also reinforce fairness by ensuring all customers receive consistent escalation treatment. When ignored, issues bounce between teams without ownership. Customers feel abandoned and unsupported. Over time, Journey Control deteriorates as escalation becomes chaotic.

When executed well, Escalation Path Diagrams strengthen Journey Control by creating predictable, accountable resolution pathways. Customers trust that their concerns will be handled efficiently and transparently. This improves satisfaction, reduces churn, and enhances operational discipline. Diagrams reinforce the principle that escalation must be structured, not improvised. They remind leaders that clarity is essential to customer confidence. Ultimately, they build an environment where problems are resolved with precision and accountability.

MORAL COMPASS SCENARIO	LEADERSHIP ACTION
Priya discovers that customers escalate issues directly to executives because the diagram is unclear. She realizes the path lacks visibility.	The leader redesigns the diagram with clear steps and publishes it across channels. They also train support teams on proper routing.
Rowan notices that escalation paths differ between mobile and desktop. He fears inconsistent user experiences.	The leader standardizes escalation flows across platforms. They also test them with real users.
Serena finds that escalation paths don't include accessibility accommodations. She recognizes exclusion.	The leader adds accessible escalation options and tests them with disabled users. They also require accessibility checks for all escalation flows.
Omar sees that after-hours escalations are missing from the diagram. He senses coverage gaps.	The leader adds 24/7 escalation rules and publishes on-call schedules. They also require after-hours testing.
Lila discovers that escalation paths rely heavily on email. She worries about slow response times.	The leader adds real-time escalation channels and updates workflows. They also track escalation latency.
Andre notices that escalation diagrams are not updated after org changes. He realizes teams are following outdated paths.	The leader refreshes diagrams and enforces version control. They also require updates after every reorg.

4. PERSONALIZATION CONTROL RULES

Journey Control requires boundaries around customization, and Personalization Control Rules define how tailored experiences are delivered without compromising fairness, privacy, or consistency. They ensure that personalization enhances the journey rather than distorting it. This matters because uncontrolled personalization can create unpredictable or inequitable experiences. Personalization Control Rules prevent overreach by setting limits on what can be customized and how. They also reinforce fairness by ensuring all customers benefit from personalization without bias. When ignored, personalization becomes intrusive, inconsistent, or discriminatory. Customers feel manipulated or miscategorized. Over time, Journey Control collapses as tailored experiences become chaotic.

When executed well, Personalization Control Rules strengthen Journey Control by balancing customization with stability. Customers trust that personalization serves their interests rather than exploiting them. This improves engagement, reduces risk, and enhances ethical alignment. Rules reinforce the principle that personalization must be governed, not left to chance. They remind leaders that tailored experiences require discipline. Ultimately, they build an environment where personalization enhances the journey without compromising control.

MORAL COMPASS SCENARIO	LEADERSHIP ACTION
Nadia sees that personalization settings are buried deep in menus. She realizes users cannot easily control their experience.	The leader surfaces controls prominently and simplifies toggles. They also require usability testing for all personalization settings.
Trevor notices that personalization rules override user opt-outs. He senses a consent violation.	The leader fixes the override logic and validates compliance. They also add opt-out enforcement checks.
Aisha discovers that personalization relies on inferred sensitive attributes. She recognizes ethical risk.	The leader bans sensitive-attribute inference and updates data-use rules. They also retrain teams on ethical personalization.
Marco finds that personalization behaves inconsistently across devices. He worries users will lose trust.	The leader synchronizes personalization logic across platforms. They also add cross-device consistency tests.
Evelyn sees that personalization rules don't include safety constraints. She fears harmful recommendations.	The leader adds safety filters and tests them with edge cases. They also require safety-aware personalization reviews.
Jonah notices that personalization is applied even to child accounts. He recognizes a COPPA risk.	The leader disables personalization for minors and updates age-screening rules. They also audit all child-facing experiences.

5. INTERACTION GOVERNANCE MAPS

Journey Control requires structured oversight of customer interactions, and Interaction Governance Maps define the rules, responsibilities, and boundaries that shape how interactions occur. They clarify who controls each part of the journey and how decisions are made. This matters because unclear governance leads to inconsistent experiences. Interaction Governance Maps prevent fragmentation by aligning teams around shared standards. They also reinforce fairness by ensuring all customers receive interactions governed by the same principles. When ignored, interactions become unpredictable and vary by team or channel. Customers experience inconsistency that erodes trust. Over time, Journey Control deteriorates as governance weakens.

When executed well, Interaction Governance Maps strengthen Journey Control by creating a unified framework for interaction quality. Customers trust that every interaction reflects the same values and standards. This improves reliability, reduces confusion, and enhances brand cohesion. Maps reinforce the principle that governance is essential to experience integrity. They remind leaders that interactions must be managed intentionally. Ultimately, they build an environment where every interaction is consistent, principled, and well controlled.

MORAL COMPASS SCENARIO	LEADERSHIP ACTION
Rhea discovers that teams are designing overlapping interaction patterns. She realizes governance is fragmented.	The leader consolidates patterns into a unified governance map. They also enforce pattern-reuse rules.
Omar notices that governance maps don't include chatbot interactions. He senses a gap in conversational oversight.	The leader adds conversational flows and tests them for safety. They also require chatbot-specific governance reviews.
Lila finds that governance maps ignore cultural differences in interaction norms. She recognizes global usability risk.	The leader adds cultural-variation layers and tests flows across regions. They also require cultural-sensitivity reviews.
Andre sees that governance maps are not updated after major UI changes. He realizes teams are working from outdated rules.	The leader refreshes maps and enforces update triggers. They also require governance reviews during design sprints.
Serena notices that governance maps don't include escalation triggers for risky interactions. She senses safety gaps.	The leader adds risk-based escalation rules. They also test high-risk flows with safety teams.
Hugo discovers that governance maps are stored in siloed documents. He worries teams lack visibility.	The leader centralizes maps in a shared repository. They also assign governance owners.

Journey Control requires continuous verification, and Journey Integrity Audits evaluate whether the actual customer journey matches the intended design. They examine alignment across channels, processes, and touchpoints. This matters because journeys degrade over time as systems evolve. Journey Integrity Audits prevent drift by identifying where execution diverges from design. They also reinforce fairness by ensuring all customers experience the journey as promised. When ignored, inconsistencies multiply and customers encounter unpredictable pathways. Trust erodes as the journey becomes unreliable. Over time, Journey Control collapses as the designed experience loses coherence.

When executed well, Journey Integrity Audits strengthen Journey Control by restoring alignment between intention and reality. Customers trust that the organization maintains the journey with discipline. This improves satisfaction, reduces operational risk, and enhances brand stability. Audits reinforce the principle that journeys must be maintained, not just designed. They remind leaders that integrity requires ongoing attention. Ultimately, they build an environment where the customer journey remains consistent, reliable, and true to its design.

MORAL COMPASS SCENARIO	LEADERSHIP ACTION
Keisha sees that users are being redirected to third-party pages without clear disclosure. She realizes journey integrity is broken.	The leader adds disclosure prompts and audits all external redirects. They also require integrity checks for partner integrations.
Damon notices that dark-pattern elements appear in a new flow. He senses manipulative design.	The leader removes the patterns and reinforces ethical-design standards. They also add dark-pattern detection to audits.
Liora discovers that users are forced into upsells before completing essential tasks. She recognizes coercive friction.	The leader removes forced upsells and redesigns the flow. They also require ethical-friction reviews.
Mateo finds that journey integrity breaks when users switch devices mid-flow. He worries about session continuity.	The leader adds cross-device continuity and tests handoff flows. They also track integrity failures.
Serena sees that error states redirect users to irrelevant pages. She senses disorientation risk.	The leader fixes routing logic and tests error-state recovery. They also require error-path audits.
Omar notices that integrity audits are performed only annually. He realizes issues linger too long.	The leader increases audit frequency and adds automated integrity checks. They also assign owners for remediation.

Ai6x6™ Transformation Blueprint

Legal Rights

"Morality in AI is simply morality — amplified."

Desmond Tutu, Nobel Peace Prize laureate and human rights leader.

Legal Rights require adherence to consumer protections, and Consumer Protection Requirements define the legal standards organizations must uphold in every customer interaction. They establish the minimum obligations for lawful and ethical treatment. This matters because violations expose customers to harm and organizations to legal risk. Consumer Protection Requirements prevent misconduct by ensuring rights are clearly defined and respected. They also reinforce fairness by ensuring all customers receive equal legal protection. When ignored, violations become more frequent and severe. Teams experience uncertainty as decisions drift outside legal boundaries. Over time, Legal Rights deteriorate as compliance weakens. Mandated Fair Treatment ensures customer interactions remain aligned with regulatory expectations.

Protecting the customer's future through rigorous adherence to rights. Ensuring FTC, CCPA, and Truth in Lending compliance keeps the relationship evolving within a secure, legally sound framework. This topic treats "Legal Rights" as the "Armor" of the customer relationship. By reviewing "Disclosure Notices" and "Accuracy Reports," the leader ensures the organization is never a "Museum" waiting for its last visitor. Legal rights are the clinical benchmarks of trust. In the AI era, respecting consumer sovereignty is the only way to avoid the regulatory sledgehammers that destroy brand value and market defensibility. Using the print series, you can sit with these legal protections away from distractions, seeing how they form the "Moral Compass" of your customer strategy. This clinical mastery ensures that your "Customer Transformation" is not just profitable, but legally unassailable and morally sound, providing the ultimate insurance policy against the market extinction that follows the failure to protect the rights of your sovereign customers.

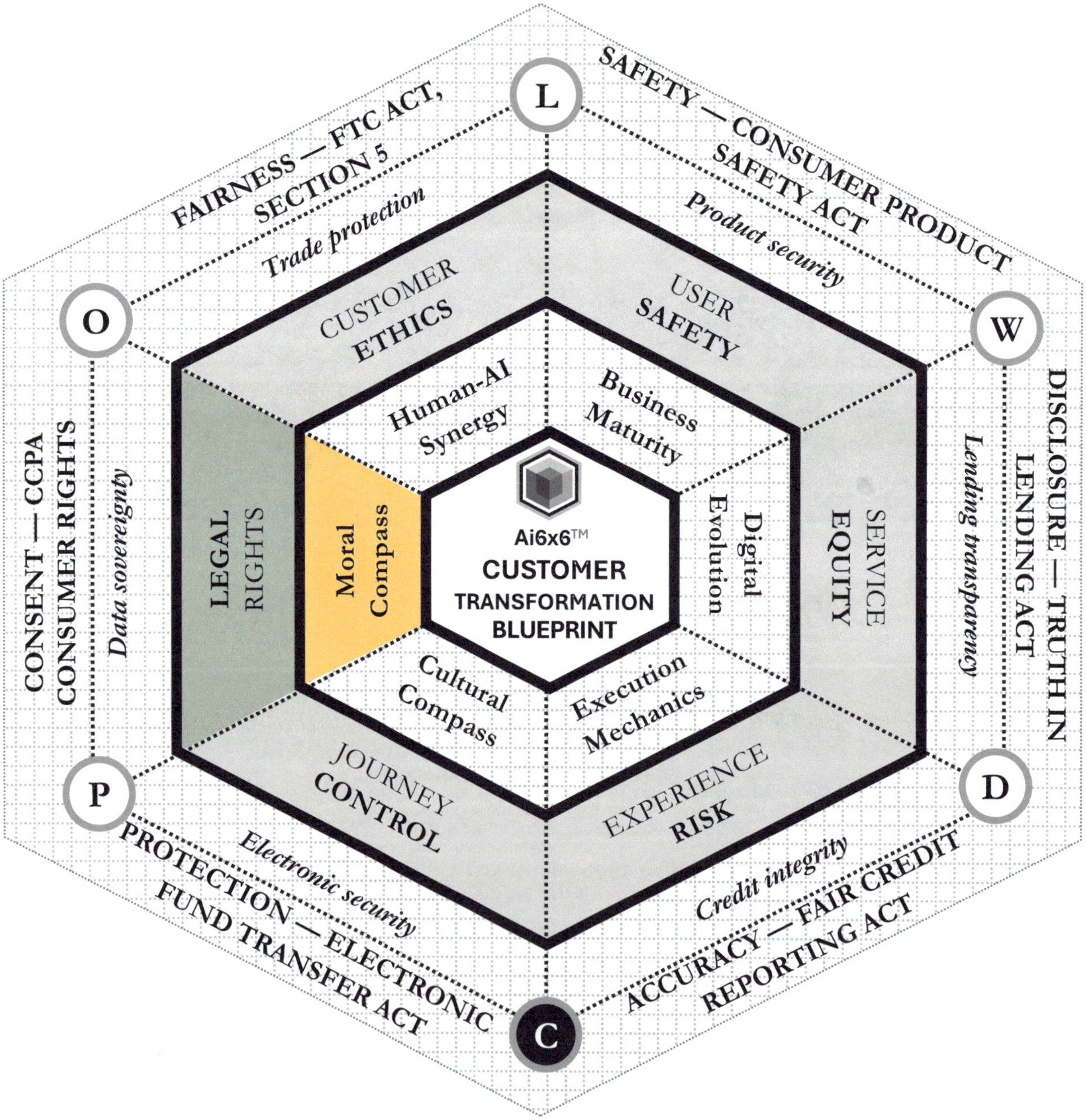

Ai6x6 LEGAL RIGHTS FOR MORAL COMPASS | CUSTOMER TRANSFORMATION

1. FAIRNESS — FTC ACT, SECTION 5

Legal Rights require protection from unfair or deceptive practices, and the FTC Act, Section 5 establishes the foundational standard for fairness in customer interactions. It prohibits misleading claims, manipulative tactics, and practices that distort customer decision making. This matters because customers rely on truthful, transparent information to evaluate products and services. The FTC Act prevents exploitation by defining clear boundaries for ethical conduct. It also reinforces fairness by ensuring all customers receive accurate representations, not selectively crafted narratives. When ignored, organizations drift into deceptive behavior that erodes trust. Customers feel misled and lose confidence in the brand. Over time, Legal Rights deteriorate as unfair practices become normalized.

When executed well, adherence to the FTC Act strengthens Legal Rights by embedding fairness into every communication and transaction. Customers trust that the organization respects their right to truthful information. This improves loyalty, reduces regulatory exposure, and enhances ethical credibility. Compliance reinforces the principle that fairness is a legal and moral obligation. It reminds leaders that customer trust is built on honesty. Ultimately, it creates an environment where fairness governs every customer interaction.

MORAL COMPASS SCENARIO	LEADERSHIP ACTION
Mara discovers that a promotional offer hides mandatory fees until the final checkout step. She realizes this could be considered a deceptive practice under Section 5.	The leader updates the flow to disclose all fees upfront. They also require deceptive-pattern screening for all promotions.
Devin notices that a comparison chart exaggerates product performance relative to competitors. He senses misleading representation.	The leader corrects the chart and adds substantiation requirements. They also implement a truth-in-advertising review.
Selah finds that a subscription auto-renews without clear notice. She recognizes an unfair billing practice.	The leader adds explicit renewal disclosures and confirmation prompts. They also audit all recurring-billing flows.
Mateo sees that refund policies are written ambiguously. He worries customers are being misled.	The leader rewrites policies in plain language and tests comprehension. They also require legal review for all customer-facing policies.
Farah notices that a "free trial" requires a credit card with unclear cancellation terms. She senses a dark-pattern risk.	The leader clarifies cancellation rules and removes manipulative elements. They also add FTC-aligned fairness checks.
Idris discovers that customer complaints about misleading claims are dismissed internally. He recognizes a compliance culture issue.	The leader escalates the issue and retrains teams on FTC fairness standards. They also track complaint-driven

Legal Rights require protection from physical harm, and the Consumer Product Safety Act establishes standards to ensure products are safe for customer use. It mandates testing, reporting, and corrective action when hazards are identified. This matters because customers expect products to function without exposing them to danger. The Act prevents harm by requiring organizations to identify and mitigate risks proactively. It also reinforces fairness by ensuring all customers receive products that meet the same safety standards. When ignored, unsafe products reach the market and cause preventable injuries. Customers lose trust in the organization's commitment to safety. Over time, Legal Rights collapse as safety failures accumulate.

When executed well, compliance with the Consumer Product Safety Act strengthens Legal Rights by embedding safety into product development and oversight. Customers trust that the organization prioritizes their well being. This improves satisfaction, reduces liability, and enhances brand integrity. Compliance reinforces the principle that safety is a non negotiable requirement. It reminds leaders that risk must be eliminated before products reach customers. Ultimately, it builds an environment where safety is consistently protected through rigorous standards.

MORAL COMPASS SCENARIO	LEADERSHIP ACTION
Nia discovers that a product defect causing overheating has been reported multiple times but not escalated. She realizes this may trigger CPSA reporting obligations.	The leader initiates a formal hazard investigation and notifies compliance teams. They also implement mandatory escalation rules for safety complaints.
Corin notices that packaging lacks age-appropriate safety warnings. He fears regulatory non-compliance.	The leader updates packaging with required warnings. They also add safety-label audits to production workflows.
Elara finds that a physical accessory breaks easily under normal use. She recognizes a potential safety hazard.	The leader halts shipments and conducts stress testing. They also implement corrective design changes.
Mateo sees that customer-reported injuries are logged inconsistently. He senses a reporting-chain failure.	The leader standardizes injury-report logging and assigns a safety-incident coordinator. They also audit historical reports.
Soren notices that safety instructions are buried in a long manual. He worries users won't see critical warnings.	The leader elevates safety instructions to the first page and adds visual cues. They also require readability testing.
Hana discovers that a product variant sold internationally lacks required safety certifications. She recognizes a compliance breach.	The leader suspends distribution and obtains proper certifications. They also enforce certification checks for all variants.

3. DISCLOSURE — TRUTH IN LENDING ACT

Legal Rights require transparency in financial transactions, and the Truth in Lending Act ensures customers receive clear, accurate information about credit terms. It mandates disclosure of interest rates, fees, and repayment obligations. This matters because customers must understand the true cost of borrowing. The Act prevents deception by standardizing how financial information is presented. It also reinforces fairness by ensuring all customers receive the same level of clarity. When ignored, customers face hidden charges, confusing terms, and unexpected financial burdens. Trust erodes as lending practices appear predatory. Over time, Legal Rights deteriorate as transparency disappears.

When executed well, compliance with the Truth in Lending Act strengthens Legal Rights by ensuring customers can make informed financial decisions. Customers trust that the organization communicates credit terms honestly. This improves financial literacy, reduces disputes, and enhances ethical alignment. Compliance reinforces the principle that transparency is essential to responsible lending. It reminds leaders that clarity protects both customers and the organization. Ultimately, it builds an environment where financial disclosures are complete, accurate, and accessible.

MORAL COMPASS SCENARIO	LEADERSHIP ACTION
Priya sees that APR disclosures are displayed in small, low-contrast text. She realizes customers cannot reasonably understand loan costs.	The leader redesigns disclosures with prominent, readable formatting. They also enforce TILA-aligned design standards.
Rowan notices that promotional rates are advertised without clear expiration dates. He senses misleading disclosure.	The leader adds explicit timeframes and updates marketing rules. They also require legal review for all lending promotions.
Serena finds that repayment schedules are hidden behind multiple clicks. She recognizes a transparency failure.	The leader surfaces schedules upfront and tests comprehension. They also add disclosure-visibility checks.
Omar sees that late-fee policies are vague and inconsistently communicated. He worries about regulatory exposure.	The leader rewrites policies with clear, standardized language. They also require policy-consistency audits.
Lila discovers that loan calculators exclude mandatory fees. She senses deceptive omission.	The leader updates calculators to include all required costs. They also add accuracy validation to financial tools.
Andre notices that disclosures differ between mobile and desktop. He recognizes cross-platform inconsistency.	The leader standardizes disclosures across platforms. They also test for parity before release.

4. ACCURACY — FAIR CREDIT REPORTING ACT

Legal Rights require accuracy in personal financial data, and the Fair Credit Reporting Act ensures that customer credit information is correct, protected, and used responsibly. It establishes standards for data accuracy, dispute resolution, and permissible use. This matters because inaccurate credit data can harm customers' financial opportunities. The Act prevents injustice by giving customers the right to review and correct their information. It also reinforces fairness by ensuring all customers are evaluated based on accurate records. When ignored, errors go uncorrected and customers face unfair denials or penalties. Trust erodes as credit systems appear biased or unreliable. Over time, Legal Rights collapse as accuracy becomes optional.

When executed well, compliance with the Fair Credit Reporting Act strengthens Legal Rights by ensuring financial decisions are based on truthful data. Customers trust that their credit information is handled with precision. This improves confidence, reduces harm, and enhances ethical responsibility. Compliance reinforces the principle that accuracy is foundational to fairness. It reminds leaders that data integrity directly affects customer livelihoods. Ultimately, it builds an environment where credit information is consistently verified and protected.

MORAL COMPASS SCENARIO	LEADERSHIP ACTION
Nadia discovers that customer credit data is being updated manually, leading to errors. She realizes this violates FCRA accuracy requirements.	The leader automates data ingestion and adds validation checks. They also audit historical entries for inaccuracies.
Trevor notices that dispute investigations exceed the required timelines. He senses compliance risk.	The leader restructures the dispute-resolution workflow and assigns SLA owners. They also track investigation timeliness.
Aisha finds that negative information is being reported without proper verification. She recognizes a high-risk violation.	The leader enforces verification protocols and retrains staff. They also require documentation for all negative entries.
Marco sees that credit-related notifications are not being sent consistently. He worries customers are uninformed.	The leader standardizes notification workflows and adds delivery tracking. They also audit notification completeness.
Evelyn discovers that credit data from third-party sources is not validated. She senses data-quality risk.	The leader adds third-party data-verification steps. They also require vendor-accuracy attestations.
Jonah notices that credit-reporting logs lack audit trails. He recognizes traceability gaps.	The leader adds immutable audit logs and restricts editing rights. They also enforce audit-trail reviews.

Legal Rights require security in financial transactions, and the Electronic Fund Transfer Act protects customers from unauthorized withdrawals, errors, and electronic fraud. It establishes rules for liability, dispute resolution, and transaction transparency. This matters because customers rely on digital payments for daily life. The Act prevents financial harm by ensuring customers are not held responsible for unauthorized activity when they act promptly. It also reinforces fairness by standardizing protections across institutions. When ignored, customers face unpredictable losses and unresolved disputes. Trust erodes as digital transactions feel unsafe. Over time, Legal Rights deteriorate as protections weaken.

When executed well, compliance with the Electronic Fund Transfer Act strengthens Legal Rights by ensuring customers can transact with confidence. Customers trust that their funds are protected and recoverable. This improves adoption of digital services, reduces fear, and enhances ethical alignment. Compliance reinforces the principle that financial security is a fundamental right. It reminds leaders that protection must be built into every transaction. Ultimately, it builds an environment where electronic payments are safe, transparent, and reliable.

MORAL COMPASS SCENARIO	LEADERSHIP ACTION
Rhea sees that unauthorized transaction reports are not being investigated within required timelines. She realizes EFTA compliance is at risk.	The leader accelerates investigation workflows and assigns case owners. They also track resolution timeliness.
Omar notices that error-resolution disclosures are missing from the app. He senses regulatory gaps.	The leader adds required disclosures and tests visibility. They also enforce disclosure checks for all financial flows.
Lila finds that customers are charged overdraft fees without explicit opt-in. She recognizes a major violation.	The leader refunds affected customers and updates opt-in workflows. They also require explicit consent for all overdraft features.
Andre sees that transaction receipts omit key information. He worries about incomplete disclosures.	The leader updates receipt templates and validates compliance. They also add receipt-accuracy audits.
Serena discovers that recurring payments continue after cancellation. She senses systemic failure.	The leader fixes cancellation logic and audits all recurring-payment flows. They also add cancellation-confirmation steps.
Hugo notices that fraud alerts are delayed during peak hours. He recognizes a safety risk.	The leader improves alert infrastructure and adds redundancy. They also track alert-delivery latency.

Legal Rights require control over personal data, and the CCPA Consumer Rights provisions give customers the ability to access, delete, and restrict the sale of their information. They establish clear requirements for consent, disclosure, and data handling. This matters because customers deserve agency over how their information is used. The CCPA prevents exploitation by requiring organizations to honor customer choices. It also reinforces fairness by ensuring all customers—not just those with technical expertise—can exercise their rights. When ignored, data practices become opaque and customers lose control. Trust erodes as privacy expectations are violated. Over time, Legal Rights collapse as consent becomes meaningless.

When executed well, compliance with CCPA Consumer Rights strengthens Legal Rights by ensuring customers maintain authority over their data. Customers trust that the organization respects their privacy and autonomy. This improves transparency, reduces risk, and enhances ethical credibility. Compliance reinforces the principle that consent must be explicit and honored. It reminds leaders that data rights are central to modern customer relationships. Ultimately, it builds an environment where customers control their information with confidence.

MORAL COMPASS SCENARIO	LEADERSHIP ACTION
Keisha discovers that "Do Not Sell My Info" requests are not honored in advertising systems. She realizes consent is being overridden.	The leader synchronizes consent flags across all systems and validates enforcement. They also add automated compliance checks.
Damon notices that opt-out flows require multiple steps and confusing language. He senses friction designed to discourage users.	The leader simplifies the flow and rewrites language in plain terms. They also require friction-free consent pathways.
Liora finds that consent logs are missing timestamps. She recognizes auditability issues.	The leader enforces timestamping and updates logging standards. They also audit historical entries.
Mateo sees that consent withdrawals are not propagated to analytics pipelines. He worries about unlawful processing.	The leader updates data-flow orchestration to enforce real-time revocation. They also add revocation-propagation tests.
Serena discovers that minors' data is being processed without parental consent. She senses a COPPA and CCPA risk.	The leader deletes improperly collected data and implements parental-consent workflows. They also add age-screening to onboarding.
Omar notices that consent banners are inconsistent across platforms. He recognizes cross-platform compliance gaps.	The leader standardizes banners and tests them across devices. They also require platform-parity checks.

BONUS CHAPTER

Ai6x6™ Transformation Blueprint

The Influence Paradox

"Buy less, choose well, make it last." -- **Vivienne Westwood**, the iconic designer and climate activist who used her global stage to champion "Slow Fashion," provides the mantra for the **Influence Paradox**. Her challenge to the customer is the new North Star for brands: moving from high-volume transaction models to a value-based relationship where the brand succeeds by helping the customer consume more intentionally and durably.

The legacy marketing model is fundamentally at odds with Net Zero; it is designed to trigger infinite consumption in a finite world. Traditional "Green Marketing" has often been a thin layer of persuasion over a volume-based engine, leading to customer skepticism and "greenwashing" fatigue. To move forward, leaders must navigate the **Influence Paradox**. This requires a shift from selling products to providing "Circular Value," where profit is decoupled from units sold. By using AI to nudge customers toward sustainable defaults while respecting their privacy and autonomy, brands can transform from vendors into value partners. Traction is achieved when the customer's desire for convenience is balanced with the planet's need for slower, more intentional, and higher-value consumption cycles.

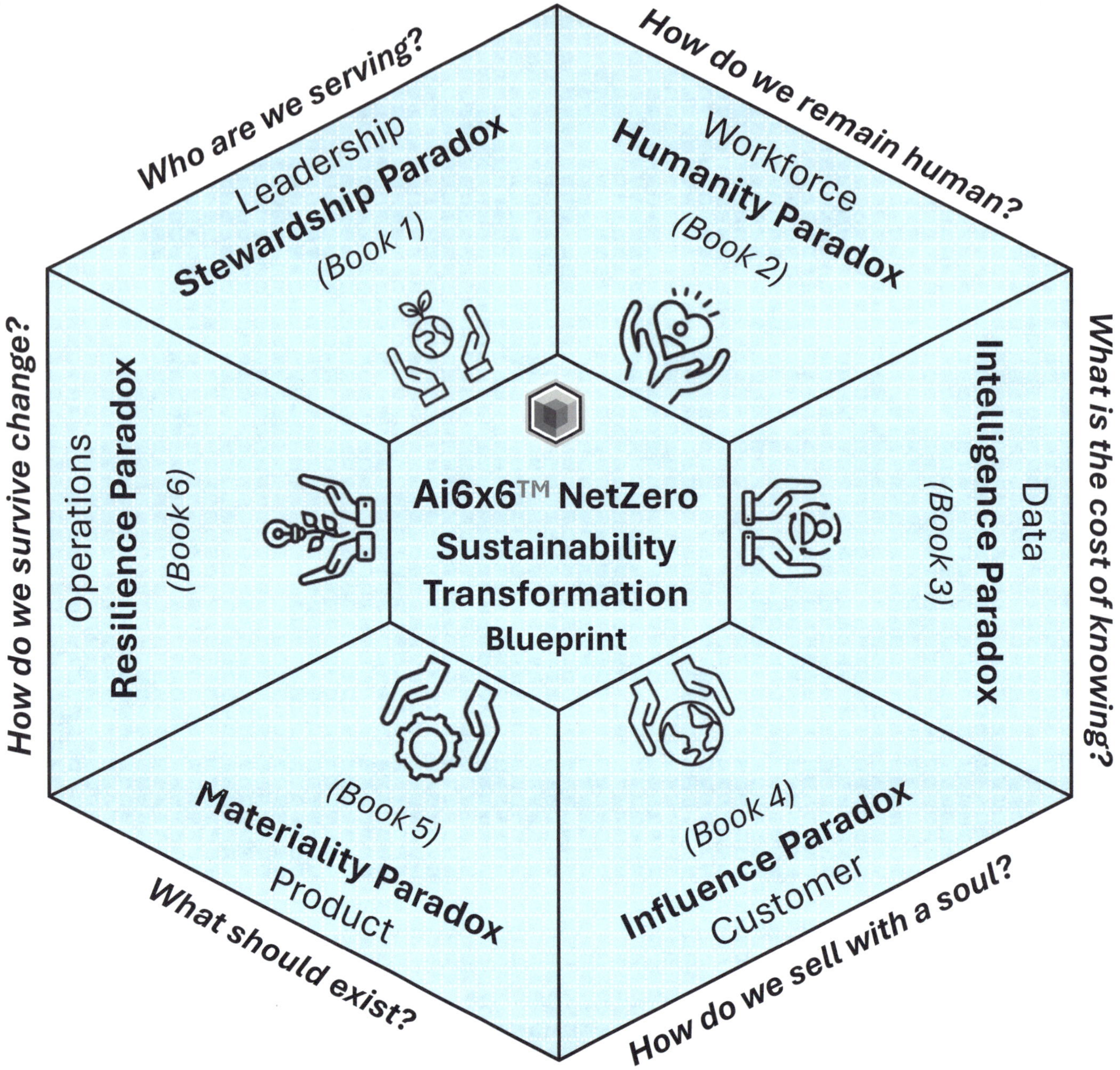

The **Ai6x6 NetZero Sustainability Transformation Blueprint** redefines the path to Net Zero by replacing the traditional "check-the-box" compliance mindset with a rigorous framework of **paradoxical leadership**. In a world accelerated by AI, sustainability is rarely a choice between a clear right and wrong; instead, it is a constant navigation of thirty-six competing tensions across six organizational lenses. By framing the journey through these paradoxes—such as the conflict between **Digital Intelligence and Energy Consumption** or **Global Targets and Local Infrastructure**—the framework prepares leaders for the messy, non-linear realities of systemic change.

Rather than viewing these tensions as obstacles to be avoided, the **Ai6x6 methodology** utilizes them as engines for innovation. It guides executives to move past the paralysis of "either/or" thinking, teaching them to embrace a "both/and" strategy that maintains fiduciary health while securing long-term planetary survival. This approach ensures that a company's moral compass remains calibrated even as technology and climate volatility shift the ground beneath them. Ultimately, the series provides a diagnostic roadmap that turns systemic friction into strategic advantage, empowering leaders to achieve radical Net Zero goals without sacrificing the operational resilience or human dignity required to thrive in the post-AI era.

THE INFLUENCE PARADOX is the central struggle of modern commerce: the tension between driving consumption to hit sales targets and guiding customers toward a lower-carbon lifestyle. For decades, the goal of customer engagement was to make buying as frictionless and frequent as possible. Now, sustainability requires us to introduce "positive friction"—encouraging customers to buy less, choose slower, or repair instead of replace. This creates a fundamental conflict for the marketing and sales engine. To transform, a business must learn to grow by providing more value and better outcomes, rather than moving more volume through the checkout.

1. THE PERSONALIZED NUDGE PARADOX (Helpful Hints vs. Personal Privacy)

Using AI to "nudge" customers toward green choices requires deep insights into their personal habits and motivations. The paradox is that the more "helpful" and personalized these suggestions become, the more they rely on invasive data harvesting that can feel like a violation of privacy. Brands must walk a thin line between being a supportive sustainability coach and being a digital "big brother." Success in this area requires radical transparency about how data is used, ensuring that the customer feels empowered by the nudge rather than surveilled by the system.

2. THE CIRCULAR-VALUE PARADOX (Using Things vs. Owning Things)

Shifting from selling physical products to providing "Product-as-a-Service" is a massive win for resource efficiency. However, the paradox lies in the deep-seated human desire for ownership as a symbol of status and security. Customers are used to "having" things rather than just "accessing" them. For a business, this means rethinking the entire customer relationship—moving from a one-time transaction to a long-term partnership. It requires convincing the market that the value lies in the experience and the result, rather than in the clutter of a physical object.

3. THE CONVENIENCE-CARBON PARADOX (Fast Delivery vs. Earth's Cost)

The digital economy has trained customers to expect everything to be delivered "instant and free." The paradox is that this culture of convenience is the enemy of carbon efficiency. Radical decarbonization of logistics requires slower, consolidated shipping and fewer individual "last-mile" trips. Brands are caught between the market pressure to deliver in hours and the moral requirement to optimize for the planet. Solving this requires a shift in the customer "reward" system, making slower, greener shipping the most attractive and socially celebrated choice for the conscious consumer.

4. THE PRICE-EQUITY PARADOX (Green Prices vs. Fair Costs)

Ethically sourced, low-carbon products often come with a higher price tag due to more expensive materials and fair labor. The paradox is that while the planet needs everyone to buy "green," the current price premium often limits these choices to the wealthy. If sustainability remains a luxury, we will never hit our global targets. Transformation means finding ways to use AI and operational efficiency to bring the cost of sustainable goods down. A brand's moral compass must aim for "Green for All," ensuring that the ethical choice is also the affordable one.

5. THE CHOICE-ARCHITECTURE PARADOX (Smart Defaults vs. Freedom to Choose)

Designers can use "choice architecture" to make the most sustainable option the default—like automatically selecting carbon-neutral shipping at checkout. The paradox is that while this is highly effective for the planet, it can feel like a manipulation of the customer's free will. There is a fine line between making the right choice easy and taking the choice away altogether. Ethical engagement requires a balance: designing interfaces that lead with sustainability while still respecting the customer's intelligence and their right to make the final decision for themselves.

6. THE SECOND-LIFE PARADOX (Fixing Old vs. Buying New)

Traditional business models rely on "planned obsolescence" to keep sales moving. The paradox of the second life is that a truly sustainable brand must help its customers buy *less* by offering repair services and resale platforms. This creates a direct conflict with internal sales teams who are measured on new units sold. To resolve this, the definition of "customer value" must change. Revenue must be decoupled from manufacturing new items, shifting toward revenue from maintenance, upgrades, and circular marketplaces that keep products in use for as long as possible.

1. THE PERSONALIZED NUDGE PARADOX (Helpful Hints vs. Personal Privacy)

- **Insurance:** A provider offers premium discounts for customers who share smart-home data, using it to send "nudge" notifications when appliances are running inefficiently.
- **Ride-Hailing:** An app defaults users to "Green/EV" rides, providing a real-time carbon-savings counter that compares the impact of their choice to a standard petrol vehicle.
- **Personal Finance:** A neobank sends proactive push notifications when a customer's monthly "Carbon Footprint" (calculated from transaction categories) exceeds their personal goal.
- **Fitness:** A wearable brand rewards users for "Active Commuting" by converting walking or cycling miles into credits that can be used for sustainable gear.
- **Utilities:** A power company sends "Peak-Shaving" alerts, nudging customers to delay heavy laundry or dishwashing until wind-power production is at its highest on the grid.
- **Beauty:** A skincare brand uses AI skin-analysis to prevent over-purchasing, nudging customers toward a minimalist, three-product routine tailored to their specific needs.

2. THE CIRCULAR-VALUE PARADOX (Using Things vs. Owning Things)

- **Outdoor Apparel:** A brand integrates a "Certified Pre-Owned" section into its main website, allowing customers to trade in used jackets for store credit.
- **Toys:** A subscription service launches a "Play Library," where parents pay a monthly fee to swap age-appropriate toys, ensuring plastic items never reach a landfill.
- **Smartphones:** A manufacturer offers a "Guaranteed Future Value" for devices, incentivizing customers to return them for professional refurbishment or material recovery.
- **Appliances:** A kitchen brand shifts to a "Performance Lease," where customers pay for "Cold Storage" as a service; the brand remains responsible for all repairs and energy efficiency.
- **Baby Goods:** A stroller company offers a rental model that grows with the child, allowing parents to swap infant bassinets for toddler seats without buying new hardware.
- **Tools:** A hardware chain installs "IoT Tool Lockers" in urban neighborhoods, allowing residents to rent high-quality power drills by the hour instead of purchasing them.

3. THE CONVENIENCE-CARBON PARADOX (Fast Delivery vs. Earth's Cost)

- **E-Commerce:** A major retailer sets "Green Delivery" (consolidating orders into a 5-day window) as the default option, requiring an explicit "Carbon Fee" for next-day shipping.
- **Food Delivery:** An app "batches" orders from the same apartment complex or office building, offering a discount to customers who agree to a shared delivery window.
- **Logistics:** A global courier implements "Slow Shipping" by rail and sea for non-urgent parcels, transparently showing customers that the cost of speed is a 10x increase in carbon.
- **Grocery:** A supermarket offers a "Packaging-Free" home delivery service, where fresh goods are delivered in reusable, returnable crates rather than plastic bags.
- **Retail:** A brand provides "In-Store Pickup" incentives, such as loyalty points or free coffee, to encourage customers to eliminate the carbon-heavy "last-mile" home delivery.
- **Pharmacy:** A health chain uses 100% reusable insulated totes for medication delivery, collecting the empty containers during the next scheduled drop-off.

4. THE PRICE-EQUITY PARADOX (Green Prices vs. Fair Costs)

- **Grocery Retail:** A supermarket uses AI to dynamic-price organic and plant-based produce nearing its "sell-by" date, making sustainable food affordable for lower-income shoppers.
- **Energy:** A utility company implements "Social Tariffs," using profits from high-usage commercial clients to subsidize 100% renewable energy for low-income households.
- **Automotive:** A car-share service offers subsidized "EV Credits" in underserved urban neighborhoods to ensure clean mobility is not a luxury-only benefit.
- **Fashion:** A brand creates a "Basics" line using recycled materials priced exactly the same as virgin cotton, cross-subsidizing the cost through its high-end luxury collections.
- **Travel:** A booking platform offers "Eco-Grants" for students and low-income travelers who choose train travel over high-carbon regional flights.
- **FMCG:** A soap brand launches a "Refill-at-Home" concentrate that is 20% cheaper than the bottled version, incentivizing sustainability through direct cost savings.

5. THE CHOICE-ARCHITECTURE PARADOX (Smart Defaults vs. Freedom to Choose)

- **Travel/Hospitality:** A flight search engine ranks results by "Lowest Carbon Footprint" by default, forcing price-only comparisons to the second page of results.
- **E-Commerce:** A checkout page uses "Transparent Friction," asking for a second confirmation if a customer chooses a high-carbon shipping method or a single-use product.
- **Dining:** A corporate cafeteria lists plant-based meals at the top of the digital menu and highlights them with "Chef's Choice" badges to nudge healthier selections.
- **Home Services:** A contractor's booking portal defaults to "Energy Audit + Repair" instead of "Full Replacement," requiring a manual override to buy new equipment.
- **Retail UI:** A clothing site allows users to set a "Green Filter" in their profile, which automatically hides any products that do not meet a minimum 70% recycled-content score.
- **Banking:** A credit card app highlights the "Planetary Cost" of a purchase (in liters of water or grams of CO2) alongside the currency price in the transaction history.

6. THE SECOND-LIFE PARADOX (Fixing Old vs. Buying New)

- **Furniture:** A brand sells official "Refurbishment Kits" (new foam, covers, and legs) for its most popular sofa models to extend their lifespan by another decade.
- **Electronics:** A laptop manufacturer provides open-source 3D-printing files for spare parts, allowing customers to print their own buttons or hinges at local maker spaces.
- **Apparel:** A brand hosts "Repair Cafes" in its flagship stores, where tailors fix any brand of garment for free to build community trust around longevity.
- **Watchmaking:** A luxury brand offers a "Lifetime Service Guarantee" that includes a guaranteed trade-in value for the watch's internal movements to be used in new models.
- **Footwear:** A sneaker brand launches a "Resole Service," where customers can ship back worn shoes to have a fresh, recycled-rubber sole bonded to the original upper.
- **Kitchenware:** A cookware company offers a "Lifetime Re-tinning" service for its copper pans, ensuring the physical product remains in use for generations.

Ai6x6 COMPREHENSIVE TRANSFORMATOION BOOK SERIES

Most comprehensive business transformation book series worldwide

Ai6x6™ Transformation Engine

The AI6x6 Transformation Engine™

Your Operating System for the AI Era

In an age of autonomous systems, fragmented strategy is a liability. The **AI6x6 Transformation Engine™** is a precision-engineered, 18-volume masterwork designed to move your enterprise from organizational inertia to scalable momentum. Built on a proprietary 6×6 geometry, this ecosystem unifies architecture, execution, and calibration into one high-performance unit.

The journey begins with the **Transformation Blueprint Series**, your **Intelligence Layer**. It codifies a "Navigation Deck" of Six Pillars and Six Lenses to anchor your strategic North Star. Next, the **Consulting Playbook Series** acts as the **Transmission Layer**, utilizing the IGNITE™ methodology to translate executive intent into high-torque, repeatable action. Finally, the **Maturity Diagnostic Series** serves as the **Sensor Layer**, providing a data-driven intake to quantify readiness and calibrate performance.

This isn't just a book collection; it is a rigid **Chassis** for institutional excellence. Secure the definitive architecture for sustained, high-stakes digital evolution. Drive the future with the Engine.

Transformation Series

To ensure this book series benefits your company holistically, we strongly recommend you buy the series and then distribute it to the appropriate transformation sponsors in each of these departments. Don't forget to browse through consulting offers at the end of the book.

WE NEED YOU!

WELCOMING OUR NEXT SET OF THOUGHT LEADERS

The remarkable success of the Ai6x6 Organization Transformation Series is built upon the **hard-won wisdom** of executive leaders who have successfully navigated the AI transformation within their industries, turning theory into a deeply practical, actionable guide. As we expand the series, we are extending a highly selective invitation to a new cohort of pioneers—individuals whose professional journey is at the **leading edge of AI strategy and execution**. We are looking for executives who possess a proven track record of rewriting business models, mastering unforeseen ethical and competitive challenges, and offering a **clear blueprint** for the next 3-5 years of the AI-Era.

Contributing to the Ai6x6 Series is a unique opportunity to **cement your professional legacy**, joining an exclusive, curated network of global authorities whose insights are distilled into executive-level knowledge cards, directly shaping the governance and innovation decisions of the world's most influential companies. Our selection process is based strictly on **pure expertise and demonstrated leadership** in the AI-Era, reinforcing our commitment to editorial integrity and authentic thought leadership—there is no monetary exchange involved. Select contributors will also receive an invitation to blog on our expert network microsite, further amplifying their platform. To confidentially inquire about contributing your expertise, please contact us via our dedicated portal at **www.Ai6x6.com/contact** or by scanning the adjacent QR code.

Contact

SUBSCRIBE TO OUR NEWSLETTERS: To stay ahead of the curve and receive continuous, executive-level intelligence on the rapid evolution of the AI-Era, we invite you to register for our **Ai6x6 Weekly Newsletter**. Each edition delivers curated insights directly from our thought leaders and highlights critical shifts in AI strategy, governance, and provides relevant excepts from our website and books. Keep your finger on the pulse of the future and ensure your company remains competitive by signing up on the same page

PLEASE REVIEW US ON AMAZON!

Customer Blueprint

We know your time is valuable, but if the Ai6x6 Organization Transformation Series delivered the real-world, actionable strategies you needed, please consider leaving a brief review on Amazon. If we missed the mark, we would also like to know, and that is the only we we can improve. Your honest feedback is crucial; it helps fellow executives discover this unique framework and benefit from the insights of our thought leaders. Thank you for supporting our mission!

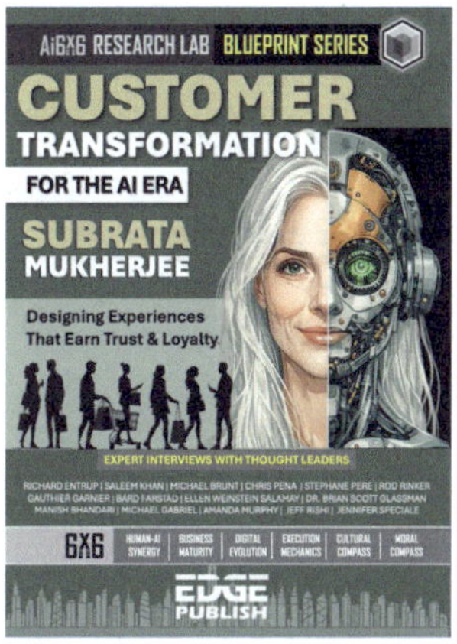

THE Ai6x6™ CERTIFIED TRANSFORMATION STRATEGIST

Certified for Thinking, Not Reciting

The **Ai6x6™ CTS (Certified Transformation Strategist)** is an elite designation for MBAs and strategic leaders with a minimum of **five years of professional experience**. This is not a test of your ability to regurgitate theory; it is a clinical evaluation of your ability to orchestrate high-stakes business transformation in real-time. It signals to the market that you are a strategist who can think on your feet, navigate ambiguity, and utilize a world-class clinical framework to deliver results.

The Non-Traditional Certification

We don't believe in "black box" testing. We believe in the "Open Book" reality of the modern boardroom. To earn your CTS, you must demonstrate mastery through a **60-minute Board-Level Defense** before a member of the Ai6x6™ Global Leadership Team.

$1000 per defense

The Preparation

Mastery of the Series: Candidates must familiarize themselves with all six volumes in the **Ai6x6™ Transformation Series.** You choose **two Retreat Exercises** from any two **Consulting Playbooks** within the series that best align with your expertise. Best practices for the presentation are provided on our website.

The 60-Minute Defense

1. **The Presentation (30 Minutes):** You have 15 minutes per exercise to present your strategic approach and expected outcomes. 60% of your score is for this exercise.
2. **The Situational Audit (30 Minutes):** Our leadership team will pose **six situational transformation scenarios**. You must demonstrate how to apply the **6 Lens Sequence™** to solve these complex frictions.
3. **The Open-Resource Advantage:** You are **encouraged** to refer to any of the six Ai6x6™ books during the exam. We test your ability to navigate and apply the framework under pressure, not your memory.

The Induction: Beyond the Credential

Passing the Board Defense is only the beginning. Successful candidates are immediately inducted into the **Ai6x6™ Executive Mentor Network**, unlocking:

- **Thought Leadership:** Opportunities to contribute to the Ai6x6™ books, knowledge articles and gain visibility among global industry leaders.
- **Engagement Pipeline:** Direct access to a network of future consulting engagements and strategic partnerships.
- **The Fast Track:** Exceptional participants may be invited to join our elite **Practitioner Program**, bypassing the standard entry requirements to lead their own white-label boutique practice.

SCHEDULE YOUR BOARD DEFENSE AT:

Ai6x6.com/certification

Consulting Certification

THE Ai6x6™ STRATEGIC ROADMAP SERVICE

Your 3-Year Vision, Stress-Tested for the AI Era

The **Ai6x6™** book series has given you the "source code"—the clinical lenses, the vulnerability audits, and the diagnostic frameworks. But strategic orchestration at the highest level requires a cold, objective eye. If you are ready to move from self-diagnosis to radical transformation, we invite you to engage the **Ai6x6™ Roadmap Service.**

The 21-Day Transformation Engine

We don't believe in "billable hours" or six-month discovery phases. We believe in high-fidelity data and rapid execution. Our domain experts take your raw diagnostic output and synthesize it into a personalized, holistic blueprint for the next three years of your organization's life.

How It Works

1. **Submit Your Data:** Upload your completed Discovery Questionnaires, Vulnerability Audits, and Radar Charts directly to our secure portal or via email or via courier.
2. **Contextualize:** Provide your current 3-year roadmap and any business documentation that can provide more context on your line of business, your mission and success KPIs.
3. **The Clinical Review:** Our experts analyze your organizational DNA against the **6 Lens Sequence**, identifying hidden frictions and untapped leverage points. Assign a single point within your organization for asking questions.
4. **The Delivery:** Within **2 to 3 weeks**, you receive a personalized **Holistic Transformation Roadmap**, including our proprietary **Self-Disruption Strategy**.

Why Self-Disruption?

The greatest threat to your organization isn't your competitor; it is your own legacy success. Our roadmap doesn't just show you how to grow; it shows you how to cannibalize your outdated models before the market does it for you.

The Commitment

- **Speed:** 14–21-day turnaround.
- **Security:** Full NDA-backed engagement.
- **Precision:** Fact-based, quantitative analysis. No "fluff".

Packages from $3000

SECURE YOUR DIAGNOSTIC REVIEW:

Ai6x6.com/roadmap

Consulting Roadmap

THE Ai6x6™ TRANSFORMATION RETREAT

One Playbook. Your Leadership. Total Orchestration.

The **Ai6x6™ Business Transformation Series** is designed to eliminate the need for traditional, slow-moving consulting teams. Using our "Clinical Blueprints," your organization can self-organize, self-diagnose, and self-disrupt. Everything you need to facilitate a high-stakes, 2.5-day executive offsite is contained within the books.

The Self-Guided Mastery (100% Internal & Free)

You have the "Source Code." Your internal Offsite Coordinator or Chief of Staff can lead the entire 2.5-day agenda end-to-end using the **Consultant Playbook (Part 2)** and our **Supplemental Workbooks**.

- **No External Fees:** Run the entire diagnostic without "billable hours."
- **Full Control:** Use the 6-book series to "mix and match" exercises that fit your specific departmental frictions.
- **Tactile Execution:** Move from whiteboards to prioritized MoSCoW cards and RACI matrices using the physical templates provided in the series. Our Advisory section has all the inspiration your team needs.

Packages from $4999

The Orchestrated Upgrade (White-Glove Support)

For leadership teams seeking institutional weight and an objective "Cold Eye" review without the overhead of travel, we offer the **Orchestrated Retreat** package. This provides three critical layers of external authority:

- **Phase 1: The 30-Minute Architect's Briefing (Pre-Retreat):** A high-velocity strategic sprint with your internal Facilitator. we help you select the exact modules from across the 6-book series to target your unique friction points and set the tactical "Rules of Engagement."
- **Phase 2: The 30-Minute Executive Keynote (Live Remote):** A virtual address delivered by the **Ai6x6™ Leadership Team** to kick off Day 1. This keynote provides the external authority required to break through internal political silos and align the room's energy.
- **Phase 3: The 14-Day Clinical Roadmap Audit (Post-Retreat) :** Submit your digital outputs—prioritized lists, RACI matrices, and visualized Roadmaps. Our team performs a rigorous review and delivers a **Comprehensive Strategic Report** within **two weeks**, stress-testing your plan for diagnostic integrity and self-disruption readiness.

UPGRADE TO AN ORCHESTRATED RETREAT:

Ai6x6.com/retreat

Consulting Retreat

"Peace is its own reward."

Mahatma Gandhi

Ai6x6 Research Lab
Publishing Team

NEW YORK OFFICE
(Editorial Hub)

Editorial Lead: Marisa Thompson
Research Lead: Talia Rosen
Line Editor: Jordan Alvarez

LONDON OFFICE
(Design Hub)

Design Lead: Keiko Tanaka
Layout Specialist: Oliver Whitmore
Illustrator: Aisha Rahman
Design Team from Sumu Arts

KOLKATA OFFICE
(Support Hub)

Project Manager: Sohini Chatterjee
Production : Arindam Ghosh
Quality Lead: Priya Deshmukh
Fact Checker: Ravi Patel
Research Assistant: Ananya Sengupta
Research Assistant: Rohan Mukherjee
Project Assistant: Madhumita Roy
Production Assistant: Sagnik Dutta
Quality Assistant: Ishita Banerjee
Series Lead (Blueprint): Amit Verma
Series Lead (Playbook): Neha Reddy
Series Lead (Diagnostic): Karan Mehta

The End

Made in the USA
Monee, IL
05 November 2026

67814276R00210